CELL SURFACE RECEPTORS

THE ELLIS HORWOOD SERIES IN
BIOCHEMISTRY IN MEDICINE AND PHARMACOLOGY
Series Editor: JACKIE de BELLEROCHE
Lecturer in Neurochemistry, Departments of Biochemistry and Neurology, Charing Cross Hospital Medical School, London

A series of up-to-date works covering a wide range of applications in biochemistry, medicine and pharmacology. The books will be of interest to industry, particularly pharmaceuticals, as well as to many areas of the biological and medical sciences. The level will vary between introductory texts, and advanced level research works.

THE BIOCHEMICAL EFFECT OF DRUGS IN PREGNANCY
Volume 1: The Nervous System, Cardiovascular and Haemopoietic Systems.
Volume 2: Diaretics, Digestive Tract Drugs, Broncho-pulmonary Drugs, Hormones, Antipyretics, Intermediary Metabolism Drugs, Antibiotics, Vaccines

A. ONNIS, Professor of Clinical Obstetrics and Gynaecology, and Director of the Obstetric and Gynaecological Clinic, University of Padova, and P. GRELLA, Professor of Prenatal Puericulture, Clinic of Obstetrics and Gynaecology, University of Padova, Italy.

CELL SURFACE RECEPTORS
Edited by P. G. STRANGE, Department of Biochemistry, The Medical School, Queen's Medical Centre, Nottingham

CELL SURFACE RECEPTORS

Editor:
P. G. STRANGE
Department of Biochemistry
The Medical School
Queen's Medical Centre, Nottingham

ELLIS HORWOOD LIMITED
Publishers · Chichester

Halsted Press: a division of
JOHN WILEY & SONS
New York · Brisbane · Chichester · Toronto

First published in 1983 by
ELLIS HORWOOD LIMITED
Market Cross House, Cooper Street, Chichester, West Sussex, PO19 1EB, England

The publisher's colophon is reproduced from James Gillison's drawing of the ancient Market Cross, Chichester.

Distributors:

Australia, New Zealand, South-east Asia:
Jacaranda-Wiley Ltd., Jacaranda Press,
JOHN WILEY & SONS INC.,
G.P.O. Box 859, Brisbane, Queensland 40001, Australia

Canada:
JOHN WILEY & SONS CANADA LIMITED
22 Worcester Road, Rexdale, Ontario, Canada.

Europe, Africa:
JOHN WILEY & SONS LIMITED
Baffins Lane, Chichester, West Sussex, England.

North and South America and the rest of the world:
Halsted Press: a division of
JOHN WILEY & SONS
605 Third Avenue, New York, N.Y. 10016, U.S.A.

© 1983 P. G. Strange/Ellis Horwood Ltd.

British Library Cataloguing in Publication Data
Strange, Philip G.
Cell surface receptors.
1. Cytochemistry
I. Title
547.87'6042 QH611

Library of Congress Card No. 82-23376

ISBN 0-85312-569-4 (Ellis Horwood Ltd., Publishers)
ISBN 0-470-27418-2 (Halsted Press)

Typeset in Press Roman by Ellis Horwood Ltd.
Printed in Great Britain by R. J. Acford, Chichester.

COPYRIGHT NOTICE –
All Rights Reserved. No part of this publication may be reproduced, stored in a retrieval system, or transmitted, in any form or by any means, electronic, mechanical, photocopying, recording or otherwise, without the permission of Ellis Horwood Limited, Market Cross House, Cooper Street, Chichester, West Sussex, England.

Table of Contents

Editor's Preface 11

Editor's Introduction 13

Chapter 1   Selective α_2-adrenoreceptor agonists and antagonists — pharmacology and therapeutic potential
J. C. Doxey, D. S. Walter, C. B. Chapleo, P. W. Dettmar, P. L. Myers, A. G. Roach, C. F. C. Smith and I. F. Tulloch, Departments of Pharmacology and Medicinal Chemistry, Reckitt and Colman, Pharmaceutical Division, Dansom Lane, Hull, U.K.
1. Introduction 21
2. Distribution of α_1- and α_2-adrenoreceptors 22
3. Therapeutic potential of α_2-adrenoreceptor agonists 24
4. Selective α_2-adrenoreceptor antagonists and their therapeutic potential 28

Chapter 2   Subtypes of the opiate receptor
H. W. Kosterlitz, S. J. Paterson and L. E. Robson, Unit for Research on Addictive Drugs, University of Aberdeen, Marischal College, Aberdeen, U.K.
1. Introduction 41
2. Precursors of opioid peptides 41
3. Receptors interacting with opioid peptides 42
4. Conclusions 47

Chapter 3   The benzodiazepines and their receptor
I. L. Martin, MRC Neurochemical Pharmacology Unit, Medical Research Council Centre, Medical School, Hills Road, Cambridge, U.K.
1. Introduction 50
2. Mechanism of action of benzodiazepines 53

Table of Contents

    3. Benzodiazepine receptors 54
    4. Summary and future 70

Chapter 4   Brain dopamine receptors: characterisation and isolation
    P. G. Strange, Department of Biochemistry, The Medical School, Queen's Medical Centre, Nottingham, U.K.
    1. Introduction 82
    2. Use of ligand-binding studies with [^{3}H]spiperone for studying D_2 receptors in the brain 85
    3. Solubilisation of D_2 receptor protein from brain 88
    4. Conclusion 96

Chapter 5   Receptors for dihydropyridine calcium channel antagonists
    P. Bellemann and A. Schade, Department of Pharmacology, Bayer AG, Wuppertal, West Germany
    1. Introduction 101
    2. Experimental procedures 101
    3. Results and conclusions 102

Chapter 6   Coexistence of neuropeptides and monamines: possible implications for receptors
    P. C. Emson and R. F. T. Gilbert, MRC Neurochemical Pharmacology Unit, Medical Research Council Centre, Medical School, Hills Road, Cambridge, U.K., and J. M. Lundberg, Department of Pharmacology, Karolinska Institute, Stockholm, Sweden
    1. Coesistence of 5-hydroxytryptamine, substance P and TRH in the bulbo-spinal neurones of the rat spinal cord 112
    2. Cholecystokinin and dopamine 116
    3. Vasoactive intestinal polypeptide and acetylcholine 118
    4. Conclusions 121

Chapter 7   Structure and function of the nicotinic acetylcholine receptor: proteolytic digestion and subunit specificity
    S. Fuchs and D. Bartfeld, Department of Chemical Immunology, The Weizmann Institute of Science, Rehovot, Israel
    1. Introduction 126
    2. Molecular dissection of AChR: denaturation and proteolytic digestion 127
    3. On the antigenic specificity of the subunits of AChR 133
    4. Concluding remarks 138

Table of Contents

Chapter 8 Adaptive changes in dopamine neuronal function in response to chronic neuroleptic administration
P. Jenner and C. D. Marsden, University Department of Neurology, Institute of Psychiatry and King's College Hospital Medical School, London, U.K.
1. Introduction...................................142
2. Time course of changes occurring in striatal dopamine receptor function in response to chronic administration of trifluoperazine...................................143
3. Functional changes in striatal acetylcholine content in response to continued neuroleptic intake...................................146
4. Alterations in characteristics of striatal dopamine receptors....149
5. Striatal dopamine function following withdrawal from chronic neuroleptic treatment...................................154
6. Regional changes in cerebral dopamine function in response to chronic neuroleptic administration...................................156
7. Conclusions...................................158

Chapter 9 Ligand-binding studies in brains of schizophrenics
F. Owen, A. J. Cross and T. J. Crow, Division of Psychiatry, Clinical Research Centre, Harrow, Middlesex, U.K.
1. Introduction...................................163
2. Materials and methods...................................164
3. Results...................................166
4. Discussion...................................178

Chapter 10 Pituitary gonadotrophin-releasing-hormone receptors: physiological regulation
R. N. Clayton, Department of Medicine, University of Birmingham, Edgbaston, Birmingham, U.K.
1. Introduction...................................184
2. Physiological regulation of GnRH receptors...................................187
3. Mechanism of physiological regulation of pituitary GnRH receptors...................................193
4. Autoregulation of GnRH-R by GnRH and agonist analogue....198
5. Conclusions...................................202

Chapter 11 The role of membrane phospholipids in receptor transducing mechanisms
M. J. Berridge, ARC Unit of Invertebrate Chemistry and Physiology, Department of Zoology, University of Cambridge, U.K.
1. Introduction...................................207
2. Methylation response...................................208

Table of Contents

    3. The phosphatidylinositol response 209
    4. Clinical significance of the PI response. 215

Chapter 12 The complex structure and regulation of adenylate cyclase
M. Rodbell, The National Institutes of Health, Bethesda, Maryland, U.S.A.
    1. The role of cations. 228
    2. Structure of multi-receptor systems 233
    3. Conclusions 236

Chapter 13 A human mutation affecting hormone-sensitive adenylate cyclase
H. R. Bourne, Division of Clinical Pharmacology, Departments of Medicine and Pharmacology and the Cardiovascular Research Institute, University of California, San Francisco, U.S.A.
    1. Introduction. 240
    2. Pseudohypoparathyroidism, Type 1 241
    3. N activity in erythrocytes of PHP-I patients 242
    4. N activity in other cells. 243
    5. Clinical endocrine studies 244
    6. The PHP-Ib phenotype. 246

Chapter 14 A study of the mRNA and genes coding for the nicotinic acetylcholine receptor
K. Sumikawa and M. Houghton, Molecular Genetics Department, Searle Research & Development, High Wycombe, Buckinghamshire, U.K.; R. Miledi, Department of Biophysics, University College London, U.K.; E. A. Barnard, Department of Biochemistry, Imperial College of Science and Technology, London University.
    1. Introduction. 249
    2. Structure of *Torpedo* nicotinic acetylcholine receptor 250
    3. Studies on the biosynthesis of the acetylcholine receptor 255
    4. Cloning of an AChR cDNA 262
    5. Conclusions 266

Chapter 15 Quantitative drug assays using radioreceptor techniques
S. R. Nahorski and D. B. Barnett, Department of Pharmacology and Therapeutics, Medical Sciences Building, University of Leicester, U.K.
    1. Introduction. 270
    2. Radioreceptor Assays: Principles and Methodology 271

3. *In vivo* and *ex vivo* RRAs 275
 4. Examples of radioreceptor assays 276
 5. Conclusions 288

Editor's Conclusions – Future Strategies 293

Index 296

Editor's Preface

The aim of this book is to provide a description of the current status of research into cell surface receptors emphasising clinical aspects where possible. The book consists of an introduction which gives a historical overview of the development of the field so that the non-specialist may appreciate the relevance of the later chapters which consist of a series of specialist reviews on a wide range of topics. Finally some attempt is made to predict the future of research in this field.

The book has arisen out of a Workshop on Cell Surface Receptors — Basic and Clinical Aspects, held in Jesus College, Cambridge. Financial support for the meeting was given by the following:

The Royal Society, Beechams, Boots, Glaxo, I.C.I., Merck, Pfizer, Reckitt and Colman, Sandoz, Schering, Smith Kline and French, Syntex, Upjohn.

I am most grateful to these organisations without whose assistance neither the meeting nor this book would have been possible.

I would also like to thank those who have helped me with the organisation of the meeting and the preparation of the book:

Tricia Frankham, Pete Golds, Jean Hall, Graeme Milligan, Sandra Simmonds, Millie Spooner, Liz Stuart, Mark Wheatley.

Finally, I would like to thank my wife Margaret for her constant support and encouragement.

Philip Strange

Editor's Introduction

This short introduction to the field is designed to set the scene for the more detailed discussions below. It is hoped that it will provide an overview of the field for the non-specialist whereas it may be superfluous for the specialist in this area.

The concept of a receptor (receptive substance) was first introduced by Langley (1905) in studies on the effect of nicotine and curare on muscle. Since then receptors have been shown to be responsible for the effects of many hormones and neurotransmitters and a broad division of receptor systems may be made into those dependent on cell surface receptors and those dependent on receptors in the cytosol (Fig. 1) but this book is concerned only with the former class. It is now clear that cell surface receptor systems form a group of multi-component information transfer systems involved in cellular transmembrane signalling but the development of Langley's idea into our present rather detailed knowledge of cell surface receptors has depended on several key concepts.

For many years receptors could be studied only by physiological or pharmacological methods, e.g. whole animal behavioural tests, organ responses, electrophysiological methods. In terms of a detailed analysis of receptor mechanism

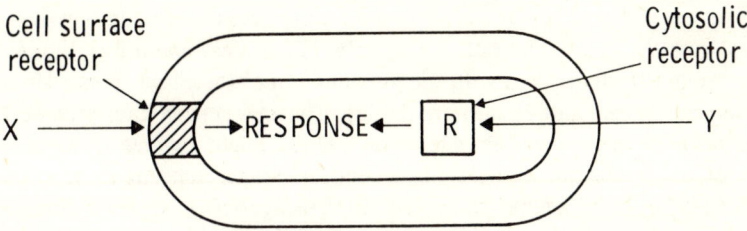

Fig. 1 — Subdivision of receptor systems into cell-surface and cytosolic categories. Responses to hormones or neurotransmitters X and Y in a cell are shown dependent on cell surface receptors (X) and cytosolic receptors (Y). The two systems are shown for convenience in one cell although this is not meant to imply a necessary coexistence.

these studies suggested that in some cases, e.g. nicotinic acetylcholine receptor (Colquhon, 1981), the receptor was closely coupled to an ion channel and that receptor activation lead to channel opening. The speed of the response precluded the intervention of many additional processes. Many receptor systems, however, did not show such rapid responses allowing the possible intervention of other processes between receptor activation and response. The discovery of second messengers, e.g. cyclic AMP by Sutherland (Fig. 2) (see Robison *et al.*, 1971) was a key step in the study of such systems. The recognition that for some receptors activation on the outside of the cell gave rise to increased levels of cyclic AMP inside the cell allowed new ideas to develop and offered an entirely new way to study some receptors.

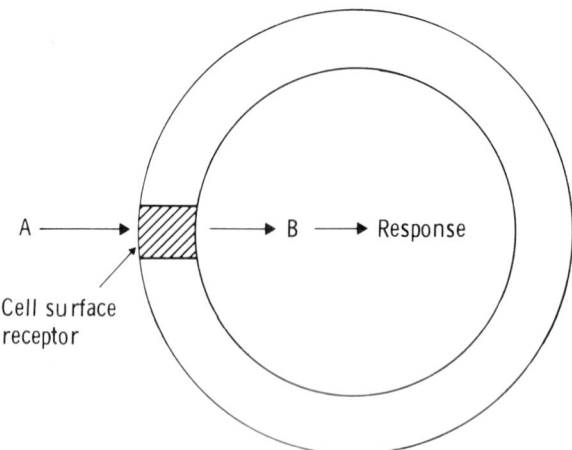

Fig. 2 — The concept of a second messenger. Hormone or neurotransmitter A is shown acting on a cell via a cell surface receptor to provoke a cellular response via second messenger compound B.

The original concept of the second messenger was essentially one of a static array across the membrane but in the early 1970s ideas about membrane structure were drastically revised with the introduction of the Fluid Mosaic Model for membranes (Singer and Nicholson, 1972). The recognition that integral membrane proteins, e.g. receptors, adenylate cyclase might be free to move in the plane of the membrane allowed the second messenger concept to be expanded so that receptor and adenylate cyclase were recognised as separate entities free to move independently and that activation of the system required coupling of the two entitites in some way (Fig. 3). Thus the receptor-stimulated adenylate cyclase could be viewed in a similar manner to soluble enzymes like aspartate transcarbamoylase where separate regulatory and catalytic sub-units contributed to the full regulatory enzyme activity (Gerhart and Schachman, 1965). This

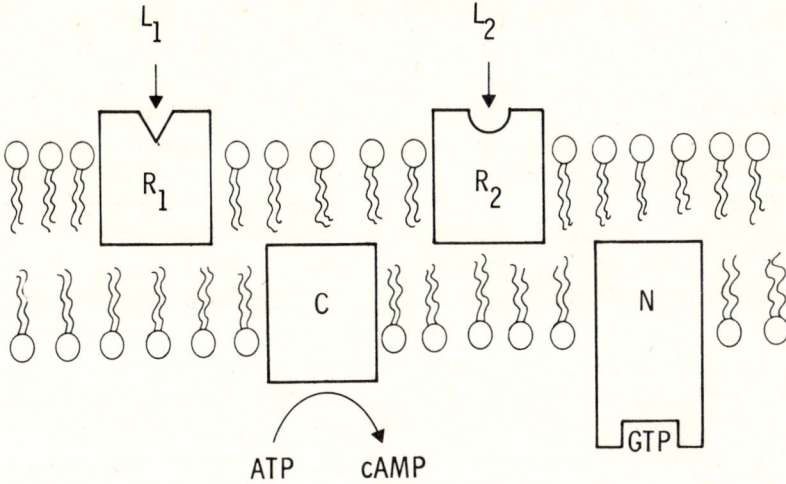

Fig. 3 — Mobile receptor hypothesis applied to receptor-sensitive adenylate cyclase. The plasma membrane is shown together with Receptors R_1, R_2, specific for substances L_1, L_2, the adenylate cyclase enzyme (C) and the guanine nucleotide regulatory protein (N-protein) (N). Receptors, when activated, facilitate the formation of an active N.C complex. All species are free to move in the plane of the membrane and either R_1 or R_2 is capable of leading to activation of cyclase.

model of adenylate cyclase, sometimes called the Mobile Receptor Hypothesis (see Cuatrecasas and Hollenberg, 1976) postulated that separate specific receptors could activate the same cyclase enzyme (Fig. 3) and this was verified in some systems.

A further understanding of receptor-linked adenylate cyclase systems came with the discovery by Rodbell of the requirement of guanine nucleotides, e.g. guanosine triphosphate (GTP) for activity in the system (see Chapters 12 and 13 of this volume). It is now clear that GTP acts via a separate guanine nucleotide regulatory protein (N-protein) accessible to the cytosol (Fig. 3) and the role of the hormone receptor complex is to facilitate formation of an active N-protein (GTP)-cyclase complex (see for example Levitzki and Helmreich, 1979) although other, as yet unidentified proteins may also be involved in this system and interactions with the cytoskeleton may be important. Systems dependent on stimulation of cyclic AMP are relatively clearly understood and have been described for many hormones and neurotransmitters. Other means of transmembrane signalling are available and it may be possible to divide receptors into groups according to their mechanism of action. Broadly speaking two groups emerge: those receptors linked to stimulation of adenylate cyclase and those where activation may lead to a variety of responses dependent on the location of the receptor, i.e. calcium ion gating, inhibition of adenylate cyclase, stimulation of phosphatidylinositol breakdown. Receptors such as the nicotinic acetylcholine receptor and possibly receptors for some amino acid transmitters may form a third group where the receptor is linked quite closely to an ionophore and no other major changes have been detected between receptor and ion flux (see also Chapter 11).

The Fluid Mosaic Model for membrane structure has also focused attention on the lipid part of the membrane. It has become clear that some membrane proteins may be critically dependent on phospholipid for activity and this consideration may apply to some receptors. Secondly, changes in certain membrane phospholipids, e.g. phosphatidylinositol and polyphosphoinositides or phosphatidylethanolamine/phosphatidylcholine have been demonstrated following receptor activation. Although these changes may be a secondary consequence of activation of the membrane proteins more specific functions have been ascribed to these changes and second messenger functions have been considered. The present situation is considered in more detail in Chapter 11.

Receptors have many features in common with enzymes and some useful ideas have emerged from considering the analogy between the two. Ideas about cooperativity in multi-subunit proteins were extended to propose a theory to account for desensitisation in receptors in terms of reversible conformational changes between active and inactive states of receptors (Heidemann and Changeux, 1978). Ideas about the use of binding energy in enzymic reactions have led to the proposal of a theory to explain some aspects of agonism and antagonism. It has been suggested that the intrinsic binding energy of enzyme-substrate interactions is partitioned into catalysis and observed binding (Jencks, 1975). Hence good substrates may often not show high affinities for binding to active sites as intrinsic binding energy has been directed into catalysis. These ideas have been taken up by Franklin (1980) and applied to the case of receptor agonists and antagonists. Whereas agonists do not undergo catalytic transformation at receptor sites they nevertheless may induce a conformational change in the receptor in order to trigger the response. Hence a part of the intrinsic binding energy of agonists may be diverted to driving this conformational change and this may account for the lower apparent affinities of many agonists for receptors compared to good antagonists.

A further aspect of receptors concerns their accessibility. By their very nature receptors must be accessible to the cell exterior and it was pointed out by Lennon and Carnegie (1971) that this might make receptors prime targets as autoantigens in autoimmune diseases. This idea has been verified subsequently for Myasthenia Gravis (see Chapter 7 of this volume) and there is strong suggestive evidence for a receptor-linked autoimmune component in several other diseases, e.g. asthma, diabetes, Graves's Disease.

There has been a dramatic increase in the amount of information obtained about receptors, particularly in the central nervous system, over the past few years. A key factor in this has been the availability of receptor ligands radioactively labelled to a high specific radioactivity and their use in ligand-binding assays. Radioactively labelled ligand (drug, hormone or neurotransmitter) is incubated with the tissue preparation bearing receptors and receptor-bound ligand determined. Owing to the apparent simplicity of the ligand-binding technique, data have been obtained on many systems and work has begun on isolation and characterisation of receptor proteins. Much useful information has been obtained in this way but it must be said that conflicting results have been obtained in some cases and this may be due to the use of slightly different conditions by different laboratories, e.g. varying ionic conditions. In addition it is likely that ligand-binding sites with no physiological relevance have been described using this technique. This raises the important question of when is a ligand-binding site a receptor site. Classically receptors have been defined in functional terms as sites linked to some physiological or pharmacological function where agonists and antagonists interact competitively, and it seems reasonable to emphasise this functional aspect. Thus it is important to establish a physiological or pharmacological function for any ligand-binding site and this should involve the demonstration of agonist and antagonist interaction at the site and the establishment of correlations between potencies for binding of ligands at the sites and ligand potencies in recognised *in vivo* or functional tests for the receptor in question.

This kind of analysis presupposes a certain knowledge of the receptor and its natural ligands and antagonists. The ultimate realisation of Langley's receptor concept came, however, with the discovery of opiate receptors (see Chapter 2) where knowledge of the receptor preceded knowledge of the ligands. The characteristics of the effects of opiates implied the existence of opiate receptors which were subsequently identified using the ligand-binding technique (see Kosterlitz, 1979). In turn the idea of an opiate receptor predicted the existence of endogenous opiates and these were subsequently discovered using opiate receptor test sytems (see Chapter 2). Whether or not further endogenous ligands remain to be identified is a matter for speculation at present (see Chapter 3).

It has also become apparent over the past few years that receptor systems are not fixed entities but may undergo short- and long-term changes in sensitivity. Short-term desensitisation of receptors is a widely observed phenomenon involving rapid loss of sensitivity to a ligand following stimulation by that ligand. The mechanism is not well understood and reversible changes in protein conformation or the loss of an essential cofactor, e.g. GTP, may be involved (see above). By analogy with enzymes, however, it is important to consider the possibility that reversible covalent modification, e.g. phosphorylation of receptors may also be occurring. Longer-term change of sensitivity is observed following prolonged treatment of receptor systems with drugs or agonists (see Chapters 8–10). Both positive and negative regulation have been described and the changes may be influenced by steroid hormones. It is not clear at present whether a general mechanism for such changes exists but it may be associated with the general turnover of receptors in the cell membrane. A model has been proposed to account for the loss of sensitivity (down-regulation) following prolonged stimulation (Pastan and Willingham, 1981) and is summarised in Fig. 4 where receptors on the cell surface cluster over coated pits following binding of ligand, e.g. hormone or neurotransmitter. Clustered receptors are internalised on smooth vesicles called receptosomes which are processed by the Golgi and lysosomes. Some receptors may be recycled but resynthesis also accounts for reappearance of receptors in the cell membrane. An increase in the concentration of ligand activates the internalisation process but does not affect the replacement processes thus leading to a loss of receptors from the surface and an apparent loss of sensitivity. Thus cell surface receptor systems show a degree of plasticity and this may be only a reflection of general homeostatic mechanisms designed to maintain a constant flow of information across the membrane through transmembrane signalling systems.

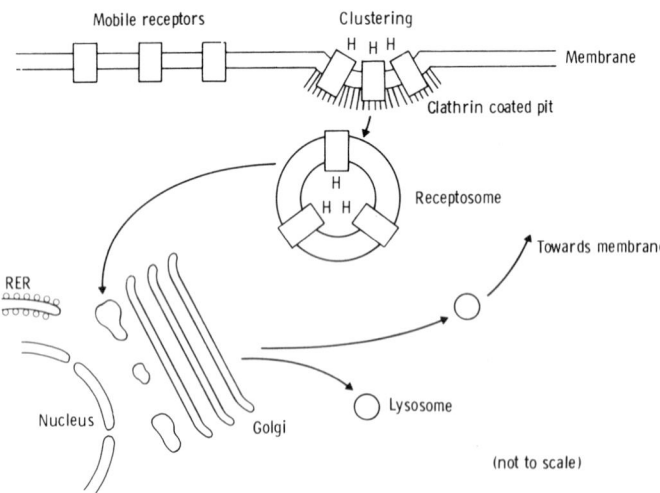

Fig. 4 — Model for internalisation and resynthesis of cell surface receptors. Cell surface receptors bearing hormone or neurotransmitter (H) cluster over clathrin coated pits and are internalised on a smooth vesicle (receptosome). This is processed by the Golgi and the contents may be destroyed in a lysosome although some recycling may occur. In addition new cell surface components synthesised on rough endoplasmic reticulum (RER) are packaged and delivered to the plasma membrane in Golgi vesicles (see Pastan and Willingham, 1981).

These ideas have then led us to our current knowledge of cell surface receptors. Many of the concepts are developed in more detail in the chapters to follow and at the end of the book some ideas will be given about future progress.

REFERENCES

Colquhon, D. (1981) How fast do drugs work? in *Towards understanding receptors,* ed J. W. Lamble. Elsevier/North Holland, pp. 16–27.

Cuatrecasas, P. and Hollenberg, M. D. (1976) Membrane receptors and hormone action. *Adv. Prot. Chem.* **30**, 252–428.

Franklin, T. J. (1980) Binding energy and the activation of hormone receptors. *Biochem. Pharmacol.* **29**, 853–856.

Gerhart, J. C. and Schachman, H. R. (1965) Distinct sub-units for the regulation and catalytic activity of aspartate transcarbamylase. *Biochemistry* **4**, 1054–1062.

Heidemann, T. and Changeux, J. P. (1978) Structural and functional properties of the acetylcholine receptor in purified and membrane-bound states. *Ann. Rev. Biochem.* **47**, 317–357.

Jencks, W. P. (1975) Binding energy, specificity and enzymic catalysis: the Circe effect. *Adv. Enzymol.* **43**, 219–410.

Kosterlitz, H. W. (1979) The best laid plans o' mice an' men gang aft agley. *Ann. Rev. Pharmacol. Toxicol.* **19**, 1–12.

Langley, J. N. (1905) On the reaction of cells and of nerve-endings to certain poisons, chiefly as regards the reaction of striated muscle to nicotine and curari. *J. Physiol.* **33**, 374–413.

Lennon, V. A. and Carnegie, P. R. (1971) Immunopharmacological disease: a break in tolerance to receptor sites. *Lancet,* i, 630–633.

Levitzki, A. and Helmreich, E. J. M. (1979) Hormone receptor-adenylate cyclase interactions. *FEBS Letts.* **101**, 213–219.

Pastan, I. H. and Willingham, M. C. (1981) Journey to the center of the cell: role of the receptosome. *Science,* **214**, 504-509.

Robison, G. A., Butcher, R. W. and Sutherland, E. W. (1971) *Cyclic AMP.* Academic Press, New York.

Singer, S. J. and Nicholson, G. L. (1972) The fluid mosaic model of the structure of cell membranes. *Science,* **175**, 720–731.

1

Selective α_2-adrenoreceptor agonists and antagonists – pharmacology and therapeutic potential

J. C. Doxey, D. S. Walter, C. B. Chapleo, P. W. Dettmar, P. L. Myers, A. G. Roach, C. F. C. Smith and **I. F. Tulloch**

Departments of Pharmacology and Medicinal Chemistry, Reckitt and Colman, Pharmaceutical Division, Dansom Lane, Hull HU8 7DS, U.K.

Abbreviations used
- ADP — adenosine diphosphate
- DOPA — dihydroxyphenylalanine
- EEG — electroencephalogram
- EMG — electromyogram
- MHPG — 3-methoxy, 4-hydroxy-phenylglycol

1. INTRODUCTION

In the early 1970s the accepted concepts of noradrenergic transmission were challenged by a hypothesis which proposed that transmitter release was regulated by prejunctional adrenoreceptors. The hypothesis suggested that α-adrenoreceptors are not only situated on postjunctional membranes but also prejunctionally on sympathetic nerve terminals. It is now widely accepted that activation of these prejunctional receptors by noradrenaline itself, or synthetic agonists such as clonidine, inhibits transmitter release. Conversely blockade of these receptors by antagonists increases transmitter release evoked by nerve stimulation. These prejunctional α-adrenoreceptors therefore regulate the release of noradrenaline through a negative feedback mechanism mediated by the neurotransmitter itself. This regulation of transmitter release can only be shown for Ca^{2+} dependent processes such as those initiated by nerve stimulation or by membrane depolarisation induced by high K^+ or veratridine (Langer, 1981). The release of noradrenaline by tyramine, which is Ca^{2+}-independent, is not influenced by α-adrenoreceptor agonists and antagonists (Langer, 1981). There is also evidence that prejunctional and postjunctional α-adrenoreceptors differ in their sensitivity to agonists (Starke, 1972; Starke *et al.*, 1974; Starke *et al.*, 1975a) and antagonists (Dubocovich and Langer, 1974; Thoenen *et al.*, 1964; Starke *et al.*, 1975a). For this reason Langer in 1974 suggested that the postjunctional α-adrenoreceptors responsible for initiating smooth muscle contraction should be designated with the prefix α_1. The prejunctional α-adrenoreceptors were given the prefix α_2. However, in 1977 Berthelsen and Pettinger reported the existence of postjunctional α_2-adrenoreceptors and therefore suggested that the terms α_1- and α_2- should be applied independently of the location of the receptor sites. Support for this view has been found in the cardiovascular system of the rat where both postjunctional α_2- and α_1-adrenoreceptors occur, each of which causes vasoconstriction when activated by appropriate agonists (McGrath, 1982).

At present therefore the subclassification of α-adrenoreceptors into α_1 and α_2-subtypes is based solely on the relative affinity of agonists and antagonists for the receptors. The terms prejunctional and postjunctional (or presynaptic and postsynaptic) only refer to the anatomical location of the receptors within the nerve terminal (Starke and Langer, 1979).

2. DISTRIBUTION OF α_1- AND α_2-ADRENORECEPTORS

Prejunctional α-adrenoreceptors located on sympathetic nerve terminals were the original examples of α_2-adrenoreceptors. There is, however, increasing evidence that α_2-adrenoreceptors are widely distributed throughout the body and that their location is not limited to sympathetic nerve terminals. Recently, for example, they have been demonstrated both at postjunctional sites and on non-neuronal structures, e.g. platelets (Grant and Scrutton, 1979). Furthermore there is evidence that prejunctional α_2-adrenoreceptors may not be the same in all tissues (Doxey and Everitt, 1977; Starke, 1977; Roach *et al.*, 1978; Dubocovich, 1979). Parasympathetic nerve terminals also possess α_2-adrenoreceptors whose activation inhibits the release of acetylcholine (Wikberg, 1978; Grundstrom *et al.*, 1981). α_1-Adrenoreceptors have been shown to be present at postjunctional sites (Starke, 1981). Non-neuronal α_1-adrenoreceptors have also been reported to exist on platelets but it is not known whether these platelet receptors have a physiological role, (Grant and Scrutton, 1979). On the basis of present knowledge there is little evidence to support the existence of prejunctionally located α_1-adrenoreceptors in the periphery.

The distribution and function of α_1- and α_2-adrenoreceptors are shown in Table 1. α-Adrenoreceptors have also been demonstrated at other sites but the subtype involved is less clearly defined. For example spinal cord α-adrenoreceptors have been implicated in cardiovascular control (Connor *et al.*, 1981) and in the noradrenergic regulation of pain transmission (Kuraishi *et al.*, 1979). In both instances it is not clear which subtype of α-adrenoreceptor is involved although Connor *et al* (1981) suggested that the spinal receptors involved in

Table 1 – Distribution and function of α_1- and α_2-adrenoreceptors.

Location	Activation response	References
α_1-Adrenoreceptors		
Central nervous system	Excitation	Pichler and Kobinger, 1981
Heart	Hypotension, bradycardia	Cavero and Roach, 1978
	Positive inotropism	Skomedal et al., 1980
Liver	Ureogenesis	Corvera and Garcia-Sainz, 1981
	Glycogenolysis	Fain and Garcia-Sainz, 1980
Smooth muscle	Contraction	McGrath, 1982
α_2-Adrenoreceptors		
Central nervous system	Decreased noradrenaline turnover	Andén et al., 1976
	Decreased MHPG levels	Sugrue, 1980
	Hypotension, bradycardia and sedation	Timmermans et al., 1981
	Increased Growth Hormone secretion	Checkley et al., 1981a
Cell bodies on central noradrenergic neurones	Hyperpolarisation	Aghajanian and Vander Maelen, 1981
Sympathetic ganglia	Hyperpolarisation	Brown and Caulfield, 1979
Peripheral sympathetic nerve terminals	Reduced noradrenaline release	Langer, 1977; Starke, 1977
Peripheral parasympathetic nerve terminals	Reduced acetylcholine release	Wikberg, 1978
Vascular smooth muscle	Contraction	McGrath, 1982
Platelets	Aggregation	Grant and Scrutton, 1979
Fat cells	Inhibition of lipolysis	Aktories et al., 1980
Pancreatic islets	Inhibition of insulin secretion	Nakaki et al., 1980
Kidney	Increased Na^+ reabsorption in proximal tubules	Scott-Young and Kuhar, 1980
	Inhibition of renin release	Pettinger et al., 1976

cardiovascular regulation may be different from peripheral α_1- and α_2-adrenoreceptors.

3. THERAPEUTIC POTENTIAL OF α_2-ADRENORECEPTOR AGONISTS

Despite the widespread distribution of α_2-adrenoreceptors throughout the body, and the many possible physiological consequences of their activation, the clinical application of selective α_2-adrenoreceptor agonists such as clonidine has been mainly restricted to the treatment of hypertension. A number of selective α_2-adrenoreceptor agonists have been introduced into the clinic and all appear to possess a similar pharmacological profile. Compounds such as clonidine and α-methyl DOPA (via its active metabolite α-methyl noradrenaline) reduce blood pressure and heart rate via activation of central α_2-adrenoreceptors; unfortunately these beneficial actions are often accompanied by unwanted α_2- effects such as sedation and dry mouth. Furthermore abrupt withdrawal of clonidine from hypertensive patients can cause a rapid and potentially dangerous rise in blood pressure towards or above pre-treatment levels. This increase in blood pressure is accompanied by a rise in plasma noradrenaline and other symptoms such as headache and anxiety (Reid et al., 1977).

Low doses of clonidine have also been used in migraine prophylaxis (for review see Brogden et al., 1975). Clarke (1976) demonstrated a beneficial prophylactic effect with clonidine but no consistent benefit was obtained when the drug was administered during migraine attacks. Clonidine has also been reported to produce a rapid attenuation of opiate withdrawal symptoms (Gold et al., 1980). These workers originally tested the effects of clonidine in opiate withdrawal in the locus coeruleus of monkeys. Electrical or pharmacological activation of this nucleus produced changes which resembled those in opiate withdrawal. Morphine and clonidine blocked these effects and this led to the suggestion that opiate withdrawal may be due partly to increased noradrenergic activity in areas such as the locus coeruleus. The similarity between the hyperpolarising effects of opiates and clonidine on locus coeruleus neurones (Aghajanian and Vander Maelen, 1981) may serve to explain the anti-withdrawal properties of clonidine: administration of clonidine effectively prevents the disinhibition of the locus coeruleus neurones that occurs on withdrawal of chronic opiate treatment. However, other workers have suggested that clonidine suppresses opiate withdrawal by an action on spinal sympathetic neurones (Franz et al., 1982).

Recent animal studies have also indicated that other therapeutic indications for clonidine-like agents should be considered. For instance clonidine has been shown to have antinociceptive effects which are equivalent to those of morphine (Fielding et al., 1978) but which are unaffected by naloxone. However, it has been suggested that a major portion of clonidine-induced antinociception appears to be at the spinal level with no input from supraspinal structures (Spaulding et. al., 1979). Furthermore studies in man have failed to demonstrate analgesic properties with clonidine on psychophysical pain (Uhde et al., 1980). Selective α_2-adrenoreceptor agonists also reduce intraocular pressure in the conscious rabbit and it has been suggested that this class of compound could represent a new group of anti-glaucoma agents (Innemee et al., 1981). It has also been reported that clonidine delays small intestinal transit in the rat. This effect appeared to be mediated via α_2-adrenoreceptors since it was blocked by yohimbine and phentolamine but not by prazosin (Ruwart et al., 1979). Other workers (Lal et al., 1981) have suggested that α_2-adrenoreceptor agonists such as clonidine and lofexidine are potential antidiarrhoal drugs of a non-narcotic nature.

Although clonidine has been an invaluable tool in investigating the pharmacological role of α_2-adrenoreceptors it is now apparent that this compound has limitations. It has been shown that clonidine is a partial agonist on the prejunctional α_2-adrenoreceptors of guinea-pig atrium (Medgett et al., 1978) and dog saphenous vein (Drew and Sullivan, 1980) and on non-neuronal platelet α_2-adrenoreceptors (Grant and Scrutton, 1979). In the guinea-pig ileum the maximum inhibitory effect of clonidine is less than that produced by noradrenaline (Wikberg, 1978); this could be due to partial agonist properties. The partial agonism of clonidine on the cholinergic nerves of the guinea-pig ileum is confirmed in Fig. 1. The maximum inhibitory effect of clonidine on the coaxially stimulated guinea-pig ileum was markedly less than that produced by noradrenaline. Furthermore, when the inhibitory effect of clonidine was removed by washing the tissue with fresh physiological solution the residual antagonist properties of clonidine were apparent since the sensitivity of the tissue to noradrenaline was reduced. In addition, in the rat perfused mesenteric artery (Chapleo et al., 1981a) and rat perfused hindquarters (Kobinger and Pichler, 1981) clonidine was devoid of agonist activity but antagonised α_1-mediated vasoconstriction.

Although clonidine acts preferentially on α_2-adrenoreceptors its selectivity is poor when compared with more recently reported compounds (Table 2). By using the rat vas deferens (α_2) and the rat anococcygeus muscle (α_1) it is apparent that compounds such as UK-14,304, B-HT 933 and B-HT 920 are extremely selective for prejunctional α_2-adrenoreceptors. Although clonidine will potentiate the platelet aggregation induced by ADP it antagonises that produced by the natural agonist, adrenaline (Grant and Scrutton, 1979). In contrast UK-14,304 can induce aggregation of the platelets of rabbits and man suggesting that this compound is a full agonist at platelet α_2-adrenoreceptors (Grant and Scrutton, 1980). These more recently introduced agonists may be more useful tools both in investigating α_2-adrenoreceptor mechanisms and in defining new therapeutic targets for compounds which interact with α_2-adrenoreceptors. However, clinical utility for α_2-adrenoreceptor agonists will only be expanded if side-effects such as sedation can be minimised either by identifying compounds which interact with specific α_2-adrenoreceptor subtypes or by restricting the agonist to peripheral sites of action.

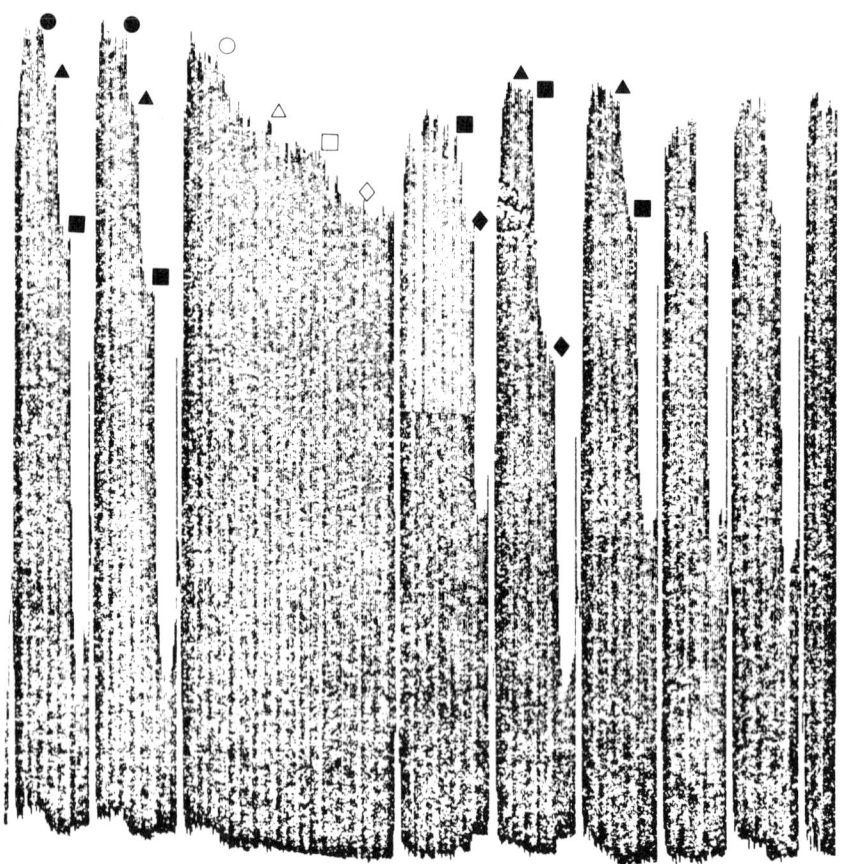

Fig. 1 — Isolated guinea-pig ileum. The inhibitory effects of noradrenaline and clonidine on the twitch responses of the transmurally stimulated guinea-pig ileum (0.1 Hz, 3 ms, 80 v). The concentrations of noradrenaline were 75 nM (●), 225 nM (▲), 675 nM (■) and 2250 nM (♦); the concentrations of clonidine were 7 nM (○), 22 nM (△) 74 nM (□) and 220 nM (◇). The tissue was washed after each concentration-response schedule. (±)-Propranolol (1.0 μM) and prazosin (70 nM) were present in the Krebs solution throughout the experiment.

Table 2 — The effects of agonists on prejunctional α_2-adrenoreceptors of the rat vas deferens and postjunctional α_1-adrenoreceptors of the rat anococcygeus muscle (for methods see Doxey *et al.*, 1981). Results are expressed as the negative logarithm of the molar concentration of agonists producing 50% maximum effect (pD$_2$ value). The selectivity ratio is the antilogarithm of the difference between the pD$_2$ values at pre-and post-junctional sites. The results are the mean of a minimum of 5 experiments.

Agonist	Postjunctional (α_2) potency pD$_2$ vs clonidine (vas deferens)	Postjunctional (α_1) potency pD$_2$ vs noradrenaline (rat anococcygeus)	Ratio α_2/α_1
clonidine	8.97	7.31	46
guanfacin	8.64	6.80	69
guanoxabenz	7.37	5.23	138
UK-14,304	9.13	5.95	1,514
B-HT 933	7.08	<4.40	> 479
B-HT 920	8.92	<4.98	>9,333

4. SELECTIVE α_2-ADRENORECEPTOR ANTAGONISTS AND THEIR THERAPEUTIC POTENTIAL

4.1 Selective α_2-adrenoreceptor antagonists

Therapeutic applications for selective α_2-adrenoreceptor antagonists are relatively unexplored due to the lack of suitable drugs. Although yohimbine and rauwolscine (Fig. 2) preferentially block α_2-adrenoreceptors (Tanaka et al., 1978) and have been extensively used in animal experiments, both have affinity for other receptor systems (Grant and Scrutton, 1979; Scatton et al., 1980; Waldmeir and Bischoff, 1981; Dedek et al., 1981). Selective α_2-adrenoreceptor antagonist properties have recently been described in two series of benzodioxans (Michel and Whiting, 1981; Chapleo et al., 1981b) and in a series of substituted benzoquinolizines (Lattimer et al., 1982); these compounds (Fig. 2) may offer advantages in evaluating the therapeutic potential of this class of compound. The pharmacological profile of a selective α_2-adrenoreceptor antagonist, as exemplified by 2-(2-(1,4-benzodioxanyl))-2-imidazoline HCl (RX 781094, Chapleo et al., 1981b), is described below.

In isolated tissue experiments RX 781904 had a pA_2 value of 8.56 against clonidine on the prejunctional α_2-adrenoreceptors of the rat vas deferens. Although RX 781094 also competitively antagonised the effects of noradrenaline on the postjunctional α_1-adrenoreceptors of the rat anococcygeus muscle its affinity (pA_2=6.10) for these receptors was much lower than for α_2-adrenoreceptors (Chapleo et al., 1981b). In these studies RX 781094 was more potent and more selective for α_2-adrenoreceptors than yohimbine. RX 781094 had extremely low affinity for other receptor systems and was only a very weak inhibitor of the uptake of biogenic amines. RX 781094 has also been shown to be an antagonist at α_2-adrenoreceptors on human platelets (Kerry and Scrutton, 1982).

The potency of RX 781094 at peripheral prejunctional α_2-adrenoreceptors has also been established *in vivo* in pithed rats (Chapleo et al., 1981b). Contractions of the vas deferens induced by electrical stimulation of the spinal sympathetic outflow were abolished by clonidine (100 µg/kg, i.v.). The cumulative intravenous doses of RX 781094 and yohimbine required to produce 50% reversal of the inhibitory effects of clonidine were 24 and 670 µg/kg respectively. Prazosin was inactive in doses up to 4.3 mg/kg, i.v. Pithed rats were also used to study the effects of RX 781094 on post-junctional α_2-adrenoreceptors (Berridge et al., 1982a). In the cardiovascular system of the rat the pressor responses induced by UK-14,304 and cirazoline are the result of the stimulation of postjunctional α_2- and α_1-adrenoreceptors respectively. RX 781094 competitively antagonised the pressor responses induced by UK-14,304 without affecting cirazoline responses (Fig. 3). In contrast, prazosin failed to displace the dose-response curve to UK-14,304 at doses which markedly antagonised cirazoline (Fig. 3). These results are consistent with the hypothesis that both postjunctional α_2- and α_1-adrenoreceptors exist in the cardiovascular system of the

α_2-ADRENOCEPTOR ANTAGONISTS

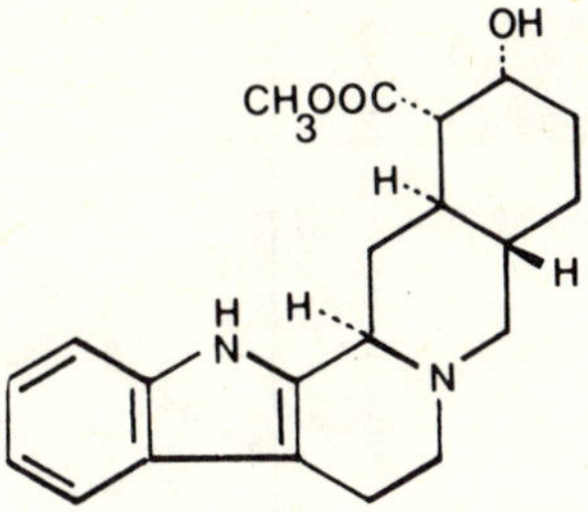

Yohimbine

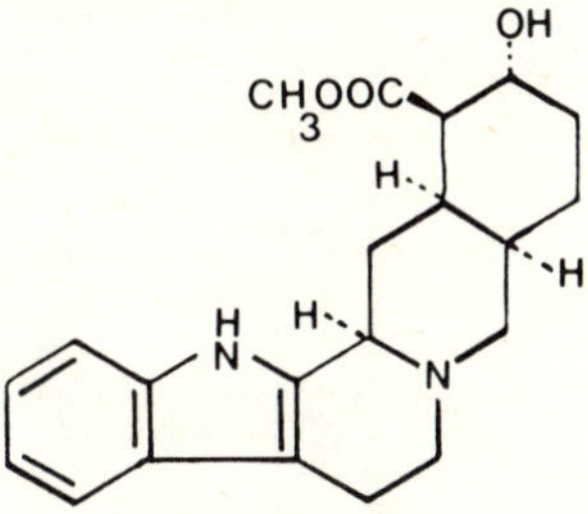

Rauwolscine

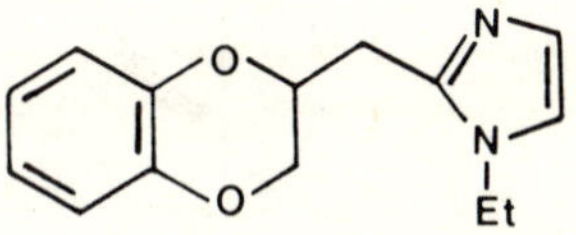

RS 21361

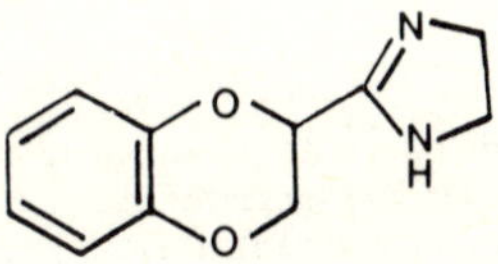

RX 781094

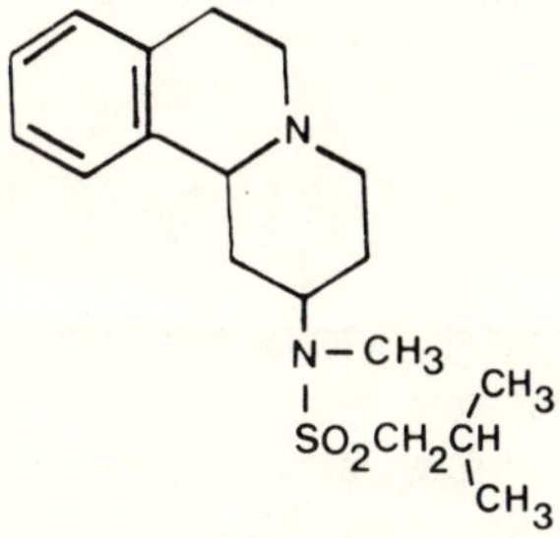

WY 26703

Fig. 2 — Chemical structures of the established α_2-adrenoreceptor antagonists, yohimbine and rauwolscine and of RS 21361 (Michel and Whiting, 1981), RX 781094 (Chapleo et al., 1981b) and WY 26703 (Lattimer et al., 1982).

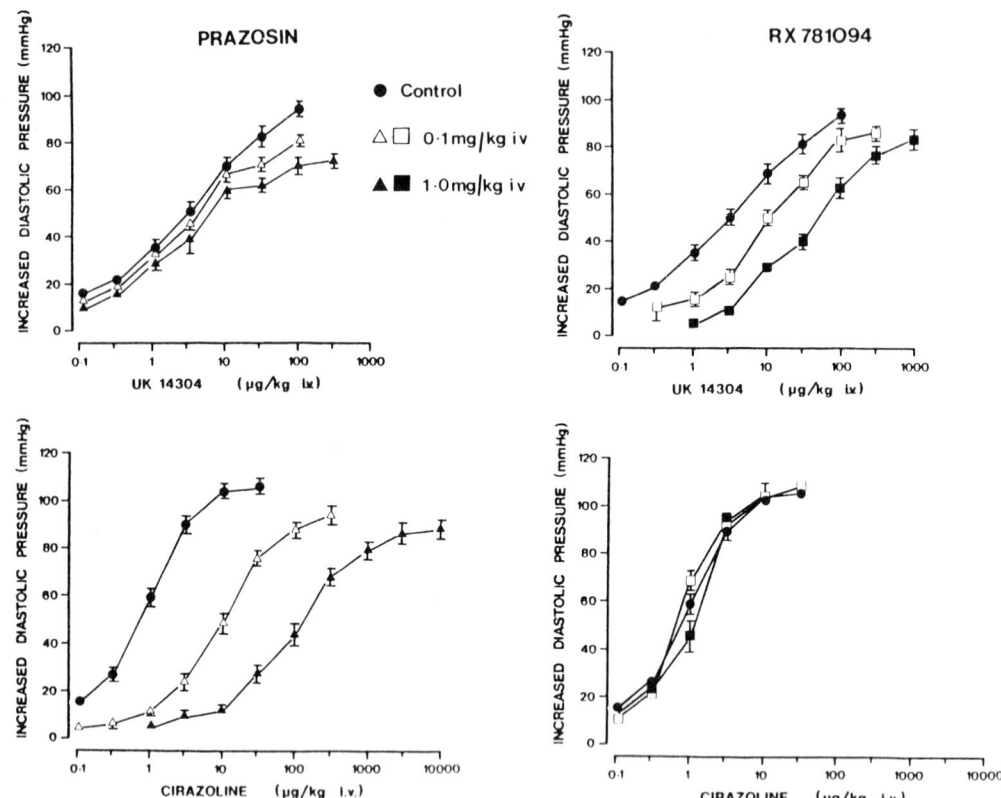

Fig. 3 – Pressor responses in pithed rats. The effects of saline (0.1 ml/kg, i.v., controls), RX 781094 and prazosin on the increases in diastolic blood pressure induced by UK 14,304 and cirazoline in pithed rats. Dose-response curves to the agonists were constructed 5 min after the administration of either antagonist or saline. All rats were vagotomised and received atropine (1.0 mg/kg, i.v.) and dl-propranolol (1.0 mg/kg, i.v.). The results are expressed as the mean of 6–8 experiments ± SEM.

rat. Furthermore, they indicate that the selectivity of RX 781094 for α_2-adrenoreceptors extends to postjunctional sites.

Radioligand binding studies using rat cerebral membranes showed that RX 781094 was about seven times more selective than yohimbine (Howlett et al., 1982). Although both compounds had similar affinities for the α_1- binding site labelled by [^{3}H] prazosin (Ki values of 547 and 544 nM for RX 781094 and yohimbine, respectively), RX 781094 had a higher affinity than yohimbine for the α_2-binding site labelled by [^{3}H]-clonidine (Ki values of 5.8 and 39 nM respectively).

The central α_2-adrenoreceptor antagonist properties of RX 781094 were confirmed in functional studies. Occupation of central α_2- adrenoreceptors by clonidine produces profound effects which manifest as hypotension and sedation (Timmermans et al., 1981). Following intravenous administration RX 781094 was approximately 3–10 times more potent than yohimbine and rauwolscine in antagonising the central cardiovascular effects of clonidine (Berridge et al., 1982b). Dettmar et al., (1981) have shown that the hypothermia, behavioural sedation and EEG synchronisation induced by clonidine are also antagonised by RX 781094 (Fig. 4). It is clear from these studies that selective α_2-adrenoreceptor antagonists block the profound pharmacological effects of α_2-adrenoreceptor agonists.

Although it is less easy to demonstrate the effects of the antagonists themselves in functional studies, biochemical changes can be readily demonstrated. In the mouse isolated vas deferens RX 781094 significantly increased the release of [^{3}H]-noradrenaline evoked by nerve stimulation (Baker and Marshall, 1982). This only occurred following trains of 10 pulses; the release evoked by single pulses was unchanged. Thus RX 781094 revealed a feedback of noradrenaline onto prejunctional α_2-adrenoreceptors in this tissue. The effects of RX 781094 on cortical noradrenaline, MHPG and striatal dopamine has been studied in rats (Flockhart et al., 1982). The turnover of catecholamines was estimated either by measuring the rate of decline of noradrenaline and dopamine after treatment with α-methyl p-tyrosine or by measuring the accumulation of the noradrenaline metabolite MHPG. RX 781094 (10–40 mg/kg, p.o.) produced a dose-related increase in the apparent rate of turnover of noradrenaline in the rat cortex (Fig. 5); the rate of turnover of dopamine in the striatum of the same rats was unchanged. Consistent with these results RX 781094 (20 mg/kg, p.o.) also produced an 100% ($P<0.01$) increase in the cortical concentration of MHPG. Relatively high oral doses of RX 781094 were needed in these studies because of a marked first pass metabolism of RX 781094 in the rat. A later communication (Dedek et al., 1982) has shown that RX 781094 produced essentially the same results following intraperitoneal injection (1–10 mg/kg). In a-methyl p-tyrosine treated rats yohimbine increases both noradrenaline and dopamine turnover (Hedler et al., 1981) and these workers have suggested that the activity of dopamine neurons is controlled by α_2-adrenoreceptors. However, the lack of effect of RX 781094 on dopamine turnover is not consistent with this suggestion. The effects produced by yohimbine are probably a reflection of a combination of α_2-adrenoreceptor and dopamine antagonist properties (Scatton et al., 1980; Dedek et al., 1982).

These results characterise RX 781094 as a potent and highly selective competitive antagonist at peripheral and central α_2-adrenoreceptors. The therapeutic potential of this compound is at present being assessed.

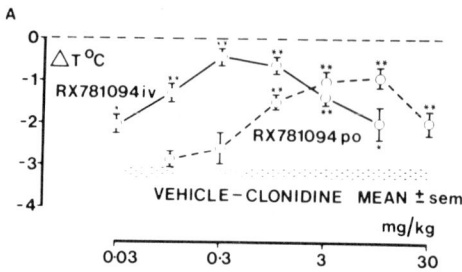
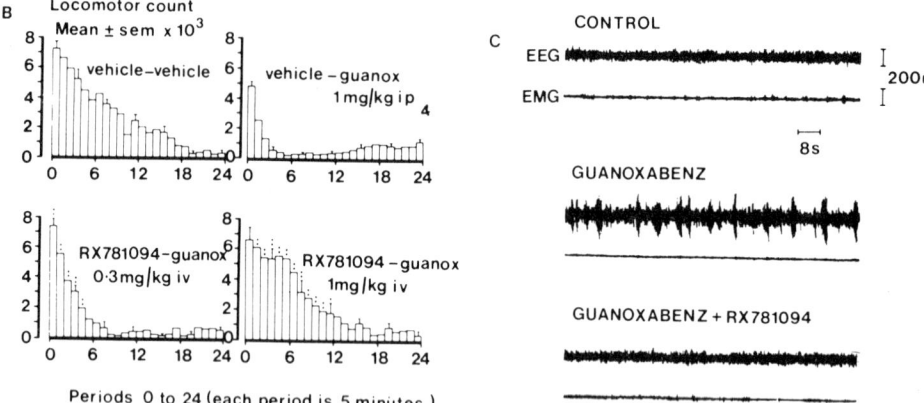

Fig. 4 — Antagonism by RX 781094 of CNS effects induced by selective α_2-adrenoreceptor agonists. (A) Hypothermia in mice. Antagonism of clonidine hypothermia was determined in groups of 8 mice by measuring oesophageal temperature prior to and at 15 min intervals after RX 781094 treatment. RX 781094 or vehicle was administered either p.o. or i.v. 15 min before clonidine (0.1 mg/kg, i.p.). The temperatures shown were measured 15 min after clonidine injection. All experiments were performed in a temperature-controlled room at 21 ± 1 °C. (B) Behavioural hypoactivity in mice. Antagonism of the profound hypoactivity induced by the selective α_2-adrenoreceptor agonist guanoxabenz was assessed by recording the activity of groups of 10 mice in cages designed to monitor locomotor activity (horizontal movement). RX 781094 or vehicle was injected i.v. 15 min before guanoxabenz (1.0 mg/kg, i.p.). (C) EEG synchronisation in rats. Drug effects on EEG activity were measured in conscious, unrestrained rats (Sprague-Dawley, 350–400 g) equipped with screw electrodes over the frontal cortex; muscle tone (EMG) was recorded from the temporalis muscles. The upper EEG trace shows desynchronised (low amplitude, higher frequency) EEG during normal awake behaviour; the middle EEG trace illustrates synchronised activity (high amplitude, lower frequency) seen 20 min after guanoxabenz administration (0.5 mg/kg, i.p.); the bottom EEG trace shows the complete reversal of guanoxabenz-induced synchronised activity by i.v. injection of RX 781094 (1.0 mg/kg). RX 781094 was administered at the time of peak agonist effect (30 min) and this trace was recorded 30 min after the injection of antagonist. (* $P < 0.05$; ** $P < 0.01$, different from vehicle-treated control (Dunnett's test).)

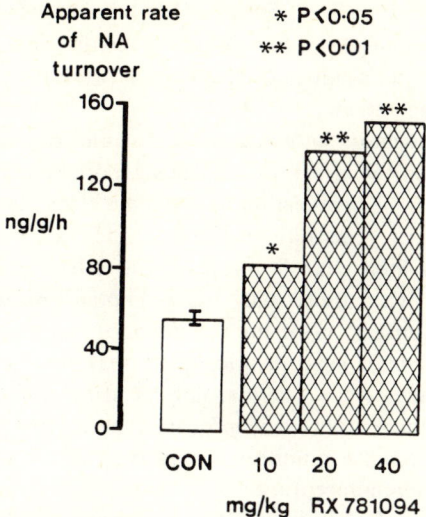

Fig. 5 — Apparent rate of turnover of noradrenaline in rat cerebral cortex. RX 781094, 10–40 mg/kg, p.o., or distilled water vehicle was dosed immediately before α-methyl-p-tyrosine methyl ester HCl, 320 mg/kg, i.p. Cortical noradrenaline concentrations were measured in groups of 4 rats every 15 min for 2 hours, from which the rates of decline (± SEM) of the noradrenaline concentration were obtained. The apparent rates of turnover were calculated from the products of the first order rates of decline and the initial noradrenaline concentrations. Control rates were compared with rates following each dose of RX 781094 using a statistical routine as incorporated in the SAS 'General linear models procedure' (*Statistical Analysis System Users Guide*, 1979, S.A.S. Institute Inc., P.O. Box 10066, Raleigh, North Carolina, 27605) from which significance levels were obtained. The control shows the mean and SEM of the three control experiments.

4.2 Central α_2-adrenoreceptors and antidepressant therapy

It has been suggested that selective α_2-adrenoreceptor antagonists may have therapeutic potential in the treatment of depression (Langer, 1978). The rationale for such clinical utility arises from the hypothesis that depressive illness is associated with a reduced availability of noradrenaline within regions of the brain (Schildkraut, 1965). Checkley *et al.* (1981a) reported that depressed patients have a specific defect in the central α-adrenoreceptors involved in the release of growth hormone. More recently, it has been postulated that supersensitive central α_2-adrenoreceptors play in important role in the etiology and maintenance of affective illness (Cohen *et al.*, 1980). An increased responsiveness of α_2-adrenoreceptors would be predicted to enhance feedback inhibition of noradrenaline release and thus reduce the synaptic concentration of noradrenaline. If this supersensitivity is a factor in the etiology of depression then compensatory antagonism of these receptors may provide an effective and novel antidepressant treatment. However, in studies in which platelet α_2-adrenoreceptors have been used as an index of receptor sensitivity contradictory results have been found. Although Garcia-Sevilla *et al.*. (1981a) found evidence of supersensitive platelet α_2-adrenoreceptors in depressed patients, other workers (Daiguji *et al.*, 1981) have found no significant difference between the platelet receptors of healthy volunteers and patients.

A number of investigators have suggested that the tricyclic antidepressant drugs produce their clinical effect by enhancing noradrenergic transmission. However, there is a temporal dissociation between the blockade of noradrenaline re-uptake produced by tricyclic antidepressants, which occurs within hours of ingestion of drugs, and their clinical antidepressant effect which usually takes 2–3 weeks to develop. This discrepancy has prompted investigations into the effects of chronic administration of uptake inhibitors on receptor sensitivity. Studies in animals and man have demonstrated that chronic treatment with uptake inhibitors attenuates the pharmacological effects of clonidine. These studies, together with more direct evidence, have been cited as evidence that chronic treatment with uptake inhibitors is associated with a gradually developing subsensitivity of α_2-adrenoreceptors.

In 1978, Crews and Smith reported a subsensitivity of presynaptic α_2-adrenoreceptors (on sympathetic nerves) in arterial tissue taken from rats treated with desipramine for 3 weeks. Other workers have shown that the ability of clonidine to produce either hypothermia in mice (Von Voigtlander *et al.*, 1978), behavioural depression in rats (Spyraki and Fibiger, 1980) or inhibition of locus coeruleus neurone firing rate in rats (Svensson and Usdin, 1978) is reduced following chronic but not acute treatment with noradrenaline uptake inhibitors. Furthermore, the number of α_2-adrenoreceptor binding sites in rat brain was decreased following two weeks' treatment with amitriptyline (Smith *et al.*, 1981). A number of studies have also shown that the hypotensive and sedative effects of clonidine are reduced in depressed patients following long-term treatment with desipramine (Charney, 1981; Checkley *et al.*, 1981b; Siever *et al.*, 1981). In addition α_2-adrenoreceptors on the platelets of depressed patients are decreased in number after chronic treatment with imipramine (Garcia-Sevilla *et al.*, 1981b).

Crews and Smith (1980) suggested that the delay in onset of antidepressant activity of tricyclic antidepressants may be due to the fact that the acute effects of uptake inhibition on the synaptic concentration of noradrenaline are counteracted by prejunctional α_2-adrenoreceptor activation which reduces the release of noradrenaline. It is only after the prejunctional adrenoreceptors have become subsensitive that noradrenaline accumulates in the synaptic cleft. Consistent with this view, chronic treatment of rats with desipramine is associated with an increase in both noradrenaline turnover and MHPG levels in the brain. In contrast acute administration of desipramine has no effect or reduces the turnover of noradrenaline and decreases MHPG levels (Sugrue, 1980). It is possible that the increase in noradrenaline turnover is responsible for the subsensitivity of cortical β-adrenoreceptors which occurs following chronic administration of most antidepressants (Sulser *et al.*, 1978). This β-adrenoreceptor down-regulation is accelerated when α_2-adrenoreceptor antagonists are co-administered with desipramine (Paul and Crews, 1980; Johnson *et al.*, 1980). Crews *et al.* (1981) have shown that the administration of either a tricyclic antidepressant (desipramine) or a monoamine oxidase inhibitor (tranylcypramine) in combination with phenoxybenzamine accelerates and intensifies the desensitisation of β-adrenoreceptors. On the basis of animal experiments Crews *et al.* (1981) suggested that the combination of an antidepressant with a presynaptic α_2-adrenoreceptor antagonist may provide a therapy with a rapid onset of action. Furthermore, the enhanced reduction in β-adrenoreceptor density produced by this combination, compared with antidepressant alone, indicates that co-administration of these two agents may be more efficacious and produce antidepressant effects in previously unresponsive patients. Whether selective α_2-adrenoreceptor antagonists are themselves antidepressant will be the subject of the initial clinical investigation of RX 781094. It should be stressed, however, that α_2-adrenoreceptors may subserve other physiological roles and that elucidation of these may lead to a wider utility for selective α_2-adrenoreceptor antagonists in a range of pathological states.

The authors wish to thank Pamela Parker for typing the manuscript.

REFERENCES

Aghajanian, G. K. and Vander Maelen, C. P. (1981) α_2-Adrenoreceptor-mediated hyperpolarisation of locus coeruleus neurons: Intracellular studies in vivo. *Science* **215**, 1394–1396.

Aktories, K., Schultz, G. and Jakobs, K. H. (1980) Regulation of adenylate cyclase activity in hamster adipocytes. *Naunyn-Schmiedeberg's Arch. Pharmac.* **312**, 167–173.

Andén, N-E, Grabowska, M. and Strombom, U. (1976) Different alpha-adrenoreceptors in the central nervous system mediating biochemical and functional effects of clonidine and receptor blocking agents. *Naunyn-Schmiedeberg's Arch. Pharmac.* **292**, 43–52.

Baker, S. and Marshall, I. (1982) The effect of RX 781094, a selective α_2-adrenoreceptor antagonist on ^{3}H-noradrenaline release in the mouse vas deferens. *Br. J. Pharmac.* **76**, Proc. Suppl. 212P.

Berridge, T. L., Doxey, J. C., Roach, A. G. and Strachan, D. A. (1982a) Comparison of the effects of RX 781094 and prazosin on the pressor responses to various α-adrenoceptor agonists. *Br. J. Pharmac.* **75**, Proc. Suppl. 140P.

Berridge, T. L., Doxey, J. C. and Roach, A. G. (1982b) Antagonism of clonidine-induced hypotension and bradycardia by RX 781094 and other α-adrenoceptor antagonists in the rat. *Br. J. Pharmac.* **75**, Proc. Suppl. 139P.

Berthelsen, S. and Pettinger, W. A. (1977) A functional basis for classification of α-adrenergic receptors. *Life Sci.* **21**, 595–606.

Brogden, R. N., Pinder, R. M., Sawyer, P. R., Speight, T. M. and Avery, G. S. (1975) Low-dose clonidine: A review of its therapeutic efficacy in migraine prophylaxis. *Drugs* **10**, 357–365.

Brown, D. A. and Caulfield, M. P. (1979) Hyperpolarising α_2-adrenoreceptors in rat sympathetic ganglia. *Br. J. Pharmac.* **65**, 435–445.

Cavero, I. and Roach, A. G. (1978) The effects of prazosin on the clonidine-induced hypotension and bradycardia in rats and sedation in chicks. *Br. J. Pharmac.* **62**, 468–469P.

Chapleo, C. B., Doxey, J. C., Myers, P. L., Roach, A. G. and Smith, S. E. (1981a) Clonidine – influence of aromatic substitution on α-adrenoceptor selectivity. *Br. J. Pharmac.* **73**, 280P.

Chapleo, C. B., Doxey, J. C., Myers, P. L. and Roach, A. G. (1981b) RX 781094, a new potent, selective antagonist of α_2-adrenoceptors. *Br. J. Pharmac.* **74**, 842P.

Charney, D. S. (1981) Presynaptic adrenergic receptor sensitivity in depression: The effect of long-term desipramine treatment. *Arch. Gen. Psychiat.* **38**, 1334–40.

Checkley, S. A., Slade, A. P. and Shur, E. (1981a) Growth hormone and other responses to clonidine in patients with endogenous depression. *Br. J. Psychiat.* **138**, 51–55.

Checkley, S. A., Slade, A. P., Shur, E. and Dawling, S. (1981b) A pilot study of the mechanism of action of desipramine. *Br. J. Psychiat.* **138**, 248–251.

Clarke, G. J. R. (1976) Use of clonidine hydrochloride (Dixarit) in migraine. A trial in general practice. *Clin. Trials J.* **13**, 137–41.

Cohen, R. M., Campbell, I. C., Cohen, M. R., Torda, T., Pickar, D., Siever, L. and Murphy, D. L. (1980) Presynaptic noradrenergic regulation during depression and antidepressant drug treatment. *Psychiat. Res.* **3**, 93–105.

Connor, H. E., Drew, G. M., Finch, L. and Hicks, P. E. (1981) Pharmacological characteristics of spinal α-adrenoreceptors in rats. *J. Aut. Pharmac.* **1**, 149–156.

Corvera, S. and Garcia-Sainz, J. A. (1981) α_1-Adrenoceptor activation stimulates ureogenesis in rat hepatocytes. *Eur J. Pharmac.* **72**, 387–390.

Crews, F. T. and Smith, C. B. (1978) Presynaptic alpha-receptor subsensitivity after long-term antidepressant treatment. *Science* **202**, 322–324.

Crews, F. T. and Smith, C. B. (1980) Potentiation of responses to adrenergic nerve stimulation in isolated rat atria during chronic tricyclic antidepressant administration. *J. Pharm. Exp. Ther.* **215**, 143–149.

Crews, F. T., Paul, S. M. and Goodwin, F. K. (1981) Acceleration of β-receptor desensitisation in combined administration of antidepressants and phenoxybenzamine. *Nature* **290**, 787–789.

Daiguji, M., Meltzer, H. Y., Tong, C., U'Prichard, D. C., Young, M. and Kravitz, H. (1981) α_2-Adrenergic receptors in platelet membranes of depressed patients: no change in number or ^{3}H-yohimbine affinity. *Life Sci.* **29**, 2059–2064.

Dedek, J., Scatton, B. and Zivcovic, B. (1982) α_2-Receptors are not involved in

the regulation of striatal dopaminergic transmission. *Br. J. Pharmac.* **77**, Proc. Suppl. 361P.
Dettmar, P. W., Lynn, A. G. and Tulloch, I. F. (1981) Neuropharmacological evaluation of RX 781094, a new selective α_2-adrenoceptor antagonist. *Br. J. Pharmac.* **74**, 843–844P.
Doxey, J. C. and Everitt, J. (1977) Inhibitory effects of clonidine on responses to sympathetic nerve simulation in the pithed rat. *Br. J. Pharmac.* **61**, 559–566.
Doxey, J. C., Frank, L. W. and Hersom, A. S. (1981) Studies in the pre-and post-junctional activities of α-adrenoreceptor agonists and their cardiovascular effects in the anaesthetised rat. *J. Aut. Pharmac.* **1**, 157–169.
Drew, G. M. and Sullivan, A. T. (1980) Effect of α-adrenoceptor agonists and antagonists on adrenergic neurotransmitter overflow from dog isolated saphenous veins. *Br. J. Pharmac.* **68**, 139–140P.
Dubocovich, M. L. (1979) Pharmacological differences between the alpha-presynaptic adrenoceptors in the peripheral and central nervous system. In *Presynaptic Receptors, Advances in the Biosciences,* Vol. 18, pp. 29–26, eds. Langer, S. Z., Starke, K. and Dubocovich, M. L. Pergamon Press, Oxford.
Dubocovich, M. L. and Langer, S. Z. (1974) Negative feed-back regulation of noradrenaline release by nerve stimulation in the perfused cats spleen: differences in potency of phenoxybenzamine in blocking the pre-and postsynaptic adrenergic receptors. *J. Physiol., Lond.* **237**, 505–519.
Fain, J. N. and Garcia-Sainz, J. A. (1980) Role of phosphatidylinositol turnover in alpha$_1$ and of adenylate cyclase inhibition in alpha$_2$ effects of catecholamines. *Life Sci.* **26**, 1183–1194.
Fielding, S., Wilker, J., Hynes, M., Szewczak, M., Novick, W. J. and Lal, H. (1978) A comparison of clonidine with morphine for antinococeptive and withdrawal actions. *J. Pharmac. Exp. Ther.* **207**, 899–905.
Flockhart, J. R., Haynes, M. J. and Walter, D. S. (1982) RX 781094 increases the α-mpt-induced rate of decline of noradrenaline of rat cortex. *Br. J. Pharmac,* **76**, Proc. Suppl. 236P.
Franz, D. N. Hare, B. D. and McCloskey, K. L. (1982). Spinal sympathetic neurons: Possible sites of opiate withdrawal suppression by clonidine. *Science* **215**, 1643–1645.
Garcia-Sevilla, J. A., Zis, A. P., Hollingsworth, P. J., Greden, J. F. and Smith, C. B. (1981a) Platelet α_2-adrenergic receptors in major depressive disorder: binding of tritiated clonidine before and after tricyclic antidepressant drug treatment. *Arch. Gen. Psychiat.* **38**, 1327–33.
Garcia-Sevilla, J. A., Zis, A. P., Zelnik, T. C. and Smith, C. B. (1981b) Tricyclic antidepressant-drug treatment decreases α_2-adrenoceptors on human platelet membranes. *Eur. J. Pharmac.* **69**, 121–123.
Gold, M. S., Pottash, A. C., Sweeney, D. R. and Kleber, H. D. (1980) Opiate withdrawal using clonidine, *J. Am. Med. Assoc.* **243**, 343–346.

Grant, J. A. and Scrutton, M. C. (1979) Novel α_2-adrenoceptors primarily responsible for inducing human platelet aggregation, *Nature* **277**, 659–661.

Grant, J. A. and Scrutton, M. C. (1980) Interaction of selective α-adrenoceptor agonists and antagonists with human and rabbit blood platelets. *Br. J. Pharmac.* **71**, 121–134.

Grundstrom, N., Andersson, R. G. G. and Wikberg, J. E. S. (1981) Prejunctional alpha$_2$-adrenoceptors inhibit contraction of tracheal smooth muscle by inhibiting cholinergic transmission. *Life Sci.* **28**, 2981–2986.

Hedler, L., Stamm, G., Weitzell, R. and Starke, K. (1981). Functional characterisation of central α-adrenoceptors by yohimbine diastereoisomers. *Eur. J. Pharmac.* **70**, 43–52.

Howlett, D. R., Taylor, P. and Walter, D. S. (1982) α-Adrenoceptor selectivity studies with RX 781094 using radioligand binding to cerebral membranes. *Br. J. Pharmac.* **76**, Proc. Suppl. 294P.

Innemee, H. C., de Jonge, A., van Meel, J. C. A., Timmermans, P. B. M. W. M. and van Zwieten, P. A. (1981) The effect of selective α_1- and α_2-adrenoceptor stimulation on intraocular pressure in the conscious rabbit. *Naunyn-Schmiedeberg's Arch. Pharmac.* **316**, 294–298.

Johnson, R. W., Reisine, T., Spotnitz, S., Wiech, N., Ursillo, R. and Yamamura, H. I. (1980) Effects of desipramine and yohimbine on α_2- and β-adrenoreceptor sensitivity. *Eur. J. Pharmac.* **67**, 123–127.

Kerry, R. and Scrutton, M. C. (1982). Interaction of RX 781094, a new selective α_2-adrenoceptor antagonist, with human blood platelets. *Br. J. Pharmac.* **75**, Proc. Suppl. 124P.

Kobinger, W. and Pichler, L. (1981) α_1-Adrenoceptor blockade by α_2-adrenoceptor agonists in the isolated perfused hindquarter of rats. *Eur. J. Pharmac.* **72**, 113–115.

Kuraishi, Y., Harada, Y. and Takagi, H. (1979) Noradrenaline regulation of pain transmission in the spinal cord mediated by α-adrenoceptors. *Brain Res.* **174**, 333–336.

Lal, H., Shearman, G. T. and Ursillo, R. C. (1981) Non-narcotic antidiarrhoeal action of clonidine and lofexidine in the rat. *J. Clin. Pharmac.* **21**, 16–19.

Langer, S. Z. (1974) Presynaptic regulation of catecholamine release. *Biochem. Pharmac.* **23**, 1793–1800.

Langer, S. Z. (1977) Presynaptic receptors and their role in the regulation of transmitter release. *Br. J. Pharmac.* **60**, 481–497.

Langer, S. Z. (1978) Presynaptic receptors. *Nature* **275**, 479–480.

Langer, S. Z. (1981) Presynaptic regulation of the release of catecholamines. *Pharmac. Rev.* **32**, 337–362.

Lattimer, N., Rhodes, K. F., Ward, T. J., Waterfall, J. F. and White, J. F. (1982) Selective α_2-adrenoceptor activity of novel substituted benzoquinolizines. *Br. J. Pharmac.* **75**, Proc Suppl. 154P.

McGrath, J. C. (1982) Evidence for more than one type of postjunctional α-adrenoceptor. *Biochem. Pharmac.* **31**, 467–484.

Medgett, I. C., McCulloch, M. W. and Rand, M. J. (1978) Partial agonist action of clonidine and prejunctional and postjunctional α-adrenoceptors. *Naunyn-Schmiedeberg's Arch. Pharmac.* **304**, 215–221.

Michel, A. D. and Whiting, R. L. (1981) 2-(2-imidazolyl methyl)-1,4-benzodioxans, a series of selective α_2-adrenoceptor antagonists. *Br. J. Pharmac.* **74**, 255–256P.

Nakaki, T., Nakadate, T. and Kato, R. (1980) α_2-Adrenoceptors modulating insulin release from isolated pancreatic islets. *Naunyn-Schmiedeberg's. Arch. Pharmac.* **313**, 151–153.

Paul, S. M. and Crews, F. T. (1980) Rapid desensitisation of cerebral cortical β-adrenergic receptors induced by desmethylimipramine and phenoxybenzamine. *Eur. J. Pharmac.* **62**, 349–350.

Pettinger, W. A., Keeton, T. K., Campbell, W. B. and Harper, D. C. (1976) Evidence for a renal alpha adrenergic receptor inhibiting renin release. *Circulation Res.* **38**, 338–346.

Pichler, L. and Kobinger, W. (1981) Modulation of motor activity by α_1- and α_2-adrenoceptor stimulation in mice. *Naunyn-Schmiedeberg's Arch. Pharmac.* **317**, 180–182.

Reid, J. L., Dargie, H. J., Davies, D. S., Wing, L. M. H., Hamilton, C. A. and Dollery, C. T. (1977) Clonidine withdrawal in hypertension. *Lancet* i, 1171–1174.

Roach, A. G., Lefèvre, F. and Cavero, I. (1978) Effects of prazosin and phentolamine on cardiac presynaptic α-adrenoceptors in the cat, dog and rat. *Clin. Exp. Hypertension* **1**, 87–101.

Ruwart, M. J., Klepper, M. S. and Rush, B. D. (1979) Clonidine delays small intestinal transit in the rat. *J. Pharmac. Exp. Ther.* **212**, 487–490.

Scatton, B., Zivkovic, B. and Dedek, J. (1980) Antidopaminergic properties of yohimbine., *J. Pharmac. Exp. Ther.* **215**, 494–499.

Schildkraut, J. J. (1965) The catecholamine hypothesis: a review of the supporting evidence. *Am. J. Psychiat.* **122**, 509–522.

Scott-Young, W. and Kuhar, M. J. (1980) α_2-adrenergic receptors are associated with renal proximal tubules. *Eur. J. Pharmac.* **67**, 493–495.

Siever, L. J., Cohen, R. M. and Murphy, D. L. (1981) Antidepressants and α_2-adrenergic autoreceptor desensitisation. *Am. J. Psychiat.* **138**, 681–682.

Skomedal, T., Osnes, J. B. and Oye, J. (1980) Competitive blockade of α-adrenergic receptors in rat heart by prazosin. *Acta Pharmac. Toxicol.* **47**, 217–222.

Smith, C. B., Garcia-Sevilla, J. A. and Hollingsworth, P. J. (1981) α_2-Adrenoceptors in rat brain are decreased after long term tricyclic antidepressant drug treatment. *Brain Res.* **210**, 413–418.

Spaulding, T. C., Venafro, J. J., Ma, M. G. and Fielding, S. (1979) The dissociation of the antinociceptive effect of clonidine from supraspinal structures. *Neuropharmacol.* **18**, 103–105.

Spyraki, C. and Fibiger, H. C. (1980) Functional evidence for subsensitivity of noradrenergic α_2-receptors after chronic desipramine treatment. *Life Sci.* **27**, 1863–1867.

Starke, K. (1972) α-Sympathomimetic inhibition of adrenergic and cholinergic transmission in the rabbit heart. *Naunyn-Schmiedeberg's Arch. Pharmac.* **274**, 18–45.

Starke, K. (1977) Regulation of noradrenaline release by presynaptic receptor systems. *Rev. Physiol. Biochem. Pharmac.* **77**, 1–124.

Starke, K. (1981) α-Adrenoceptor subclassification. *Rev. Physiol. Biochem. Pharmac.* **88**, 199–236.

Starke, K. and Langer, S. Z. (1979). A note on terminology for presynaptic receptors. In *Presynaptic receptors. Advances in the Biosciences,* Vol. 18, pp. 1–3, eds. Langer, S. Z., Starke, K. and Dubocovich, M. L. Pergamon Press, Oxford.

Starke, K., Montel, H., Gayk, W. and Merker, R. (1974) Comparison of the effects of clonidine on pre- and postsynaptic adrenoceptors in the rabbit pulmonary artery. *Naunyn-Schmiedeberg's Arch. Pharmac.* **285**, 133–150.

Starke, K., Borowski, E. and Endo, T. (1975a) Preferential blockade of presynaptic α-adrenoceptors by yohimbine. *Eur. J. Pharmac.* **34**, 385–388.

Starke, K., Endo, T. and Taube, H. D. (1975b). Pre- and postsynaptic components in effect of drugs with α-adrenoceptor affinity. *Nature, Lond.* **254**, 440–441.

Sugrue, M. F. (1980) Changes in rat brain monoamine turnover following chronic antidepressant administration. *Life Sci.* **26**, 423–429.

Sulser, F., Vetulani, J. and Mobley, P. L. (1978) Mode of action of antidepressant drugs. *Biochem. Pharmac.* **27**, 257–261.

Svensson, T. H. and Usdin, T. (1978) Feedback inhibition of brain noradrenaline neurons by tricyclic antidepressants: α-receptor mediation. *Science* **202**, 1089–1091.

Tanaka, T., Weitzell, R. and Starke, K. (1978) High selectivity of rauwolscine for presynaptic α-adrenoceptors. *Eur. J. Pharmac.* **52**, 239–240.

Thoenen, H., Hurlimann, A. and Haefely, W. (1964) Dual site of action of phenoxybenzamine in the cat's spleen. *Experientia* **20**, 272–273.

Timmermans, P. B. M. W. M., Schoop, A. M. C., Kwa, H. Y. and van Zwieten, P. A. (1981) Characterisation of α-adrenoceptors participating in the central hypotensive and sedative effects of clonidine using yohimbine, rauwolscine and corynanthine. *Eur. J. Pharmac.* **70**, 7–15.

Uhde, T. W., Post, R. M., Siever, L. J. and Buchsbaum, M. S. (1980) Clonidine and psychophysical pain. *Lancet* ii, 1375.

Von Voigtlander, P. F., Triezenberg, H. J. and Losey, E. G. (1978) Interactions between clonidine and antidepressant drugs: A method for identifying antidepressant-like agents. *Neuropharmac.* **17**, 375 381.

Waldmeir, P. C. and Bischoff, S. (1981). Interaction of α-agonists and antagonists with dopaminergic transmission. *Experientia* **37**, 677.

Wikberg, J. E. S. (1978) Pharmacological classification of adrenergic α-receptors in the guinea-pig. *Nature, Lond.* **273**, 164–166.

2

Subtypes of the opiate receptor

Hans W. Kosterlitz, Stewart J. Paterson and **Linda E. Robson**
Unit for Research on Addictive Drugs, University of Aberdeen, Marischal College, Aberdeen AB9 1AS, U.K

Abbreviations used
- ACTH — corticotropin
- CLIP — corticotropin-like intermediate peptide
- MSH — melanotropin

1. INTRODUCTION

This brief presentation will deal mainly with the basic aspects of our present knowledge of the subtypes of the opiate receptor. In particular, we shall deal with the precursors of the ligands acting as neurotransmitters or neuromodulators and with their interaction with the binding sites and also with some of the more global aspects of the receptors.

2. PRECURSORS OF OPIOID PEPTIDES

Pre-proopiocortin from bovine pituitary, consisting of 265 amino acids, was the first precursor for which the full nucleotide sequence of the cloned DNA complementary to mRNA was established in *E. coli* plasmids (Nakanishi *et al.*, 1979). It contains the sequences of ACTH and β-lipotropin as contiguous peptides at the C-terminal end. The presence of paired amino acids, Lys-Lys or Lys-Arg, within these peptides suggests the formation α-MSH, β-MSH, CLIP, and in addition the sequence of γ-MSH from the N-terminal region of pre-proopiocortin.

Quite recently, the sequence of the second precursor, pre-proenkephalin has been established (Noda *et al.*, 1982; Gubler *et al.*, 1982) by a similar method. This precursor protein has been obtained from bovine adrenal medulla and contains 263 amino acids. Pre-proenkephalin is quite different from pre-proopiocortin in that it contains four copies of [Met^5]enkephalin and one copy of [Leu^5]enkephalin. In addition, there is one copy each of the enkephalin derivatives, [Met^5]enkephalyl-Arg-Gly-Leu and [Met^5]enkephalyl-Arg-Phe. The first of these sequences is situated between the third and fourth sequences of [Met^5] enkephalin; this is followed by [Leu^5] enkephalin and, finally, at the C-terminal end of the peptide by [Met^5] enkephalyl-Arg-Phe. The enkephalins and their derivatives are each bounded by paired basic residues, Lys-Arg, Lys-Lys or Arg-Arg, as already described for pre-proopiocortin. No equivocal evidence has so far been obtained for the possible physiological functions of these peptides.

An interesting observation is the absence in these two precursor proteins of the sequence of α-neo-endorphin and dynorphin, which are extended [Leu^5] enkephalins and may be contained in one or more precursor as yet unknown. α-Neo-endorphin from porcine hypothalamus is [Leu^5] enkephalyl-Arg-Lys-Tyr-Pro-Lys (Kangawa et al., 1981) while dynorphin from porcine pituitary is [Leu^5] enkephalyl-Arg-Arg-Ile-Arg-Pro-Lys-Leu-Lys-Trp-Asp-Asn-Gln (Goldstein et al., 1981). It is of considerable interest that dynorphin has high affinities not only to μ- and δ-binding sites but also to the κ-binding site (see Section 3) (Pfeiffer et al., 1981; Chavkin et al, 1982; S. J. Paterson, Unpublished observations).

Although our knowledge of the possible physiological functions of the precursors of the opioid peptides and their processed fragments is still limited, the existence of at least two and probably three different precursors is of great interest. In pre-proopiocortin the fragment exhibiting opioid activity, β-endorphin, is closely associated with ACTH and the melanotropins; furthermore, in each precursor molecule there is only one sequence of β-endorphin. In contrast, each molecule of pre-proenkephalin contains four sequences of [Met^5] enkephalin, one sequence of [Leu^5] enkephalin, in addition to the two [Met^5] enkephalins extended at the C-terminal. Thus, mole for mole, the pre-proenkephalins are much richer in opioid activity than pre-proopiocortin. It is obviously of importance for their physiological function that, in contrast to β-endorphin, all fragments of pre-proenkephalin are readily inactivated by peptidases.

3. RECEPTORS INTERACTING WITH OPIOID PEPTIDES

The concept that there are subtypes of the opiate receptor was originally based on pharmacological observations and electrophysiological experiments on the chronic spinal dog (Gilbert and Martin, 1976; Martin, 1967) and was supported by observations on isolated preparations of the guinea-pig ileum and mouse vas deferens (Hutchinson et al., 1975). The two types of opiate receptor have been designated μ- and κ-receptors, the prototype ligands being morphine and ketazocine, respectively.

Parallel observations in two pharmacological assays, the guinea-pig ileum and the mouse vas deferens, and in assays measuring the inhibition of the specific binding of [^{3}H] [Leu^5] enkephalin and [^{3}H] naloxone in homogenates of guinea-pig brain have led to the conclusion that the opioid peptides interact with the μ-receptor and, in addition, with a receptor different in binding characteristics from the κ-receptor (Lord et al., 1977). It has been shown that β-endorphin is equipotent in the guinea-pig ileum and mouse vas deferens assays and also in its potency to inhibit the binding of [^{3}H] [Leu^5] enkephalin and [^{3}H] naloxone in brain homogenates. [Leu^5] enkephalin is 50 times more potent in the mouse vas deferens than in the guinea-pig ileum and its ability to inhibit [^{3}H] [Leu^5] enkephalin binding is 25 times greater than to inhibit [^{3}H] naloxone binding. As a first approximation, the major part of [^{3}H] naloxone binding may be due to the μ-binding site while the major part of the [^{3}H] [Leu^5] enkephalin binding would be assumed to be to a site different from that of the μ- or κ-receptor and has been assigned to the δ-receptor (Lord et al., 1977; Kosterlitz et al., 1980; Gillan et al., 1980). Thus, the opiate receptor has at least three subclasses; it has, however, not yet been established which physiological functions are subserved by them.

This concept that the opioid peptides interact differentially with the μ- and δ-receptors is supported by experiments in which pre-treatment with selective unlabelled ligands protects the binding of the μ-ligand, [^{3}H] dihydromorphine, or the δ-ligand, [^{3}H] [D-Ala^2,D-Leu^5] enkephalin against the alkylating action of phenoxybenzamine in a highly specific manner (Robson and Kosterlitz, 1979). It was predicted that a ligand with a high affinity for the δ-receptor, such as unlabelled [D-Ala^2, D-Leu^5] enkephalin, should protect the binding of [^{3}H]-[D-Ala^2,D-Leu^5] enkephalin more readily than the binding of [^{3}H] dihydromorphine which has a high affinity for the μ-receptor; conversely, unlabelled dihydromorphine should protect [^{3}H] dihydromorphine binding sites more readily than [^{3}H] [D-Ala^2,D-Leu^5] enkephalin binding sites. These predictions were borne out by the experimental results.

As has already been pointed out, the enkephalins, in contrast to β-endorphin, are very readily degraded by aminopeptidases, carboxypeptidases and enkephalinases; it is therefore necessary to use stable analogues for experiments in animals and observations in man. As [Met^5] enkephalin and particularly [Leu^5]-enkephalin have a much higher affinity for the δ-receptor than for the μ-receptor, it is of importance to establish whether or not the binding and pharmacological patterns of analogues are similar to those of the natural enkephalins (Kosterlitz et al., 1980). The substitution in [Leu^5] enkephalin of Gly^2 by D-Ala and L-Leu^5 by D-Leu leads to [D-Ala^2, D-Leu^5] enkephalin which has a binding pattern similar to that of the parent compound. However, its potency in the guinea-pig ileum is increased 10-fold and in the mouse vas deferens 16-fold and therefore it may be concluded that this increase in the pharmacological assays is at least partly due to an increased resistance to enzyme action. This interpretation has been confirmed by the use of degradative enzyme inhibitors when the potency of labile [Leu^5] enkephalin is raised to that of the almost completely stable [D-Ala^2, D-Leu^5] enkephalin. On the other hand, another analogue, Tyr-D-Ala^2-Gly-MePhe-Met(O)-ol, FK 33-824 (Römer et al., 1977), has a very different pattern. Compared with [Met^5] enkephalin, its parent compound, the affinity to the μ-binding site is unchanged but that to the δ-binding site is decreased to 6.4%; its activity is increased 20-fold in the guinea-pig ileum but unchanged in the mouse vas deferens. FK 33-824 is a much more potent antinociceptive agent than [D-Ala^2, D-Leu^5]enkephalin, a fact which suggests that for this action the μ-receptor may be more important than the δ-receptor.

To obtain further information on the biochemistry and pharmacology of the binding sites, highly selective ligands are required. This goal has been attained so far only for ligands which bind to the μ-site. For instance, the Gly-ol^5 analogue of FK 33-824, Tyr-D-Ala-Gly-MePhe-Gly-ol, is such a compound (Handa et al., 1981; Kosterlitz and Paterson, 1981; Kosterlitz et al., 1981). The most commonly used but less discerning δ-ligand is [D-Ala^2, D-Leu^5]enkephalin which to a minor degree is surpassed by [D-Ser^2, L-Leu^5] enkephalyl-Thr (Gacel et al., 1980).

Binding assays in which saturation is achieved give information on the maximum number of binding sites, provided cross-reactivity between binding sites is low. The enkephalins bind preferentially with δ-sites but interact also with μ-sites; the maximal binding of either of the two enkephalins in guinea-pig brain is between 5 and 6 pmol/g brain tissue and that of dihydromorphine 4 pmol/g. It is of interest that amidation of the C-terminal carboxyl group in both [D-Ala^2, Met^5]enkephalin or [D-Ala^2, Leu^5]enkephalin increases the maximal number of binding sites to 13–14 pmol/g, a value similar to that found for one of the most potent narcotic analgesic drugs, etorphine. This finding would indicate that these compounds interact with both the μ- and δ-receptors and, as will be shown in the following paragraphs, also with κ-binding sites.

The κ-agonists are of particular interest because they have a unique pharmacological pattern: they are potent antinociceptive agents in rodents but do not suppress signs of withdrawal in morphine-dependent monkeys and have no antagonist activity in this species. Although in *in vitro* tests in the guinea-pig ileum and the mouse vas deferens the κ-compounds are pure agonists, they are pure antagonists in the rat vas deferens (Gillan et al., 1981). Since in the rat vas deferens the benzomorphans antagonise the selective μ-agonists more readily than the δ-agonists, it is possible that their antagonist action is due to interaction at the μ-site.

The κ-agonist-like compounds, e.g. ethylketazocine and bremazocine, which are available at present, have binding characteristics which indicate a very high degree of cross-reactivity to the μ-binding sites and, to a lesser degree, also to the δ-binding sites. In contrast to these findings, unlabelled highly selective μ-ligands or δ-ligands exhibit a two-phase inhibition of the binding of [^{3}H]-ethylketazocine. The first phase is due to inhibition at the μ-binding site and to a lesser extent also at the δ-binding site; the second phase occurs at a very much higher concentration, displacing [^{3}H]ethylketazocine from the κ-binding site.

An observation of particular importance is the finding that the endogenous opioid peptides, [Met⁵]enkephalin and [Leu⁵]enkephalin have only a low affinity to the κ-binding site. While the K_I value for the first phase of the inhibition of [³H]ethylketazocine binding is 25 nM for [Met⁵]enkephalin and 57 nM for [Leu⁵]enkephalin, the values for the second phase are 3,500 nM and 14,800 nM, respectively. Thus, the two pentapeptides are not the endogenous ligands for the κ-binding site. The status of β-endorphin in this respect is not quite clear yet. In contrast, dynorphin has a very high affinity to the κ-binding site but it also has high affinities to the μ- and δ-binding sites (Pfeiffer *et al.*, 1981; Chavkin *et al.*, 1982; S. J. Paterson, unpublished observations).

The views presented here are corroborated by experiments investigating the selective protection against alkylation by phenoxybenzamine. While unlabelled ethylketazocine readily protects the binding of [³H]ethylketazocine against alkylation, the μ-ligand dihydromorphine is less effective; the protection by the δ-ligand [D-Ala², D-Leu⁵]enkephalin does not reach 40% at 5000 nM (Kosterlitz *et al.*, 1981).

The present situation may be summarised as follows. The μ-binding site is selectively activated by the μ-ligand, [D-Ala², MePhe⁴, Gly-ol⁵]enkephalin which affects the δ- and κ-binding sites only to a low degree. The κ-binding site responds only to κ-ligands, the δ-binding site to δ-ligands and to ethylketazocine and the μ-binding site to μ-ligands, to ethylketazocine and, to a smaller extent, to [D-Ala², D-Leu⁵]enkephalin. It cannot be decided whether, as far as structure is concerned, the μ-binding site is less demanding than the κ-binding site and also the δ-binding site, alternatively, the available κ-ligands are not selective and thus cross-react with the μ-binding site. It is important to appreciate that there are species differences in the density and distribution of the κ-binding sites. In brains of the rat and guinea-pig it has been found by all laboratories that the putative κ-ligands have high affinities for the μ- and κ-binding sites with lower affinities for the δ-binding site (Chang *et al.*, 1980; Harris and Sethy, 1980; Hiller and Simon, 1979, 1980; Römer *et al.*, 1980; Snyder and Goodman, 1980). In contrast, the observations obtained in rat brain for the inhibition of [³H]ethylketazocine binding by unlabelled μ- and δ-ligands are inconsistent. However, the density of the κ-binding sites is greater in guinea-pig brain than in rat brain. In guinea-pig brain the estimate is 21% μ-binding sites, 38% δ-binding sites and 41% κ-binding sites, of a total 13 pmol/g wet brain tissue (Kosterlitz *et al.*, 1981) while the corresponding values in rat brain are 40%, 46% and 14% (M. G. C. Gillan, unpublished observations).

The regional distribution of the opiate receptors has been examined repeatedly but many of the data available at present have been obtained without taking into account the significance of their heterogeneity. In the last two years papers have been published which give a certain amount of information on the regional distribution of the subtypes of the opiate receptor. One approach to this problem was the use of [¹²⁵I][D-Ala², MePhe⁴, Met(O)-ol⁵]enkephalin as a probe for the 'morphine receptor' and [¹²⁵I][D-Ala², D-Leu⁵]enkephalin for the 'enkephalin receptor' (Chang et al., 1979). The maximum binding capacities of the two ligands were calculated indirectly from their dissociation constants and from the concentrations of free receptor and free and bound ligand at a single concentration of the labelled ligand. The maximum binding to the 'enkephalin receptor' occurred in the frontal cortex of rat brain with progressively decreasing binding in the sensomotor cortex, the limbic system, the hippocampus, brainstem, thalamus and finally hypothalamus having the lowest binding. The distribution of the 'morphine receptor' was more uniform with highest levels in frontal cortex, striatum, sensomotor cortex and thalamus and lowest in the brainstem; the ratio of 'morphine' binding sites to 'enkephalin' binding sites was about 1 in the frontal cortex and between 3 and 5 in the thalamus and hypothalamus.

Support for this finding was obtained from the observation that one group of [^{3}H] diprenorphine binding sites is depressed by GTP (type 1) while the other group is GTP-resistant (type 2) (Pert and Taylor, 1980). GTP-sensitive sites predominate in areas which are associated with antinociception, e.g. the periaqueductal grey, nucleus gigantocellularis, reticular formation and thalamus. GTP-insensitive receptors predominate in areas of the limbic system, e.g. the amygdala, hypothalamus, nucleus accumbens and frontal cortex and the pituitary. The type of distribution suggests that GTP-sensitive receptors may correspond to μ-receptors and GTP-insensitive receptors to δ-receptors.

Determination of specific binding of 0.5 nM [^{3}H] morphine (μ-binding site) and 0.5 nM [^{3}H] [D-Ala^2, D-Leu^5] enkephalin (δ-binding site) was carried out in homogenates obtained from frozen slices of different regions of bovine brain (Ninkovic et al., 1981). The ratios of the binding of the two ligands showed large differences, varying between 3.9 in the substantia nigra and 0.40 in the hippocampus. While certain areas, such as the periaqueductal grey, the raphe nucleus and the dorsal horn of the spinal cord have high ratios of μ-binding to δ-binding and are probably associated with antinociceptive mechanisms, no firm correlation can be made at present for the substantia nigra and the thalamus. It would appear that the limbic system is relatively rich in δ-binding sites.

Differences in the densities of the μ- and δ-binding sites have also been demonstrated by autoradiographic techniques using slices of rat brain. The results agree with those obtained with binding assays but yield greater detail of the distribution. For instance, lamina IV of the cortex is rich in μ-receptors but contains only a few δ-receptors (Goodman et al., 1980). In general, μ-receptors occur in distinct patches while δ-receptors are diffuse and more widespread (Herkenham and Pert, 1981).

No data have been published concerning the detailed distribution of the κ-binding sites but there is evidence for species and regional variations (M. G. C. Gillan and L. E. Robson, unpublished observations). At the present state of our knowledge, we cannot speculate on the physiological role played by these binding sites.

4. CONCLUSIONS

Although there is a considerable amount of information regarding the precursors of the opioid peptides and their interaction with the three subtypes of receptor, several problems are still outstanding. Firstly, by the nature of the underlying mechanisms, binding assays cannot determine whether a ligand-site interaction leads to an agonist, antagonist or non-event response. For this purpose, pharmacological models are required which are of a nature more complex than those so far developed in standard *in vitro* bioassays. Secondly, there is a need for highly selective agonists or antagonists. While µ-agonists and, to a lesser extent, δ-agonists are available, the available compounds binding to κ-sites have a very low degree of selectivity. None of the antagonists fulfils the requirements of a compound suitable for either *in vitro* or *in vivo* experimentation.

ACKNOWLEDGEMENTS

Supported by grants provided by the Medical Research Council and the U.S. National Institute on Drug Abuse (DA 00662).

REFERENCES

Chang, K. J., Cooper, B. R., Hazum, E. and Cuatrecasas, P. (1979) Multiple opiate receptors: different regional distribution in the brain and differential binding of opiates and opioid peptides. *Mol. Pharmacol.* **16**, 91–104.

Chang, K. J., Hazum, E. and Cuatrecasas, P. (1980). Possible role of distinct morphine and enkephalin receptors in mediating actions of benzomorphan drugs (putative κ and δ agonists). *Proc. Natl. Acad. Sci. USA* **77**, 4469–4473.

Chavkin, C., James, I. F. and Goldstein, A. (1982) Dynorphin is a specific endogenous ligand of the κ opioid receptor. *Science* **215**, 413–415.

Gacel, G., Fournie-Zaluski, M.-C. and Roques, B. P. (1980) Tyr-D-Ser-Gly-Phe-Leu-Thr, a highly preferential ligand for δ-opiate receptors. *FEBS Lett.* **118**, 245–247.

Gilbert, P. E. and Martin, W. R. (1976) The effect of morphine- and nalorphine-like drugs in the non-dependent, morphine-dependent and cyclazocine-dependent chronic spinal dog. *J. Pharmacol. Exp. Ther.* **198**, 66–82.

Gillan, M. G. C., Kosterlitz, H. W. and Paterson, S. J. (1980) Comparison of the binding characteristics of tritiated opiates and opioid peptides. *Br. J. Pharmacol.* **70**, 481–490.

Gillan, M. G. C., Kosterlitz, H. W. and Magnan, J. (1981) Unexpected antagonism in the rat vas deferens by benzomorphans which are agonists in other phamacological tests. *Br. J. Pharmacol.* **72**, 13–15.

Goldstein, A., Fischli, W., Lowney, L. I., Hunkapiller, M. and Hood, L. (1981) Porcine pituitary dynorphin: complete amino acid sequence of the biologically active heptadecapeptide. *Proc. Natl. Acad. Sci. USA* **78**, 7219-7223.

Goodman, R. R., Snyder, S. H., Kuhar, M. J. and Young, W. S. (1980) Differentiation of delta and mu receptor localisations by light microscopic autoradiography. *Proc. Natl. Acad. Sci. USA* **77**, 6239–6243.

Gubler, U., Seeburg, P., Hoffman, B. J., Gage, L. P. and Udenfriend, S (1982) Molecular cloning establishes proenkephalin as precursor of enkephalin-containing peptides. *Nature* **295**, 206–208.

Handa, B. K., Lane, A. C., Lord, J. A. H., Morgan, B. A., Rance, M. J. and Smith, C. F. C. (1981) Analogues of β-LPH$_{61-64}$ possessing selective agonist activity at μ-opiate receptors. *Eur. J. Pharmacol.* **70**, 531–540.

Harris, D. W. and Sethy, V. H. (1980) High affinity binding of [^{3}H]ethylketazocine to rat brain homogenate. *Eur. J. Pharmacol.* **66**, 121–123.

Herkenham. M. and Pert, C. B. (1981) Mosaic distribution of opiate receptors, parafascicular projections and acetylcholinesterase in rat striatum. *Nature* **291**, 415–418.

Hiller, J. M. and Simon, E. J. (1979) ^{3}H-Ethylketazocine binding: lack of evidence for a separate κ receptor in rats CNS. *Eur. J. Pharmacol.* **60**, 389–390.

Hiller, J. M. and Simon, E. J. (1980) Specific, high affinity [^{3}H]ethylketazocine binding in rat central nervous system: lack of evidence for κ receptors. *J. Pharmacol. Exp. Ther.* **214**, 516–519.

Hutchinson, M., Kosterlitz, H. W., Leslie, F. M., Waterfield, A. A. and Terenius, L. (1975) Assessment in the guinea-pig ileum and mouse vas deferens of benzomorphans which have strong antinococeptive activity but do not substitute for morphine in the dependent monkey. *Br. J. Pharmacol.* **55**, 541–546.

Kangawa, K., Minamino, N., Chino, N., Sakakibara, S. and Matsuo, H. (1981) The complete amino acid sequence of α-neo-endorphin. *Biochem. Biophys. Res. Commun.* **99**, 871–873.

Kosterlitz, H. W. and Paterson, S. J. (1981) Tyr-D-Ala-Gly-MePhe-NH(CH$_2$)$_2$OH is a selective ligand for the μ-opiate binding site. *Br. J. Pharmacol.* **73**, 299P.

Kosterlitz, H. W., Lord, J. A. H., Paterson, S. J. and Waterfield, A. A. (1980) Effects of changes in the structure of enkephalins and narcotic analgesic drugs on their interactions with μ-receptors and δ-receptors. *Br. J. Pharmacol.* **68**, 333–342.

Kosterlitz, H. W. Paterson, S. J. and Robson, L. E. (1981) Characterization of the κ-subtype of the opiate receptor in the guinea-pig brain. *Br. J. Pharmacol.* **73**, 939–949.

Lord, J. A. H., Waterfield, A. A., Hughes, J. and Kosterlitz, H. W. (1977) Endogenous opioid peptides: multiple agonists and receptors. *Nature* **267**, 495–499.

References

Martin, W. R. (1967) Opioid antagonists. *Pharmacol. Rev.* **19**, 463–521.

Nakanishi, S., Inoue, A., Kita, T., Nakamura, M., Chang, A. C. Y., Cohen, S. N. and Numa, S. (1979) Nucleotide sequence of cloned cDNA for bovine corticotropin-β-lipotropin precursor. *Nature* **278**, 423–427.

Ninkovic, M., Hunt, S. P., Emson, P. C. and Iversen, L. L. (1981) The distribution of multiple opiate receptors in bovine brain. *Brain Res.* **214**, 163–167.

Noda, M., Furutani, Y., Takahashi, H., Toyosato, M., Hirose, T., Inayama, S., Nakanishi, S. and Numa, S. (1982) Cloning and sequence analysis of cDNA for bovine adrenal preproenkephalin. *Nature* **295**, 202–206.

Pert, C. B. and Taylor, D. (1980) Type 1 and Type 2 opiate receptors: a subclassification scheme based upon GTP's differential effects on binding. In E. L. Way (ed.), *Endogenous and Exogenous Opiate Agonists and Antagonists*. Pergamon, New York, pp. 87–90.

Pfeiffer, A., Pasi, A., Mehraein, P. and Herz, A. (1981) A subclassification of κ-sites in the human brain by use of dynorphin 1-17. *Neuropeptides* **2**, 89–97.

Robson, L. E. and Kosterlitz, H. W. (1979) Specific protection of the binding sites of D-Ala^2-D-Leu^5-enkephalin (δ-receptors) and dihydromorphine (μ-receptors). *Proc. R. Soc. Lond. (Biol.)* **205**, 425–432.

Römer, D., Büscher, H., Hill, R. C., Maurer, R., Petcher, T. J., Welle, H. B. A., Bakel, H. C. C. and Akkerman, A. M. (1980) Bremazocine: a potent long-acting opiate kappa-agonist. *Life Sci.* **27**, 971–978.

Römer, D. Büscher, H. H., Hill, R. C., Pless, J., Bauer, W., Cardinaux, F., Closse, A., Hauser, D. and Huguenin, R. (1977) A synthetic enkephalin analogue with prolonged parenteral and oral analgesic activity. *Nature* **268**, 547–549.

Snyder, S. H. and Goodman, R. R. (1980) Multiple neurotransmitter receptors. *J. Neurochem* **35**, 5–15.

3

The benzodiazepines and their receptor

I. L. Martin

MRC Neurochemical Pharmacology Unit, Medical Research Council Centre, Medical School, Hills Road, Cambridge CB2 2QH, U.K.

Abbreviations used

β-CCE	—	ethyl-β-carboline-3-carboxylate
CNS	—	central nervous system
EEG	—	electroencephalogram
GABA	—	γ-amino butyric acid
GMP	—	guanosine monophosphate

1. INTRODUCTION

Since the introduction of chlordiazepoxide (Librium) into clinical practice in 1960 members of the benzodiazepine class have become the most frequently prescribed of all psychotropic drugs. The reasons for this popularity are undoubtedly multiple, but their wide spectrum of activity as anxiolytics, sedative/hypnotics, anti-convulsants and muscle relaxants together with their essential lack of disturbing peripheral side effects and large therapeutic ratio have certainly played their part. It is, however, worthy of note that the age which has seen the success of the benzodiazepines has also seen a steady increase in the concern of society with stress, in all its many aspects, and such a parallel makes it tempting to suggest that these drugs have contributed to the development of such concern. However, that question remains the prerogative of the social scientist and this chapter will review the development of the benzodiazepines and our present understanding of the receptors through which they are thought to mediate their effects.

1.1 History and development

Like many successful drug development programmes that of the benzodiazepines owes much to serendipity and is recounted in some detail by Sternbach (1979). In 1957 Leo Sternbach, leading a synthetic program at Hoffman–La

Roche submitted what was then thought to be a quinazoline N-oxide for testing by the pharmacologists led by Randall within the same company, though he expressed little hope in a positive outcome. The results of these and later tests indicated that the compound had a novel spectrum of pharmacological activity. It was of equal muscle relaxant potency to chlorpromazine in the cat but was more efficacious as an anti-convulsant and appeared less hypnotic than the barbiturates; it showed no action on the autonomic nervous system and exhibited a low toxicity. These results encouraged the chemists to study this compound further, and many of its analogues, with the result that it became clear that the initial structural assignment was incorrect; the compound contained a seven-membered ring and was in fact 7-chloro-2-(methylamino)-5-phenyl-3H-1, 4-benzodiazepine 4-oxide, now known by the generic name of chlordiazepoxide. So intense was the subsequent toxicological and clinical testing of this compound that the company were able to market it under the trade name Librium in 1960, less than three years after the first pharmacological studies were carried out. During the subsequent years over 3000 analogues of chlordiazepoxide were synthesised and subjected to pharmacological investigation along with 4,000 intermediates and by-products. The further exploration of this chemical structure by numerous other groups has led to decreases in the minimally effective doses and a considerable increase in the number of such compounds available to the clinician. Currently 22 benzodiazepines are available throughout the world from which the clinician must make his selection, though the guidance available to aid such a selection is limited (see Section 1.3).

1.2 Clinical uses and abuses

The most frequent use of the benzodiazepines is in the treatment of anxiety. This condition is a normal response to stressful situations in our daily life, indeed it is an integral part of developing an appropriate response effectively to reduce that particular stress. However, occasionally this response may become excessive or chronic and thus lead to what has been termed 'pathological anxiety', it is at this stage that some type of clinical intervention normally takes place, and to date the most cost-effective way is the prescription of anti-anxiety agents, inevitably nowadays the benzodiazepines.

The benzodiazepines are also used as anti-convulsants and diazepam remains the drug of choice in the treatment of status epilepticus, while clonazepam has been used in the alleviation of several seizure disorders (Browne, 1978). Tolerance to the use of clonazepam develops on chronic usage both in animal models (Rossi *et al.*, 1973, Killam *et al.*, 1973), and in man, suggesting that the therapeutic benefit as an anticonvulsant is lost within a period of months (Browne, 1978).

The hypnotic and sedative effects of this group of compounds have both advantages and disadvantages. While there is considerable individual variation in sensitivity to the actions of the benzodiazepines, one of the major disadvantages of these drugs in the treatment of anxiety is the sedation frequently produced when the medication is first adopted. However, these initial sedative effects appear to subside after the first few days of treatment, a phenomenon which has been observed both in animal models of this behaviour (Margules and Stein, 1968) and in man (Warner, 1965), while the anxiolytic effects do not. There are, however, a number of compounds within the group which are marketed specifically as hypnotics such as flurazepam, nitrazepam, flunitrazepam and more recently triazolam (see Section 1.3 for further discussion.).

Higher doses of the benzodiazepines are frequently used to produce muscle relaxation in such conditions as cerebral palsy, certain degenerative neurological disorders, muscle pain and tetanus (Greenblatt and Shader, 1974).

The benzodiazepines have also been used with increasing success in preanaesthetics where the muscle-relaxant, anxiolytic and sedative effects have considerable advantage. A further property of these compounds, that of anterograde amnesia, has also made the compounds increasingly popular especially in the area of dental anaesthesia though the additive depressive effects of the compounds with both the barbiturates (Chambers and Jefferson, 1977) and halothane (Chambers *et al.*, 1978) require the careful manipulation of dosage regimes in this particular area.

It is clear from the pharmacological profile described above that the benzodiazepines are of considerable value in the clinical armoury, indeed we have previously alluded to their widespread usage. Figures suggest that 1 in 10 Canadians receive a prescription for a benzodiazepine each year while in excess of 30% of all hospitalised patients are given one of these drugs (Ban *et al.,* 1981), and there is no reason to believe that such levels of usage are not representative of other developed countries. Such figures have recently led to much concern about the over-prescription of drugs of this class (see Marks, 1978). Psychological dependence on the compounds has been reported; many general practitioners report difficulties in persuading patients to decrease or stop the use of these compounds, though physical withdrawal symptoms would appear to be uncommon except following the abrupt cessation of long-term, high dose treatment. However, the major population at risk is that receiving treatment for anxiety of a chronic nature. Recent evidence suggests that the benzodiazepines actually *decrease* the ability of animals to adjust to stressful situations, that is the normal adaptation of the animals is impaired by treatment with the benzodiazepines, and evidence would suggest that this may also apply to man (Davis *et al.,* 1981; J. A. Gray personal communication). Such evidence would seem to imply that such drug therapy should be used for the treatment of acute conditions but not extended. If future experiments support such an hypothesis the arguments for limited term therapy with benzodiazepines would receive considerable support.

1.3 Pharmacodynamic and pharmacokinetic considerations

It has been pointed out earlier that over twenty benzodiazepines are available to the clinician and we have delineated the most important conditions against which these compounds are used. While these drugs have been available for over twenty years there remains little but anecdotal evidence to suggest that the pharmacodynamic profile of any compound differs significantly from any other so far marketed; they all appear to possess anxiolytic, anti-convulsant, sedative/ hypnotic and muscle relaxant activity in approximately the same ratio. Indeed there is a similar lack of such evidence for compounds of the class which are used only as experimental tools, their pharmacodynamic spectrum of activity appears to be invariate, though it should be pointed out that most studies so far reported have not addressed this problem specifically (but see Babbini et al., 1979).

There are, however, very marked differences with regard to the pharmacokinetic profiles of different benzodiazepines. The mean biological half lives of these compounds vary from 2 h for triazolam to in excess of 30 h for diazepam. The interpretation of such data, however, is complicated by the fact that metabolic breakdown leads, in many cases, to the formation of active metabolites with long half lives (Shader and Greenblatt, 1977). Much attention has focused on the differences in pharmacokinetic profiles of the currently marketed benzodiazepines, and, together with the lack of differences in the pharmacodynamic profiles of the compounds has led to the suggestion that the congeners with short biological half lives should be used for night-time sedation while those with extended half-lives are more appropriate for the treatment of anxiety (see Committee on the Review of Medicines, 1980).

2. MECHANISM OF ACTION OF BENZODIAZEPINES
2.1 Electrophysiological studies
The first investigations into the possible mechanisms of action of the benzodiazepines were directed towards the classical neurotransmitter systems (see Costa and Greengard, 1975) and almost without exception effects of the benzodiazepines were noted. However, the seminal observations were made by Schmidt et al. (1967) who observed that diazepam was able to potentiate presynaptic inhibition in the cat spinal cord. Later when it became clear that the mediator of this response was probably GABA, Polc et al. (1974) were able to demonstrate that other active benzodiazepines shared this property with diazepam, and that the effect of the benzodiazepines was dependent on a functional GABA system, the benzodiazepines having no GABA-mimetic action in their own right (Polc and Haefely, 1977). Since that time many experiments have been carried out to suggest that the benzodiazepines facilitate GABAergic transmission in many regions of the CNS (see Haefely, 1978).

While it was clear by the mid-1970s that the benzodiazepines had a major action in facilitating GABAergic transmission in the CNS it was apprent that this action was not due to direct activation of GABA receptors or to the inhibition of the uptake or inactivation of this major inhibitory neurotransmitter. Studies by Mitchell and Martin (1978) indicated that both flurazepam and diazepam were able to increase the potassium-induced release of GABA from rat cortical tissue *in vitro* at concentrations compatible with those known to inhibit pentylenetetrazol induced convulsions in this species, though it soon became clear that the structure–activity relationship in this series did not support the hypothesis that this increase in stimulus-secretion coupling for GABA was the mechanism by which the benzodiazepines produced their effects. Attention was therefore becoming focused on the possibility that the facilitatory actions of these compounds were mediated at the post-synaptic GABA receptor and perhaps occurred by an increase in the affinity of this receptor for GABA or by an increased coupling between the GABA receptor and the chloride conductance of the post-synaptic membrane, the ionic mechanism by which GABA is thought to produce its effects (Choi *et al.*, 1977, MacDonald and Barker, 1978).

2.2 Biochemical evidence for the GABA hypothesis

Biochemical evidence to support the contention that the benzodiazepines act post-synaptically at GABA receptor complexes has been admirably reviewed by Costa *et al.* (1978), the laboratory which was responsible for much of the early work with these compounds. This group gained evidence that the net activity of the Purkinje cells of the cerebellum could be monitored by their content of cyclic GMP, an activity which was controlled by the balance between excitatory inputs to these cells and GABAergic inhibitory influences. Activation of GABA receptors on the Purkinje cells resulted in a decrease in the cyclic GMP content of the cells and it was shown that both muscimol, as a direct GABA agonist, and diazepam were able to produce such a decrease; however, the effects of diazepam were only demonstrable in the presence of a functional GABA input. It was further shown that GABA turnover, in a number of brain areas, was decreased by diazepam, an observation incompatible with an increased release of the neurotransmitter, suggesting that whatever actions diazepam was producing in the GABA system were mediated by an increased post-synaptic efficacy of GABA itself.

At this stage evidence had accumulated, therefore, from a number of sources, to suggest that the site of action of the benzodiazepines was associated with the post-synaptically located GABA receptor system.

3. BENZODIAZEPINE RECEPTORS

3.1 Initial studies

Without doubt the most significant impetus to the research effort to elucidate the mechanism of action of the benzodiazepines came from the simultaneous reports by Squires and Braestrup (1977) and Mohler and Okada (1977) of specific high affinity binding sites for the benzodiazepines in CNS tissue. These sites were identified by equilibrium binding techniques, similar to those previously used in the biochemical identification of neurotransmitter receptors in brain (Synder and Bennett, 1976). These studies demonstrated that [^{3}H] diazepam bound to membranes obtained from the CNS in a saturable manner and with high affinity. It was further shown that the [^{3}H] diazepam so bound was displaceable specifically by benzodiazepines, and not by other psychoactive drugs, or by any of the known neurotransmitter candidates including GABA, indicating that the recognition or binding sites for GABA and benzodiazepines were different. The interaction showed stereospecificity and in a series of benzodiazepines used in these displacement experiments there was a significant correlation between the affinity of a particular compound for the diazepam binding site and its efficacy in producing muscle relaxation *in vivo* in the cat. This excellent correlation provided evidence that the recognition site for the benzodiazepines in the CNS membranes was in fact part of the pharmacological receptor through which the benzodiazepines produced their effects. The subcellular distribution of the receptor indicated that it was localised primarily to the synaptic membrane fraction and it also exhibited a heterogeneous topographical distribution in the CNS, with high densities in cortical structures and lowest in the brain stem and spinal cord (Squires and Braestrup, 1977; Mohler and Okada, 1977; Müller *et al.*, 1978; Mackerer *et al.*, 1978). Further studies indicated that benzodiazepine binding sites with similar characteristics were demonstrable in human brain (Braestrup *et al.*, 1977, Mohler *et al.*, 1978b, Speth *et al.*, 1978). Although some binding of [^{3}H] diazepam occurs in certain peripheral organs such as the kidney, lung and liver the recognition characteristics of these binding sites are different to the so-called 'brain specific' site (Braestrup and Squires, 1977).

The detailed analysis of equilibrium saturation data in a number of regions of both rat (Squires and Braestrup, 1977; Mohler and Okada, 1977) and human brain (Braestrup *et al.*, 1977; Mohler *et al.*, 1978b) indicated that the benzodiazepine receptor population in the CNS was homogeneous, i.e. only one type of benzodiazepine receptor existed. This conclusion also received some support from the good correlations between the efficacy of a number of benzodiazepines in various pharmacological tests and their affinity for the receptor (see Mohler and Okada, 1978, Speth *et al.*, 1980). The differences which were noted, e.g. the fact that medazepam was more active in animal tests than suggested by its affinity for the receptor, could be adequately explained by metabolic factors, which, in the case of medazepam is due to the *in vivo* production of N-desmethyldiazepam, a compound with a higher affinity for the receptor than the parent drug (see Braestrup and Nielsen, 1980).

Phylogenetically the benzodiazepine receptor appears to have developed rather late as it cannot be identified in invertebrate species and its binding characteristics in lower vertebrate species are atypical (Nielsen *et al.*, 1978). However, ontogenically the receptor develops very early, significant levels of [^{3}H] diazepam binding being found 8 days before birth in the rat (Braestrup and

Nielsen, 1978) and at birth the level was already about half of that found in the adult in both the cerebral cortex and cerebellum (Candy and Martin, 1979a). This situation is in marked contrast to the development of other receptor systems, e.g. GABA (Coyle and Enna, 1976) where development is somewhat delayed.

While this work was able to characterise, in some detail, the benzodiazepine receptor there was little information concerning its neuronal localisation or its relationship to other neurotransmitter receptor populations. However, work reported by Mohler and Okada (1978) indicated that there was a significant decrease in the density of benzodiazepine binding sites in the putamen and caudate nucleus of the brains of people dying with Huntington's disease a condition known to be associated with a diffuse neuronal cell loss, notably from the striatum. There is a loss particularly of GABAergic cells from putamen, where the activities of glutamic acid decarboxylase and choline acetyltransferase are markedly reduced (Bird and Iversen, 1974). Similar observations have been made by Reisine *et al.* (1980). This supported the notion that benzodiazepine receptors were localised, at least to a considerable extent on neuronal elements. Such a neuronal localisation has been supported by observations made by Braestrup *et al.* (1979a) that intracerebellar injections of kainic acid, a neurotoxin, produce a large decrease in the number of benzodiazepine receptors. The slight increase in flunitrazepam binding in the mutant mouse *jimpy*, with an almost complete loss of oligodendroglia, indicate that these receptors are not associated with this cell type (Fry and Hanwell, 1982). The suggestion that the benzodiazepine receptors are localised to some considerable extent on glial cells (Henn and Henke, 1978) has received little support and the inability to identify brain specific benzodiazepine receptors on mouse primary astroglial cell cultures (Braestrup *et al.*, 1978) would appear to conclude the evidence. However, recent reports (Schoemaker *et al.*, 1981) have shown that [^{3}H] Ro5-4864 exhibits saturable binding to rat cerebral cortex membranes; this ligand, (4$'$-chlorodiazepam) exhibits a marked preference for the peripheral type of receptor, found in kidney and liver. The Bmax value obtained in brain, however, represents only 10–20% of that obtained using flunitrazepam as the ligand. While the physiological significance of this binding site is unclear it is possible that these 'peripheral type' sites may be located on glial elements in the CNS, and it is apparent that the pharmacological specificity of this site is markedly different from that of the 'CNS' type of benzodiazepine receptor. Further studies are undoubtedly required to elucidate this current uncertainty, though from present information it is unlikely that the results already obtained with [^{3}H] flunitrazepam and [^{3}H] diazepam will suffer significant changes and our current interpretations of the data collected will remain unchanged.

While a neuronal localisation of benzodiazepine receptors has now received almost universal acceptance it has been possible to define further certain cell populations upon which these receptors are localised. This can be accomplished by using different strains of mutant mouse which are known to suffer distinct losses of certain cell types and has suggested that at least a proportion of the benzodiazepine receptors in cerebellum are localised on Purkinje cells (Lippa et al., 1978b; Braestrup et al., 1979a, Biscoe et al., 1982).

The brain-specific benzodiazepine receptor had been reasonably well characterised by the middle of 1979 and attention was then increasingly focused on its interaction with other neurotransmitter systems.

3.2 Interaction with GABA

Initial studies in which attempts were made to determine the specificity of [^{3}H] diazepam binding failed to reveal that GABA did not decrease but increased the binding of diazepam. However, the effects are quite pronounced and were later demonstrated by a number of groups (Tallman et al., 1979; Wastek et al., 1978; Martin and Candy, 1978; Karobath and Sperk, 1979). The presence of GABA appears to increase the affinity of the benzodiazepine receptor for its ligand, without significantly affecting the total number of receptors involved in the interaction. The effect is produced by a number of GABA agonists (Karobath et al., 1979; Supavilai and Karobath, 1980a) and the effect is blocked by bicuculline (Tallman et al., 1979). The mechanism of the interaction is still not understood but the previous evidence for facilitation of GABA mediated effects by the benzodiazepines may suggest that the interaction between these two receptors is a two-way system each mutually modifying the other by some allosteric mechanism.

The presence of chloride ions in the incubation medium was shown to increase the affinity of the benzodiazepine receptor for [^{3}H] diazepam still further and chloride acts synergistically with GABA rather than being merely additive (Martin and Candy, 1978). This led to the speculation that the anion effects were due to interactions at the chloride ionophore, known to be involved in GABA mediated transmission, and therefore that the benzodiazepine receptor was part of a GABA benzodiazepine receptor—chloride ionophore complex. This hypothesis received support from observations that the ability of a number of other anions to facilitate benzodiazepine binding were the same ions which were most easily able to penetrate the activated inhibitory post-synaptic membrane of cat motoneurones (Costa, Rodbard and Pert, 1979; Candy and Martin, 1979b; Rodbard, Costa and Pert, 1979). This anion and GABA modulation appears to be partially antagonised by the anion channel blockers 4,4'-diisothiocyano-2,2'-disulphonic acid stilbene (DIDS) and 4-acetamido-4'-isothiocyano-2,2' disulphonic acid stilbene (SITS) (Costa et al., 1981), which lends support to the original concept. Additional corroborative evidence for the existence of such a complex has come from studies in which a novel group of anxiolytic agents, the pyrazolopyridines, have been shown to increase the affinity of the benzodiazepine receptors for their ligands (Williams and Risley,

1979) only in the presence of chloride, bromide or iodide (Supavilai and Karobath, 1979, 1980b). Furthermore, the facilitatory effects of these compounds were different in various brain regions, perhaps indicating a variability in the modulatory action within the receptor ionophore complex in different brain regions (see also Supavilai and Karobath, 1980c).

While the modulation of benzodiazepine binding by GABA has been relatively easy to demonstrate, the reverse interaction, the modulation of GABA binding by the benzodiazepines was initially reported by Guidotti *et al.* (1978) but has proved much more ephemeral. It is, however, now quite clear that in a number of different membrane preparations it is possible to show small but consistent increases in GABA or muscimol binding in the presence of micromolar concentrations of a number of benzodiazepines.

It would appear, therefore, that neurochemical evidence is consonant with the electrophysiological evidence in the hypothesis that the benzodiazepine receptor is part of the GABA receptor−chloride ionophore complex, and that these compounds produce their effects by interactions with this complex.

3.3 The receptor *in vivo*

One criticism of the initial studies with [^{3}H] diazepam came from the observation that the binding of this ligand to its receptor was extremely temperature-dependent and could not be demonstrated at 37 °C *in vitro* but only 4 °C, an observation which initially cast doubt on the physiological relevance of such a receptor system. However, later studies with the alternative ligand [^{3}H]-flunitrazepam, exhibiting higher affinity for the benzodiazepine receptor, were able to demonstrate significant levels of receptor binding at physiological temperatures *in vitro* (Braestrup and Squires, 1978; Speth *et al.*, 1978), and evidence was accumulated to suggest that these two ligands bound to the same population of receptors. It was suggested that the inability to demonstrate [^{3}H] diazepam binding at 37 °C was due to the fact that the dissociation rate constant for this ligand at this temperature was so rapid that dissociation occurred during the separation and washing of the membrane preparation subsequent to equilibrium binding (Braestrup and Squires, 1978).

However, the fact that much of the *in vitro* binding experimentation used in the characterisation of this receptor was carried out at 4 °C made it imperative to demonstrate similar properties of the receptor system *in vivo*. Such validation came initially from the work of Chang and Snyder (1978) who were able to show that after intravenous injection of [^{3}H] flunitrazepam to rats, radioactivity was associated with brain particulate fractions, and the majority of this was unchanged flunitrazepam; they also showed that the *in vivo* potencies of four benzodiazepines in preventing [^{3}H] flunitrazepam binding paralleled, in rank order, their affinities determined *in vitro*, and sodium pentobarbitone failed to displace [^{3}H] flunitrazepam binding either *in vivo* or *in vitro*. This ligand, therefore, appeared to bind to benzodiazepine receptors *in vivo* and the doses of drugs required to reduce this binding by 50% were similar to their ED_{50} values against mouse pentylenetetrazole induced convulsions *in vivo*. Similar experiments were also carried out with [^{3}H] diazepam by Williamson *et al.* (1978). They were able to show a regional distribution of binding in brain similar to that previously found *in vitro*, together with a stereospecificity of the displacable binding only for the CNS receptor but not the peripheral (kidney etc.) type, again in concordance with the *in vitro* results. Tallman *et al.* (1979) also found an increase in the *in vivo* binding of [^{3}H] diazepam in animals pre-treated with the GABA agonist muscimol or the inhibitor of GABA metabolism aminooxyacetic acid. The above results gave confidence that the *in vitro* investigations of benzodiazepine binding interactions were an adequate representation of the situation obtained *in vivo*.

A criticism which can be offered to the *in vivo* approach to binding, however, is that the measurement is actually carried out after the animal has been killed, and this may significantly affect the results. The alternative approach of positron emission tomography has been used by Comar *et al.* (1979). In this method [^{11}C] flunitrazepam was injected into baboons and the positrons and associated annihilation gamma rays emitted on the decay of the short-half life ^{11}C were monitored by positron emission tomography. Using this approach, therefore, the *in vivo* binding of flunitrazepam can be investigated non-invasively. The results showed a concentration of flunitrazepam in brain tissue and its partial displacement after the subsequent injection of lorazepam, results which are in accord with previous *in vivo* binding experiments. However, there are considerable technical difficulties, not least of which is the limited resolution of the equipment in relation to the size of the baboon brain. The results have since been extended to man in which a clear differential topographical localisation of the flunitrazepam could be seen, though displacement experiments were not carried out (Maziere *et al.*, 1981).

3.4 Distribution as revealed by autoradiography

While *in vitro* neurochemical techniques have been used to reveal the relatively gross distribution of the benzodiazepine receptor in various mammalian species, much greater detail has been revealed by light microscopic autoradiographic techniques (see Kuhar, 1978). These studies showed high densities of the receptor in the limbic system including the amygdala, hippocampus and hypothalamus and in the cortex, thalamus and cerebellum, with a complete absence of binding in the white matter, in agreement with previous biochemical determinations. However, marked sub-regional variations were found for example in the cervical spinal cord where laminae II, III, IV and X contained relatively high concentrations of the receptor compared to lamina IX which contains the motoneurons (Young and Kuhar, 1979, 1980). The ability to obtain such detailed mapping of receptor localisation has revealed very considerable differences in the distribution of benzodiazepine and GABA receptors. The benzodiazepine receptors, for example, are present in approximately equal densities in the molecular and granule cell layers of the cerebellum (though variations were seen in different species) while the GABA receptors are more concentrated in the granule cell layer. However, facilitation of benzodiazepine binding by GABA was equal in all areas (Young and Kuhar, 1979; Palacious et al., 1980; see also Murrin, 1981). This is in agreement with the results obtained by microdissection of the bovine cerebellum and subsequent determination of benzodiazepine and GABA binding sites by neurochemical methods (Candy and Martin, unpublished).

The recent development of [^{3}H] sensitive film has allowed much more information to be gained from such experiments. One of the main difficulties with the autoradiographic technique using photosensitive emulsions is that quantification can only be carried out by counting silver grains; with [^{3}H]-sensitive film computer assisted densitometry can be used for the quantification. Though some resolution is lost with this film the greater ease with which measurements can be accomplished will allow quantitative receptor pharmacology to be carried out in areas too small to study easily with conventional techniques (Penney et al., 1981). In the case of [^{3}H] flunitrazepam binding it has been shown that both techniques provide comparable results (Palacios et al., 1981a).

Another extremely elegant approach has been used by Mohler et al. (1980) in which they have made use of their observation that [^{3}H] flunitrazepam could be irreversibly bound to the benzodiazepine receptor by exposure to UV light. This group was able to fix tissue subsequently to allow autoradiography at the electron microscopic level. They were thus able to localise specific benzodiazepine binding to regions of synaptic contacts, and in the cerebellum specifically bound silver grains were found most frequently on axodendritic contacts close to mossy fibres but rarely on the mossy fibres themselves, suggesting that the benzodiazepine binding sites were localised on the synaptic contacts of Golgi cells, which are thought to be GABAergic (McLaughlin et al., 1974; Wilkin et al., 1974).

3.5 Plasticity

If the benzodiazepine receptor serves some purpose in the normal functioning of the animal then it is reasonable to expect that it will respond to perturbations of the organism, and various means have been sought to demonstrate such responses.

There have been numerous reports that chronic exposure of a number of neurotransmitter receptors to their agonists or antagonists can lead to a change in receptor density. It has also been observed that both the sedative and anticonvulsant actions of the benzodiazepines show tolerance while the anxiolytic effects do not (Goldberg et al., 1967; Margules and Stein, 1968; Killam et al., 1973, Sepinwall et al. 1978). Studies in which rats were given daily injections of diazepam for 30 days showed no change in [^{3}H] diazepam binding in 8 brain regions subsequently studied, while [^{3}H] GABA binding was significantly decreased in striatum and cholinergic muscarinic binding ([^{3}H] quinuclidinylbenzilate) was decreased in the cerebellum, both changes being in receptor number rather than affinity (Mohler et al., 1978a). Chronic treatment of rats for 8 weeks with diazepam or lorazepam were also without gross effects on [^{3}H] diazepam binding 5 or 11 days after the last dose, these time intervals being selected as abstinence sysndromes have been noted over the 1–10 day period (Braestrup et al., 1979c; see Hollister, 1977). Even with extremely high doses of benzodiazepines for 10 days only small (15%) decreases in benzodiazepine receptors were found (Chiu and Rosenberg, 1978) indicating that the system is extremely resistant to modification under these conditions, and that the cause of withdrawal or tolerance effects of these compounds must be sought elsewhere.

The benzodiazepine receptor does appear to respond to certain pharmacological insults. Rapid increases in the number of receptors, demonstrable only under certain conditions, are seen after acute administration of diazepam (Speth et al., 1979) while multiple sub-convulsive doses of pentylenetetrazole cause marked increases in the B_{max} for this receptor (Syapin and Rickman, 1981). It has also been shown that electroconvulsive shock leads to an increase in the number of benzodiazepine receptors demonstrable in rat cortex, though the change was very short-lived being maximal at 15–30 min but returned to control 60 min after the seizure (Paul and Skolnick, 1978). However, recent attempts to reproduce this effect have been unsuccessful. Using similar conditions no change in K_d or B_{max} values for benzodiazepine binding were found, though pronounced increases in GABA concentrations were noted in certain brain areas (Bowdler and Green, 1981), the time course of the effect, however, being short-lived. This work has since been extended to show that chronic seizures (once per day for 10 days) again elicit no change in benzodiazepine binding, although a much more prolonged rise in striatal GABA levels was found (Bowdler and Green, 1982a, 1982b) and it may be this increase which is responsible for the enhanced anticonvulsant effects of the benzodiazepines after repeated electroconvulsive shock (Cowen and Nutt, 1982).

The benzodiazepine receptor system does, therefore, appear to respond to several pharmacological manipulations in the whole animal, though the interpretation of the results is difficult because of the complexity of the interactions involved (see Section 3.7 for further discussion).

Attempts have also been made to investigate the effects of various stressful or anxiety provoking reactions on the benzodiazepine receptor, in the belief that the system may respond to such situations which the benzodiazepines are known to alleviate. Lippa et al. (1978a) found that experimental stress produced a significant decrease in [^{3}H] diazepam binding in rat cortex, a result in contrast to the observation of Soubrie et al. (1980) where an increased benzodiazepine binding was found after swim stress. The confusion would appear to be supported by Braestrup *et al.* (1979b) in which the effects on benzodiazepine binding of various stressful stimuli were small and not unidirectional. In retrospect it may be unreasonable to expect a complex animal behaviour to produce a simple change in a single receptor system. Alternatively comparisons have been made between various animal strains which have been bred from a common stock to accentuate certain behavioural characteristics. Robertson *et al.* (1978) found that in two rat strains bred for high and low fearfulness there were marked differences in benzodiazepine receptor density, the former showing a lower density and similar results were found in four strains of mice (Robertson, 1979). However, in Roman high and low avoidance animals, in this case differing in certain conventional emotionality measures, the differences between the strains did not prove significant with respect to benzodiazepine receptors (Shephard *et al.*, 1982). While the attempts to correlate complex animal behaviours with simple neurochemical measures are fraught with very considerable difficulties at our present level of understanding, rather more fruitful results may be expected from the investigation of certain mutant animal strains where specific genetic deficiencies may be expected to lead to more circumscribed neurochemical abnormalities.

The mutant mouse *spastic* is characterised by muscle rigidity, a condition which is relieved by a variety of drugs, and is extremely sensitive to the benzodiazepines. Light microscopic investigation of this mutant has failed to reveal any gross structural abnormalities, though when compared with their unaffected littermates the spastic mouse showed significant increases in both diazepam and muscimol binding in the spinal cord (Biscoe *et al.*, 1981). In this case the mutant showed significant physiological malfunction, muscle spasticity, and also marked changes in the GABA-benzodiazepine receptor systems indicating that the two may be associated.

It is clear, therefore, that the benzodiazepine receptor is responsive to certain types of pharmacological manipulation though in some contrast to other receptor systems where changes take place over days or even weeks, here we find a number of examples of extremely rapid changes measured in minutes. Evidence has also been gleaned from some inbred animal strains and the mutant mouse *spastic* that longer-term changes have taken place, presumably under genetic control, which suggest correlations with behaviours which this group of pharmacological agents are known to modify. These investigations are, however, simply pieces of data at present, though further detailed investigations may allow a more precise understanding of the functional role of the receptor systems involved.

3.6 Endogenous ligands
Following the identification of the opioid peptides as the natural ligands for the opiate receptor (see Chapter 2 and Hughes *et al.*, 1975) it was inevitable that attempts would be made to identify the endogenous substance in brain which

served the same role at the benzodiazepine receptor. While this area of research has seen enormous effort expended (see Martin, 1980), arguments have been put forward concerning the necessity to even envisage the existence of such a compound (Mohler, 1981).

The techniques used to identify putative endogenous ligands for this receptor have all relied on the subfractionation of the brain, by one means or another, and the subsequent testing of these fractions for their ability to displace the benzodiazepines from their binding sites on brain membranes. Three of the endogenous compounds identified in this way, inosine, hypoxanthine and nicotinamide, exhibit a degree of specificity for the benzodiazepine receptor, though little affinity for it, with concentrations in the millimolar range required to displace [^{3}H] diazepam binding. Considering the concentrations of these substances normally found in brain it seems unlikely that they could occupy a significant number of benzodiazepine receptors under normal physiological conditions. However, nicotinamide shows a remarkably similar pharmacological profile to that of the benzodiazepines when administered to animals (Mohler *et al.*, 1979). While these compounds can be easily dismissed as putative endogenous ligands, there is the possibility that close structural analogues may have a greater affinity for the receptor, as it is clear that minor modifications in molecular structure can produce marked changes in affinity. One of the most remarkable findings to emerge from this search has been the compound ethyl β-carboline-3-carboxylate, exhibiting a very high affinity for the receptor (4–7 nM). This compound was isolated by Braestrup *et al.* (1980) from human urine and brain tissue, though it is now clear that the compound was an artefact produced in the extraction procedure. The compound has, nevertheless, proved to be a valuable experimental tool (see Section 3.8) though the precursor from which it was derived in the extraction procedure has still not been identified.

A number of high molecular weight materials have also been isolated, three of which appear promising, though work on them has not yet come to a firm conclusion.

The first studies of Guidotti *et al.* (1978) suggested that a single thermostable protein isolated from rat brain was able to inhibit the binding of both [^{3}H] GABA and [^{3}H] diazepam to their receptor populations. Further work has shown, however, that the GABA and benzodiazepine displacing activities can be separated by further purification, suggesting that the two proteins involved are separate entities. The first, GABA-modulin, appears to displace GABA non-competitively from its binding sites and has a molecular weight of 16–17,000 daltons (Mazzari *et al.*, 1981). The second, the benzodiazepine displacing material, is much smaller with a molecular weight of about 8,000 daltons and displaces the benzodiazepines competitively from their binding sites (Massotti and Guidotti, 1980; Guidotti *et al.*, 1981). The material is thermostable, pronase-sensitive and shows a differential topographical distribution in the brain (Massotti *et al.*, 1981).

Another material isolated by Davies and Cohen (1980) is able to displace the benzodiazepines competitively from their binding sites and produces EEG changes similar to those of diazepam in animals. The molecular weight of this material is about 3,000 daltons and its amino acid composition is known, though further information has not been published.

The most recent report from Woolf and Nixon (1981) concerns a third material, named nepenthin, which has been purified to apparent homogeneity and appears to have a molecular weight of around 16,000 daltons though the stability of its displacing activity after treatment with proteases suggests that a fragment of lower molecular weight retains activity. Little is so far known about the compound, but antibodies have been successfully raised to the purified material and these have been used in double antibody labelling experiments to reveal immunologically similar material in some cells of deep cortical regions of the rat forebrain.

There is, therefore, no lack of choice of hypothesis concerning the identity of the endogenous ligand for the benzodiazepine receptor, though very considerable efforts will need to be made before the relevance of any of these compounds can be judged.

3.7 The receptor and its effector complex

The majority of evidence currently available suggests that the benzodiazepines produce their pharmacological effects by an interaction with the neurotransmitter GABA. The ubiquitous occurrence of this neurotransmitter system makes it inevitable that the benzodiazepines will influence other neurotransmitter systems via the GABA system, and indeed many such interactions have been demonstrated, though none of these have been shown to take place without the influence of the GABAergic system. Until more evidence becomes available, therefore, it would seem sensible to limit our discussion to GABA-benzodiazepine interactions.

It is clear that although the benzodiazepines influence GABA mediated transmission (see Section 2.1), the ratio between the number of receptors for GABA and for the benzodiazepines is not constant in different regions of the central nervous system, and the influence of GABA on the interaction of the benzodiazepine receptor with its ligand is variable (see Palacios et al., 1981b). The simplest explanation of these observations is that not all GABA and benzodiazepine receptors are coupled to each other in the same way; it is therefore possible that in some situations they each operate in isolation. It is clear, however, that reciprocal interactions can be demonstrated and as such are worthy of further discussion.

GABA exerts its effects through increased postsynaptic membrane permeability to chloride ions (McBurney and Barker, 1978; Nistri et al., 1980); the GABA recognition site and the effector system linking it to the chloride ion channel appear to be part of a receptor complex, at least some of which contain recognition sites for the benzodiazepines. The affinity of the benzodiazepine receptor for its ligand is influenced by both GABA and certain anions which are capable of passing through the activated chloride channel, and in turn the GABA receptor is subject to modulation by the benzodiazepines (see Section 3.2). It must be made clear that the observations of the reciprocal interaction have been made with receptor-ligand binding techniques on disrupted membranes in non-physiological systems. It would of course be invaluable if evidence could be obtained on the effector system itself and the obvious system to monitor would be the chloride flux. However, at the present time it has proved impossible to monitor chloride flux neurochemically in systems which contain both GABA and benzodiazepine receptors. Attempts have been made in various clonal cell lines (Baraldi *et al.*, 1979; Syapin and Skolnick, 1979) but it would appear that the benzodiazepine receptors found in these cells exhibit different characteristics to those in the mammalian CNS, making interpretation of the data difficult (but see McCarthy and Harden, 1981). Such investigations can be carried out electrophysiologically though the experimental techniques are complex. It has, therefore, so far only been feasible to seek information about the interaction of the benzodiazepine receptor with other receptors and their effector complexes using *in vitro* binding techniques. Interesting information has become available, though great care must be taken not to over-interpret this information and continually to make reference to data obtained by different techniques.

Pentobarbitone also appears to enhance GABA mediated inhibition in the mammalian CNS (Evans, 1979; Lodge and Curtis, 1978; Ransom and Barker, 1976). It has recently been shown that pentobarbitone is able to increase the Na^+-independent GABA binding in crude synaptosomal membrane preparations. The effect is antagonised by picrotoxin (Willow and Johnston, 1980), and the relative enhancement of GABA binding amongst 6 barbiturates was correlated with their ability to induce anaesthesia in mice or to reverse the bicuculline antagonism of GABA responses in the rat superior cervical ganglion (Asano and Ogasawara, 1982). It is not surprising, therefore, to find that pentobarbitone is also able to facilitate the binding of the benzodiazepines to their receptor in the presence of GABA (Skolnick, *et al.*, 1980); the effects of GABA and pentobarbitone were synergistic, and were seen at sub-anaesthetic concentrations of pentobarbitone. It is clear, however, that this facilitation is dependent upon the presence of certain anions and a remarkable correlation is seen between these anions and those previously shown to have direct facilitatory actions on benzodiazepine binding itself (Leeb-Lundberg *et al.*, 1980; see Section 3.2) A number of pyrazolopyridines, compounds with apparent anxiolytic activity, have also been shown to facilitate benzodiazepine binding in a chloride dependent manner, (Beer *et al.*, 1978; Williams and Risley, 1979; Salama and Meiners, 1980) which is antagonised by picrotoxin (Supavilai and Karobath, 1979, 1980b, 1981).

Picrotoxin, a convulsant which blocks GABA mediated transmission without blocking GABA binding to its recognition site (Olsen et al., 1978a, 1978b) appears to exert its actions through pharmacological receptors which can be effectively labelled using [^{3}H]α-dihydropicrotoxinin (see Olsen, 1981). This ligand can be competitively displaced from its binding sites with the barbiturates (Ticku and Olsen, 1978) and the anxiolytic pyrazolopyridines (Olsen and Leeb-Lundberg, 1981), suggesting but these compounds produce their effects at the benzodiazepine receptor by interaction with the picrotoxin recognition site, the two receptors being linked in some way.

The receptor complex with which the recognition site for the benzodiazepines appears to be engaged, at least in some situations, appears also to contain recognition sites for GABA and picrotoxin together with the appropriate effector machinery which is presumably responsible for the cross-talk between the recognition sites (Simmonds, 1980). The chloride ionophore itself is the functional unit which they modulate but it is at present unclear whether or not the linkage to the ionophore from each site is direct. The situation is rendered more complex by the fact that each recognition site may also be under the influence of endogenous ligands which may exert some tone on the whole complex. Many attempts are in progress to solubilise the separate entities in order that they may be studied in isolation. Indeed there is considerable information already available but it is at present difficult to draw a consensus opinion as the solubilisation procedures used by different groups vary. The molecular weight determinations subsequently carried out indicate that while some procedures isolate a monomeric benzodiazepine recognition site essentially stripped of its effector machinery and associated sites, others appear to solubilise more complex units with some of the interactions intact (Chang et al., 1981; Doble and Iversen, 1982; see Olsen (1981) for summary of earlier work).

However, electrophysiological investigations are beginning to add further information about the interactions within this complex. Study and Barker (1981) have recently shown by fluctuation analysis in voltage clamped mouse spinal neurones grown in culture that while both pentobarbital and the benzodiazepines facilitate GABA mediated transmission, diazepam appears to produce the effect mainly by increasing the frequency of ion channel opening in the membrane compared with GABA alone, while pentobarbital decreased the frequency of channel opening but increased the channel open lifetime. This would suggest that different parts of the receptor complex are coupled to the ion channel in different ways and it is only with these sophisticated analyses that we can hope to unravel the mechanisms of action of these compounds.

3.8 The case for multiple receptors

The diverse pharmacological spectrum of the benzodiazepines has for some time raised the possibility that there may be multiple receptors for these compounds, each responsible for a different aspect of their activity profile. Such a situation would, of course, be of considerable value as the medicinal chemist may hope to devise compounds such that they interact with just one subpopulation and therefore exhibit only certain aspects of the behavioural profile of the benzodiazepines.

Evidence that this may be the case initially came from studies with a series of triazolopyridazines which were reported to possess anticonflict and anticonvulsant activities but lack the sedative actions of the benzodiazepines (Lippa et al., 1979) and although this work has not since been confirmed, the biochemical evidence with one of this series, CL 218,872, suggested that receptor multiplicity may in fact occur (Squires et al., 1979). These workers demonstrated that the benzodiazepine receptor population in rat brain exhibited polyphasic thermolability and dissociation kinetics, in accord with the hypothesis of a heterogeneous receptor population. They were unable to show this heterogeneity with the classical benzodiazepines in equilibrium binding experiments but the compound CL 218,872 appeared to possess a greater affinity for one subpopulation of the receptors than the other, evidenced by the shallow slopes of displacement against [^{3}H] flunitrazepam in rat brain membranes.

Similar biochemical results were obtained when ethyl-β-carboline-3-carboxylate (β-CCE), the compound isolated in attempts to find the endogenous ligand for the benzodiazepine receptor (see Section 3.6), was used to displace [^{3}H]-flunitrazepam from rat brain hippocampal membranes, which again exhibited shallow displacement curves (Nielsen and Braestrup, 1980). However, the same authors showed that if similar experiments were carried out in the cerebellum, no such deviation from mass action occurred. The apparent IC$_{50}$ for β-CCE in the two rat brain regions were significantly different, the compound being apparently more potent in the cerebellum by a factor of about 4. These authors suggested that the cerebellum contained largely one subtype of receptor, so called Bz$_1$ which exhibited a high affinity for β-CCE, while the hippocampus contained also a receptor with lower affinity for β-CCE, the Bz$_2$ subtype. The reason that previous evidence for heterogeneity in the receptor population had not been found when using classical 1,4-benzodiazepines as ligands was that these compounds were unable to differentiate between the two receptor populations, they exhibited the same affinity for both Bz$_1$ and Bz$_2$.

The suggestion of benzodiazepine receptor heterogeneity received further support from the work of Sieghart and Karobath (1980) who used [^{3}H] flunitrazepam as a photoaffinity label for the benzodiazepine receptor (see Section 4.4) and subsequently separated the labelled receptor proteins by SDS-gel electrophoresis and identified at least two protein bands in certain brain areas, suggesting that the different subtypes of benzodiazepine receptor were physically separable and possessed different apparent molecular weights. These experiments have been repeated with only minor technical differences but only a single labelled protein band was found and the explanations for this discrepancy are not yet clear (Thomas and Tallman, 1981). However, Braestrup and Nielsen (1981a) have extended their original findings and those of Sieghart and Karobath (1980) with the use of [^{3}H] propyl β-carboline-3-carboxylate which appeared to differentiate between the two subpopulations of binding sites in a similar manner to β-CCE. The sites appear to exhibit a differential topographical distribution in the rat brain, the cerebellum being largely Bz_1 while the hippocampal sites are almost equally distributed between the Bz_1 and Bz_2 subtypes. Recently Lo et al. (1982) have found it possible to differentiate two types of benzodiazepine receptor by preferential solubilisation in various detergents. The receptors that resist solubilisation, called type I, were most highly concentrated in the cerebellum, and showed a higher affinity for the β-carboline esters, and here again several classical benzodiazepines were unable to differentiate between the two.

Detailed analysis of the binding interaction of the benzodiazepines with the receptor population in rat brain membranes has shown that both the 'on' and 'off' rates of the ligands do not obey simple mass action kinetics. The data can, however, be adequately analysed assuming biphasic kinetics, again consistent with two separate benzodiazepine receptor populations (Quast and Mählmann, 1982; Chiu et al., 1982). However, these data, while being consistent with a multiple receptor subtype model are capable of different interpretation and Chiu et al. (1982) prefer to view their data in terms of a single benzodiazepine receptor which can exist in two states.

The shallow displacement curves produced by β-CCE on [^{3}H] flunitrazepam binding in the hippocampus can only be interpreted as multiple receptors if the ligand interaction can be shown *not* to exhibit cooperativity, as pointed out by Nielsen and Braestrup (1980). Re-investigation of the phenomenon of cooperativity in the binding of flunitrazepam has shown quite clearly that cooperativity does exist in this system, and furthermore that the sites from which the dissociation of the ligand can be shown to occur vary in proportion according to the level of occupancy of the whole population (Doble, Iversen and Martin, 1982). This evidence suggests that the two sites from which dissociation occurs are not distinct and immutable, but exist in some sort of equilibrium, that is two states of the same receptor as proposed by Chiu et al. (1982). The equilibrium which appears to govern what site is the more energetically favourable is obviously controlled by the occupancy of the sites with benzodiazepine but is controlled in a different way by occupancy with β-CCE, i.e. the receptor appears to be capable of monitoring not only its state of occupation but also the class of ligand involved.

It has now been shown that β-CCE has completely different behavioural actions to the benzodiazepines. It is a proconvulsant and can antagonise the anticonvulsant and sedative effects of the benzodiazepines *in vivo* (Tenen and Hirsch, 1980; Oakley and Jones, 1980; Cowen et al., 1981) and is also able to antagonise the action of diazepam *in vitro* (Mitchell and Martin, 1980). It has been possible to differentiate those compounds with behavioural actions like the benzodiazepines from those which have opposite actions by the fact that the former group exhibit an increased affinity for the receptor in the presence of GABA and sodium chloride while the latter group do not (Braestrup and Nielsen, 1981b; Doble, Martin and Richards, 1982; Fujimoto et al., 1982; Skolnick et al., 1982; see also Mohler and Richards, 1981). This would suggest that β-CCE is simply acting as an antagonist at the benzodiazepine receptor. The situation would appear to be rather more complex than this as recently two 'benzodiazepine antagonists' have been produced, Ro 15-1788 and CGS 8216, which are able to rapidly reverse the actions of the benzodiazepines but produce no overt behavioural actions in their own right at similar doses (Hunkeler et al., 1981; Czernick et al., 1981, 1982). The compound Ro 15-1788 has also been shown to reverse the effect of β-CCE on seizure thresholds *in vivo* and its actions on the cervical sympathetic ganglion *in vitro* (Nutt et al., 1982). Before any attempt is made to interpret these data we must consider some further evidence which is relevant.

It has been shown previously that flunitrazepam can be used to photo-affinity label benzodiazepine receptors, a procedure which effectively occludes the binding site for the benzodiazepines as evidenced by the fact that the B_{max} for the receptor population is markedly decreased in membranes so treated if this value is determined using [^{3}H] flunitrazepam (Battersby et al., 1979). We have found that under such conditions neither the affinity nor the B_{max} value for [^{3}H]-β-CCE in hippocampal membranes is markedly affected (Brown and Martin, 1982) and similar findings have recently been reported by Hirsch et al. (1982) using [^{3}H] propyl β-carboline-3-carboxylate. It would, therefore, appear that the recognition properties for β-CCE are different from those of flunitrazepam. The binding of each ligand, however, is mutually exclusive in competition experiments and from this we can conclude that the photoaffinity labelling of the benzodiazepine receptor does not occupy the binding site but may simply occlude it. The simplest, but not the only, interpretation of the unabated binding of β-CCE in such membranes is that the β-CCE binding site is sufficiently different not to be so occluded. This suggests that the two binding sites have different recognition properties but share, to some degree at least, an effector system. This may be unexpected but what is of equal significance is the fact that if photoaffinity labelled membranes are allowed to bind [^{3}H]β-CCE it is only the classical 1,4-benzodiazepines which exhibit an apparently decreased ability to displace this ligand when compared with control membranes. The two antagonists Ro 15-1788 and CGS 8216 along with all the β-carboline 3-carboxylates so far investigated show no apparent affinity decreases in the photoaffinity labelled compared with the control membranes. The most parsimonious explanation of this is that the benzodiazepine and β-CCE recognition sites are different and that the two antagonists mentioned bind to the β-CCE site rather than to the benzodiazepine site; they produce their antagonist actions via the linkage mechanism which clearly operates between the two recognition sites (Brown and Martin, 1982).

In summary, the current evidence would appear to suggest that only a single population of benzodiazepine recognition sites are present in the CNS and that they can exist in two different energetic conformations controlled by some equilibrium, the influential factors upon which are at present unclear but may include endogenous regulators of some type. The recognition sites for the β-carboline-3-carboxylates are different from those of the benzodiazepines but may merely represent the distinct domain on the benzodiazepine receptor protein, a concept which must await the collection of further information.

4. SUMMARY AND FUTURE

It has not been possible within the limits of this review to annotate all the many and significant contributions to our present understanding of the mechanism of action of the benzodiazepines that have been made over the past twenty years. Our knowledge is by no measure complete and the complexity of the effector systems revealed continues to provide considerable impetus to research.

The benzodiazepine recognition site in the mammalian CNS represents part of the receptor complex through which these drugs produce their pharmacological effects. It is intimately linked with the GABA-chloride ionophore effector system which in turn appears to suffer modulation by machinery involved in the actions of the barbiturates which are thought to produce their effects through recognition of the picrotoxinin binding site. There is also some evidence to suggest that the β-carboline-3-carboxylates possess a separate recognition site, which may simply be a domain of the benzodiazepine site, and is able to feed information in some way into this complex. However, much of the data so far obtained has been accumulated using simple binding techniques, the results of which have frequently been the subject of over-interpretation. We do not possess, at present, a simple pharmacologically relevant system to test the hypothesis built upon our test-tube measurements, and such a system is very much required. The electrophysiologists started the hypothesis building with their suggestion that these compounds facilitate GABA mediated transmission and the neurochemists must continually return to them to test rationally the ideas that have since developed.

Much store was set by the idea that the different pharmacological actions of the benzodiazepines were mediated by different subtypes of the benzodiazepine receptor. At present we have insufficient data to support the notion of multiplicity. We do have an extremely complex series of events within the action of these compounds, and it is only with a greater understanding of these events coupled with a more precise structure–activity understanding with respect to the behavioural actions of these compounds, that we can hope rationally to design new drugs with a more restricted spectrum of activity.

ACKNOWLEDGEMENTS

I would like to thank Dr. L. L. Iversen for reading the manuscript and for the many helpful suggestions that he made, Mrs. M. Wynn for typing the manuscript and Dr. A. Doble and Miss C. L. Brown for many helpful discussions.

REFERENCES

Asano, T. and Ogasawara, N. (1982) Stimulation of GABA receptor binding by barbiturates. *Eur. J. Pharmacol.* **77**, 355–357.

Babbini, M., Gaiardi, M. and Bartoletti, M. (1979) Anxiolytic versus sedative properties in the benzodiazepine series: differences in structure activity relationships. *Life Sci.* **25**, 15–22.

Ban, T. A., Brown, W. T., Da Silva, T., Gagnon, M. A., Lamont, C. T., Lemann, H. E., Lowy, F. W., Ruedy, J. and Sellers, E. M. (1981) Therapeutic monograph on anxiolytic-sedative drugs. *Can. Pharm. J.* 301–308.

Baraldi, M., Guidotti, A., Schwartz, J. P. and Costa, E. (1979) GABA receptors in clonal cell lines: A model for study of benzodiazepine action at molecular level. *Science* **205**, 821–823.

Battersby, M. K., Richards, J. G. and Mohler, H. (1979) Benzodiazepine receptor: photoaffinity labelling and localisation. *Eur. J. Pharmacol.* **57**, 277–278.

Beer, B., Klepner, C. A., Lippa, A. S. and Squires, R. F. (1978) Enhancement of ^{3}H-diazepam binding by SQ 65396: A novel antianxiety agent. *Pharmacol. Biochem. Behav.* **9**, 849–851.

Bird, E. D. and Iversen, L. L. (1974) Huntington's chorea: post-mortem measurement of glutamic acid decarboxylase, choline acetyltransferase and dopamine in the basal ganglia. *Brain Res.* **97**, 457–472.

Biscoe, T. S., Fry, J. P., Martin, I. L. and Rickets, C. (1981) Binding of GABA and benzodiazepine receptor ligands in the spinal cord of the spastic mouse. *J. Physiol.* **317**, 32–33P.

Biscoe, T. J., Fry, J. P. and Rickets, C. (1982) GABA and benzodiazepine receptors in the cerebellum of the mutant mouse Lurcher. *J. Physiol.* **328**, 12P.

Bowdler, J. M. and Green, A. R. (1981) Rat brain benzodiazepine receptor number and GABA concentration following a seizure. *Brit. J. Pharmacol.* **74**, 814–815P.

Bowdler, J. M. and Green, A. R., (1982a) Regional rat brain GABA concentrations following repeated seizures. *Brit. J. Pharmacol.* **75**, 89P.

Bowdler, J. M. and Green, A. R. (1982b) Regional rat brain benzodiazepine receptor number and γ-aminobutyric acid concentration following a convulsion. Brit. J. Pharmacol. **76**, 291–298.

Braestrup, C., Albrechsten, R. and Squires, R. F. (1977) High densities of benzodiazepine receptors in human cortical areas. *Nature* **269**, 702–704.

Braestrup, C. and Nielsen, M. (1978) Ontogenetic development of benzodiazepine receptors in the rat brain. *Brain Res.* **147**, 170–173.

Braestrup, C. and Nielsen, M. (1980) Benzodiazepine receptors. *Arzneim. Forschung.* **5a**, 852–857.

Braestrup, C. and Nielsen, M. (1981a) ^{3}H-propyl β-carboline-3-carboxylate as a selective radioligand for the Bz$_1$ benzodiazepine receptor subclass. *J. Neurochem.* **37**, 333–341.

Braestrup, C. and Nielsen, M. (1981b) GABA reduces binding of ^{3}H-methyl β-carboline-3-carboxylate to brain benzodiazepine receptors. *Nature* **294**, 472–474.

Braestrup, C., Nielsen, M., Biggio, G. and Squires, R. F. (1979a) Neuronal localisation of benzodiazepine receptors in cerebellum. *Neurosci. Lett.* **13**, 219–224.

Braestrup, C., Nielsen, M., Nielsen, B. and Lyon, M. (1979b) Benzodiazepine receptors in brain as affected by different experimental stresses: the changes are small and not unidirectional. *Psychopharmacology* **65**, 273–277.

Braestrup, C., Nielsen, M. and Squires, R. F. (1979c) No changes in rat brain benzodiazepine receptors after withdrawal from continuous treatment with lorazepam or diazepam. *Life Sci.* **24**, 347–350.

Braestrup, C., Nielsen, M. and Olsen, C. E. (1980) Urinary and brain β-carboline-3-carboxylates as potent inhibitors of brain benzodiazepine receptors. *Proc. Natl. Acad. Sci. USA* **77**, 2288–2292.

Braestrup, C., Nissen, C., Squires, R. F. and Schousboe, A. (1978) Lack of brain specific benzodiazepine receptors on mouse primary astroglial cultures. *Neurosci. Lett.* **9**, 45–49.

Braestrup, C. and Squires, R. F. (1977) Specific benzodiazepine receptors in rat brain characterised by high affinity ^{3}H-diazepam binding. *Proc. Natl. Acad. Sci. USA* **74**, 3805–3809.

Braestrup, C. and Squires, R. F. (1978) Brain specific benzodiazepine receptors. *Brit. J. Psychiat.* **133**, 249–260.

Brown, C. L. and Martin, I. L. (1982) Photoaffinity labelling of benzodiazepine receptors does not occlude the β-CCE binding site. *Br. J. Pharmacol.* **77**, 312P.

Browne, T. R. (1978) Clonazepam. *New Engl. J. Med.* **299**, 812–816.

Candy, J. M. and Martin, I. L. (1979a) The postnatal development of the benzodiazepine receptor in the cerebral cortex and cerebellum of the rat. *J. Neurochem.* **32**, 655–658.

Candy, J. M. and Martin, I. L. (1979b) Is the benzodiazepine receptor coupled to a chloride anion channel? *Nature* **280**, 172–173.

Chambers, D. M. and Jefferson, G. C. (1977) Some observations on the mechanism of benzodiazepine-barbiturate interactions in the mouse. *Brit. J. Pharmacol* **60**, 393–399.

References

Chambers, D. M., Jefferson, G. C. and Ruddick, C. A. (1978) Halothane induced sleeping time in the mouse: Its modification by benzodiazepines. *Eur. J. Pharmcol.* **50**, 103–112.

Chang, L. R., Barnard, E. A., Lo, M. S. and Dolly, J. O. (1981) Molecular sizes of benzodiazepine receptors and the interacting GABA receptors in the membrane are identical. *FEBS Letts.* **126**, 309–312.

Chang, R. S. L. and Snyder, S. H. (1978) Benzodiazepine receptors: labelling in intact animals with ^{3}H-flunitrazepam. *Eur. J. Pharmacol.* **48**, 213–218.

Chiu, T. H., Dryden, D. M. and Rosenberg, H. C. (1982) Kinetics of ^{3}H-flunitrazepam binding to membrane bound benzodiazepine receptors. *Mol. Pharmacol.* **21**, 57–65.

Chiu, T. H. and Rosenberg, H. C. (1978) Reduced diazepam binding following chronic benzodiazepine treatment. *Life Sci.* **23**, 1153–1158.

Choi, D. W., Farb, D. H. and Fischbach, G. D. (1977) Chlordiazepoxide selectively augments GABA action in spinal cord cell cultures. *Nature* **269**, 342–344.

Comar, D., Maziere, M., Godot, J. M., Berger, G., Soussaline, F., Menini, Ch., Arfel, G. and Naquet, R. (1979) Visualisation of ^{11}C-flunitrazepam displacement in the brain of the live baboon. *Nature* **280**, 329–331.

Committee on the Review of Medicines (1980) CRM report on the benzodiazepines. *Pharmaceutical J.* **224**, 384–386.

Costa, E. and Greengard, P. (Eds.) (1975) *Mechanism of Action of Benzodiazepines*. Raven Press, New York.

Costa, E., Guidotti, A. and Toffano, F. (1978) Molecular mechanisms mediating the action of diazepam on GABA receptors. *Brit J. Psychiat.* **133**, 239–248.

Costa, T. Rodbard, D. and Pert, C. (1979) Is the benzodiazepine receptor coupled to a chloride anion channel? *Nature* **277**, 315–317.

Costa, T., Russel, L., Pert, C. B. and Rodbard, D. (1981) Halide and γ-aminobutyric acid-induced enhancement of diazepam receptors in rat brain. Reversal by disulfonic acid stilbene blockers of anion channels. *Mol. Pharmacol.* **20**, 470–476.

Cowen, P. J., Green, A. R., Nutt, D. J. and Martin, I. L. (1981) Ethyl β-carboline carboxylate lowers seizure threshold and antagonises flurazepam induced sedation in rats. *Nature* **290**, 54–55.

Cowen, P. J. and Nutt, D. J. (1982) Repeated electroconvulsive shock enhances the anticonvulsant effects of benzodiazepines. *Brit. J. Phamacol.* **75**, 44P.

Coyle, J. T. and Enna, S. J. (1976) Neurochemical aspects of the ontogenesis of Gabaergic neurons in the rat brain. *Brain Res.* **111**, 119–133.

Czernik, A. J., Petrack, B., Tsai, C., Granat, R. F., Rinehart, R. K., Kalinsky, H. J., Lovell, R. A. and Cash, W. D. (1981) High affinity occupancy of benzodiazepine receptors by the novel antagonist CGS-8216. *Pharmacologist* **23**, 160.

Czernik, A. J., Petrack, B., Kalinsky, H. J., Psychoyos, S., Cash, W. D., Tsai, C., Rinehart, R. K., Grawat, F. R., Lovel, R. A., Brundish, D. E. and Wade, R. (1982) CGS 8216: receptor binding characteristics of a potent benzodiazepine antagonist. *Life Sci.* **30**, 363–372.

Davies, L. G. and Cohen, R. K. (1980) Identification of an endogenous peptide ligand for the benzodiazepine receptor. *Biochem. Biophys. Res. Commun.* **92**, 141–148.

Davis, N. M., Brookes, S., Gray, J. A. and Rawlins, J. N. P. (1981) Chlordiazepoxide and resistance to punishment. *Quart. J. Exp. Psychol.* **33B**, 227–239.

Doble, A. and Iversen, L. L. (1982) Molecular size of benzodiazepine receptor in rat brain in situ: evidence for a functional dimer. *Nature* **295**, 522–523.

Doble, A., Iversen, L. L. and Martin, I. L. (1982) The benzodiazepine binding site: one receptor or two. *Brit. J. Pharmacol.* **75**, 42P.

Doble, A., Martin, I. L. and Richards, D. A. (1982) GABA modulation predicts biological activity of ligands for the benzodiazepine receptor. *Brit. J. Pharmacol.* **76**, 238P.

Evans, R. H. (1979) Potentiation of the effects of GABA by pentobarbitone. *Brain Res.* **171**, 113–120.

Fry, J. P. and Hanwell, S. M. (1982) Elevated binding of GABA and benzodiazepine receptor ligands in the forebrain of the mutant mouse jimpy. *J. Physiol.* **325**, 32P.

Fujimoto, M., Hirai, K. and Okabayashi, T. (1982) Comparison of the effects of GABA and chloride ion on the affinities of ligands for the benzodiazepine receptor. *Life Sci.* **30**, 51–57.

Goldberg, M. E., Manian, A. A. and Efron, D. H. (1967) A comparative study of certain pharmacological responses following acute and chronic administration of chlordiazepoxide. *Life Sci.* **6**, 481–491.

Greenblatt, D. J. and Shader, R. I. (1974) in *Benzodiazepines in Clinical Practice*. Raven Press, New York.

Guidotti, A., Toffano, G. and Costa, E. (1978) An endogenous protein modulates the affinity of GABA and benzodiazepine receptors in rat brain. *Nature* **275**, 553–555.

Guidotti, A., Ebstein, B. and Costa, E. (1981) Purification and characterisation of an endogenous brain peptide that competes with ^{3}H-diazepam binding. *Soc. Neurosci. Abs.* 634.

Haefely, W. E. (1978) Central actions of the benzodiazepines: general introduction. *Brit. J. Psychiat.* **133**, 231–238.

Henn, F. A. and Henke, D. J. (1978) Cellular localisation of ^{3}H-diazepam receptors. *Neuropharmacology* **17**, 985–988.

Hirsch, J. D., Kochman, R. L. and Sumner, P. R. (1982) Heterogeneity of brain benzodiazepine receptors demonstrated by [^{3}H]propyl β-carboline-3-carboxylate binding. *Molec. Pharmacol.* **21**, 618–628.

Hollister, L. H. (1977) Withdrawal from benzodiazepine therapy. *J. Amer. Med. Assoc.* **237**, 1432–1434.

Hughes, J., Smith, T. W., Kosterlitz, H. W., Fothergill, L. A. and Morris, H. R. (1975) Identification of two related pentapeptides from the brain with potent opiate agonist activity. *Nature* **258**, 577–579.

Hunkeler, W., Mohler, H., Pieri, L., Polc, P., Bonetti, E. P., Cumin, R., Schaffner, R. and Haefely, W. (1981) Selective antagonists of benzodiazepines. *Nature* **290**, 514–516.

Johnson, R. W. and Yamamura, H. I. (1979) Photoaffinity labelling of the benzodiazepine receptor in bovine cerebral cortex. *Life Sci.* **25**, 1613–1620.

Karobath, M., Placheta, P., Lippitsch, M. and Krogsgaard-Larsen (1979) Is stimulation of benzodiazepine receptor binding mediated by a novel GABA receptor? *Nature* **278**, 748–749.

Karobath, M. and Sperk, G. (1979) Stimulation of benzodiazepine receptor binding by γ-aminobutyric acid. *Proc. Natl. Acad. Sci. USA* **76**, 1004–1006.

Killam, E. K., Matsuzaki, M. and Killam, K. F. (1973) Effects of chronic administration of benzodiazepines on epileptic seizures and brain electrical activity in Papio papio, in *The Benzodiazepines* (S. Garattini, E. Musini and L. O. Randall, eds.) pp. 443–460. Raven Press, New York.

Kuhar, M. J. (1978) Histochemical localisation of neurotransmitter receptors, in *Neurotransmitter Receptor Binding* (H. I. Yamamura, S. J. Enna and M. J. Kuhar, eds.) pp. 113–126. Raven Press, New York.

Leeb-Lundberg, F., Snowman, A. and Olsen, R. W. (1980) Barbiturate receptors are coupled to benzodiazepine receptors. *Proc. Natl. Acad. Sci. USA* **77**, 7468–7472.

Lippa, A. S., Klepner, C. A., Yunger, L., Sano, M. C., Smith, W. V. and Beer, B. (1978a) Relationship between benzodiazepine receptors and experimental anxiety in rats. *Pharmacol. Biochem. Behav.* **9**, 853–856.

Lippa, A. S., Sano, M. C., Coupet, J., Klepner, C. A., and Beer, B. (1978b) Evidence that benzodiazepine receptors reside on cerebellar Purkinje cells: studies with 'nervous' mutant mice. *Life Sci.* **23**, 2213–2218.

Lippa, A. S., Critchett, D. J., Sano, M. C., Klepner, C. A., Greenblatt, F. N., Coupet, J. and Beer, B. (1979) Benzodiazepine receptors: cellular and behavioural characteristics. *Pharmacol. Biochem. Behav.* **10**, 831–843.

Lo, M. M. S., Strittmatter, S. M. and Snyder, S. H. (1982) Physical separation and characterisation of two types of benzodiazepine receptors. *Proc. Natl. Acad. Sci. USA* **79**, 680–684.

Lodge, D. and Curtis, D. R. (1978) Time course of GABA and glycine actions on cat spinal neurones: effect of pentabarbitone. *Neurosci. Lett.* **8**, 125–129.

Macdonald, R. and Barker, J. L. (1978) Benzodiazepines specifically modulate GABA-mediated postsynaptic inhibition in cultured mammalian neurones. *Nature*, **271**, 563–564.

McBurney, R. N. and Barker, J. L. (1978) GABA induced conductance fluctuations in cultured spinal neurones. *Nature* **274**, 596–597.

McCarthy, K. D. and Harden, T. K. (1981) Identification of two benzodiazepine binding sites on cells cultured from rat cerebral cortex. *J. Pharmacol. Exp. Ther.* **216**, 183–191.

McLaughlin, B. J., Wood, J. G., Saito, K., Barber, R., Vaughn, J. E., Roberts, E. and Wu, J. E. (1974) The fine structural localisation of glutamate decarboxylase in synaptic terminals of rodent cerebellum. *Brain Res.* **76**, 377–391.

Mackerer, C. R., Kochman, R. L., Bierschenk, B. A. and Bremner, S. S. (1978) The binding of ^{3}H-diazepam to rat brain homogenates. *J. Pharmacol. Exp. Ther.* **206**, 405–413.

Margules, D. L. and Stein, L. (1968) Increase of 'antianxiety' activity and tolerance of behavioural depression during chronic administration of oxazepam. *Psychopharmacologia* **13**, 74–80.

Marks, J. (1978) *The Benzodiazepines: Use, Overuse, Misuse and Abuse.* MTP, Lancaster.

Martin, I. L. (1980) Endogenous ligands for benzodiazepine receptors. *Trends in Neurosciences* **3**, 299–301.

Martin, I. L. and Candy, J. M. (1978) Facilitation of benzodiazepine binding by sodium chloride and GABA. *Neuropharmacology* **17**, 993–998.

Massotti, M. and Guidotti, A. (1980) Endogenous regulators of benzodiazepine recognition sites. *Life Sci.* **27**, 847–854.

Massotti, M., Guidotti, A. and Costa, E. (1981) The regulation of benzodiazepine recognition sites by endogenous modulators, in *GABA and Benzodiazepine Receptors* (E. Costa, G. Di Chiara and G. L. Gessa, eds.) pp. 19–26. Raven Press, New York.

Maziere, M., Godot, J. M., Berger, G., Baron, J. C., Comar, D., Cepeda, C., Menini, Ch. and Naquet, R. (1981) Positron tomography. A new method for in vivo brain studies of benzodiazepine in animal and man, in *GABA and Benzodiazepine Receptors* (E. Costa, G. Di Chiara and G. L. Gessa, eds.) pp. 273–286. Raven Press, New York.

Mazzari, S., Massotti, M., Guidotti, A. and Costa, E. (1981) GABA receptors as supramolecular units, in *GABA and Benzodiazepine Receptors* (E. Costa, G. Di Chiara and G. L. Gessa eds.) pp. 1–8. Raven Press, New York.

Mitchell, P. R. and Martin, I. L. (1978) The effect of benzodiazepines on K$^+$-stimulated release of GABA. *Neuropharmacology* **17**, 317–320.

Mitchell, P. R. and Martin, I. L. (1980) Ethyl β-carboline-3-carboxylate antagonises the effect of diazepam on a functional GABA receptor. *Eur. J. Pharmacol.* **68**, 513–514.

Mohler, H. (1981) Benzodiazepine receptors: are there endogenous ligands in brain? *Trends in Pharmacological Sciences* **1**, 116–119.

References

Mohler, H., Battersby, M. K., Richards, J. G. (1980) Benzodiazepine receptor protein identified and visualised in brain tissue by a photoaffinity label. *Proc. Natl. Acad. Sci. USA* **77**, 1666–1670.

Mohler, H. and Okada, T. (1977) Benzodiazepine receptor: demonstration in the central nervous system. *Science* **198**, 849–851.

Mohler, H. and Okada, T. (1978) The benzodiazepine receptor in normal and pathological human brain. *Brit. J. Psychiat.* **133**, 261–268.

Mohler, H. Okada, T. and Enna, S. J. (1978a) Benzodiazepine and neurotransmitter receptor binding in rat brain after chronic administration of diazepam or phenobaribital. *Brain Res.* **156**, 391–395.

Mohler, H., Okada, T., Ulrich, J. and Heitz, P. H. (1978b) Biochemical identification of the site of action of benzodiazepines in human brain by ^{3}H-diazepam binding. *Life Sci.* **22**, 985–996.

Mohler, H., Polc, P., Cumin, R., Pieri, L. and Kettler, R. (1979) Nicotinamide is a brain constituent with benzodiazepine like actions. *Nature* **278**, 563–565.

Mohler, H. and Richards, J. G. (1981) Agonist antagonist benzodiazepine receptor interaction in vitro. *Nature* **294**, 763–764.

Müller, W. E., Schläfer, and Wollert, U. (1978) Benzodiazepine receptor binding in rat spinal membrane. *Neurosci. Lett.* **9**, 239–243.

Murrin, L. C. (1981) Neurotransmitter receptors: Neuroanatomical localisation through autoradiography. *Int. Rev. Neurobiol.* **22**, 111–171.

Nielsen, M. and Braestrup, C. (1980) Ethyl β-carboline-3-carboxylite shows differential benzodiazepine receptor interaction. *Nature* **286**, 606–607.

Nielsen, M., Braestrup, C. and Squires, R. F. (1978) Evidence for late evolutionary appearance of brain-specific benzodiazepine receptors: an investigation of 18 vertebrate and 5 invertebrate species. *Brain Res.* **141**, 342–346.

Nistri, A., Constanti, A. and Krnjevic, K. (1980) Electrophysiological studies on the mode of action of GABA on vertebrate central neurones. *Adv. Biochem. Psychopharmacol.* **21**, pp. 81–90; *Receptors for Neurotransmitters and Peptide Hormones* (G. Pepeu, M. J. Kuhar and S. J. Enna, eds.). Raven Press, New York.

Nutt, D. J., Cowen, P. J. and Little, H. J. (1982) Unusual interactions of benzodiazepine receptor antagonists. *Nature* **295**, 436–438.

Oakley, N. R. and Jones, B. J. (1980) The proconvulsant and diazepam reversing effects of ethyl β-carboline-3-carboxylate. *Eur. J. Pharmacol.* **68**, 381–382.

Olsen, R. W. (1981) GABA-benzodiazepine-barbiturate receptor interactions. *J. Neurochem.* **37**, 1–13.

Olsen, R. W. and Leeb-Lundberg, F. (1981) Convulsant and anticonvulsant drug binding sites related to GABA regulated chloride ion channels, in *GABA and Benzodiazepine Receptors* (E. Costa, G. Di Chiara and G. L. Gessa eds.) pp. 93–102. Raven Press, New York.

Olsen, R. W., Ticku, M. K. and Van Ness, P. C. (1978a) Effects of drugs on γ-aminobutyric acid receptors, uptake, release and synthesis in vitro. *Brain Res.* **139**, 277–294.

Olsen, R. W., Ticku, M. K. and Miller, T. (1978b) Dihydropicrotoxinin binding to γ-aminobutyric acid receptor ionophores. *Mol. Pharmacol.* **14**, 381–390.

Palacios, J. M., Niehoff, D. L. and Kuhar, M. J. (1981a) Receptor autoradiography with tritium sensitive film: potential for computerised densitometry. *Neurosci. Lett.* **25**, 101–105.

Palacios, J. M., Unnerstall, J. R., Young, W. S. III and Kuhar, M. J. (1981b) Radiohistochemical studies of benzodiazepines and GABA receptors and their interactions, in *GABA and Benzodiazepine Receptors* (E. Costa, G. Di Chiara and G. L. Gessa eds.) pp. 53–60. Raven Press, New York.

Palacios, J. M., Young, W. S. III and Kuhar, M. J. (1980) Autoradiographic localisation of γ-aminobutyric acid (GABA) receptors in rat cerebellum. *Proc. Natl. Acad. Sci. USA* **77**, 670–679.

Paul, S. M. and Skolnick, P. (1978) Rapid changes in brain benzodiazepine receptors after experimental seizures. *Science* **202**, 892–894.

Penney, J. B., Grey, K. and Young, A. B. (1981) Quantitative autoradiography of neurotransmitter receptors using tritium sensitive film. *Eur. J. Pharmacol.* **72**, 421–422.

Polc, P. and Haefely, W. (1977) Effects of systemic muscimol and GABA in the spinal cord and superior cervical ganglion of the cat. *Experientia* **33**, 809.

Polc, P. Möhler, H. and Haefely, W. (1974) The effects of diazepam on spinal cord activities: possible sites and mechanisms of action. *Naunyn-Schmiederberg's Arch. Pharmacol.* **284**, 319–337.

Quast, U. and Mählmann, H. (1982) Interaction of [^{3}H] flunitrazepam with the benzodiazepine receptor: evidence for a ligand-induced conformational change. *Biochem. Pharmacol.,* **31**, 2761–2768.

Ransom, B. R. and Barker, J. L. (1976) Pentobarbital selectively enhances GABA-mediated post synaptic inhibition in tissue cultured mouse spinal neurones. *Brain Res.* **114**, 530–535.

Reisine, T. D., Overstreet, D., Gale, K., Rossor, M., Iversen, L. L. and Yamamura, H. I. (1980) Benzodiazepine receptors: The effect of GABA on their characteristics in human brain and their alteration in Huntingtons's disease. *Brain Res.* **199**, 79–88.

Robertson, H. A. (1979) Benzodiazepine receptors in 'emotional' and 'non-emotional' mice: Comparison of four strains. *Eur. J. Pharmacol.* **56**, 163–166.

Robertson, H. A., Martin, I. L. and Candy, J. M. (1978) Differences in benzodiazepine binding in Maudsley reactive and Maudsley non-reactive rats. *Eur. J. Pharmacol.* **50**, 455–457.

Rodbard, D., Costa, T. and Pert, C. (1979) In reply to Candy and Martin (1979) *Nature* **280**, 173–174.

References

Rossi, G. F., Dirocco, C., Maira, G. and Meglio, M. (1973) Experimental and clinical studies of the anticonvulsant properties of a benzodiazepine derivative, clonazepam (Ro-54023), in *The Benzodiazepines* (S. Garattini, E. Mussini and L. O. Randall, eds.) pp. 461–488. Raven Press, New York.

Salama, A. I. and Meiners, B. A. (1980) Enhancement of benzodiazepine binding by the novel anxiolytic ICI 136,753. *Soc. Neurosci. Abs* **6**, 190.

Schmidt, R. F., Vogel, E. and Zimmerman, M. (1967) Die Wirkung von Diazepam auf die präsynaptische Hemmung und andere Rükenmarks reflexe. *Naunyn-Schmiedeberg's Arch. Pharmacol.* **258**, 69–82.

Schoemaker, H., Bliss, M. and Yamamura, H. I. (1981) Specific high affinity saturable binding of [^{3}H]-Ro 5-4864 to benzodiazepine binding sites in the rat cerebral cortex. *Eur. J. Pharmacol.* **71**, 173–175.

Sepinwall, J., Grodsky, F. S. and Cook, L. (1978) Conflict behaviour in the squirrel monkey: Effects of chlordiazepoxide, diazepam and N-desmethyldiazepam. *J. Pharmacol. Exp. Ther.* **204**, 88–103.

Shader, R. I. and Greenblat, D. J. (1977) Clinical implications of benzodiazepine pharmacokinetics. *Am. J. Psychiat.*, **134**, 652–656.

Shephard, R. A., Nielsen, E. B. and Broadhurst, P. L. (1982) Sex and strain differences in benzodiazepine receptor binding in Roman rat strains. *Eur. J. Pharmacol.* **77**, 327–330.

Sieghart, W. and Karobath, M. (1980) Molecular heterogeneity of benzodiazepine receptors. *Nature* **286**, 285–287.

Simmonds, M. A. (1980) A site for the potentiation of GABA mediated responses by benzodiazepines. *Nature* **284**, 558–559.

Skolnick, P., Paul, S. M. and Barker, J. L. (1980) Pentobarbital potentiates GABA enhanced [^{3}H]-diazepam binding to benzodiazepine receptors. *Eur. J. Pharmacol.* **65**, 125–127.

Skolnick, P., Schweri, M. M., Williams, E. F., Moncada, V. Y. and Paul, S. M. (1982) An in vitro binding assay which differentiates benzodiazepine agonists and antagonists. *Eur. J. Pharmacol.* **78**, 133–136.

Snyder, S. H. and Bennett, J. P. (1976) Neurotransmitter receptors in brain: biochemical identification. *Ann. Rev. Physiol.* **38**, 153–175.

Soubrie, P., Thiebot, M. H., Jobert, A., Montastruc, J. L., Hery, F. and Hamon, M. (1980) Decreased convulsant potency of picrotoxin and pentylenetetrazol and enhanced ^{3}H-flunitrazepam cortical binding following stressful manipulations in rats. *Brain Res.* **189**, 505–517.

Speth, R. C., Wastek, C. J., Johnson, P. C. and Yamamura, H. I. (1978) Benzodiazepine binding in human brain: demonstration using ^{3}H-flunitrazepam. *Life Sci.* **22**, 859–866.

Speth, R. C., Bresolin, N. and Yamamura, H. I. (1979) Acute diazepam administration produces rapid increases in brain benzodiazepine receptor density. *Eur. J. Pharmacol.* **59**, 159–160.

Speth, R. C., Johnson, R. W., Regan, J., Reisine, T., Kobayashi, R. M. Bresolin, N., Roeske, W. R. and Yamamura, H. I. (1980) The benzodiazepine receptor of mammalian brain. *Fed Proc.* **39**, 3032–3038.

Squires, R. F. and Braestrup, C. (1977) Benzodiazepine receptors in rat brain. *Nature* **266**, 732–734.

Squires, R. F., Benson, D. I., Braestrup, C., Coupet, J., Klepner, C. A., Myers, V. and Beer, B. (1979) Some properties of brain specific benzodiazepine receptors: New evidence for multiple receptors. *Pharmacol. Biochem. Behav.* **10**, 825–830.

Sternbach, L. H. (1979) the benzodiazepine story. *J. Med. Chem.* **22**, 1–7.

Study, R. E. and Barker, J. L. (1981) Diazepam and (−)pentobarbital: fluctuation analysis reveals different mechanisms for potentiation of γ-aminobutyric acid responses in cultured central neurons. *Proc. Natl. Acad. Sci. USA* **78**, 7180–7184.

Supavilai, P. and Karobath, M. (1979) Stimulation of benzodiazepine receptor binding by SQ 20009 is chloride dependent and picrotoxin sensitive. *Eur. J. Pharmacol.* **60**, 111–113.

Supavilai, P. and Karobath, M. (1980a) The effect of temperature and chloride ions on the stimulation of ^{3}H-flunitrazepam binding by the muscimol analogues THIP and piperidine-4-sulphonic acid. *Neurosci. Lett.* **19**, 337–341.

Supavilai, P. and Karobath, M. (1980b) Interaction of SQ 20009 and GABA-like drugs as modulators of benzodiazepine receptor binding. *Eur. J. Pharmacol.* **62**, 229–233.

Supavilai, P. and Karobath, M. (1980c) Heterogeneity of benzodiazepine receptors in rat cerebellum and hippocampus. *Eur. J. Pharmacol.* **64**, 91–93.

Supavilai, P. and Kardbath, M. (1981) Action of pyrazolopyridines as modulators of ^{3}H-flunitrazepam binding to the GABA/benzodiazepine receptor complex in the cerebellum. *Eur. J. Pharmacol.* **70**, 183–193.

Syapin, P. J. and Rickman, D. W. (1981) Benzodiazepine receptor increase following repeated pentylenetetrazole injections. *Eur. J. Pharmacol.* **72**, 117–120.

Syapin, P. J. and Skolnick, P. (1979) Characterisation of benzodiazepine sites in cultured cells of neural origin. *J. Neurochem.* **32**, 1047–1051.

Tallman, J. F., Thomas, J. W. and Gallager, D. W. (1979) Identification of diazepam binding in intact animals. *Life Sci.* **24**, 873–880.

Tenen, S. S. and Hirsch, J. D. (1980) Antagonism of diazepam activity by β-carboline-3-carboxylic acid ethyl ester. *Nature* **288**, 609–610.

Thomas, J. W. and Tallman, J. F. (1981) Characterisation of photoaffinity labelling of benzodiazepine binding sites. *J. Biol. Chem.* **256**, 9838–9842.

Ticku, M. K. and Olsen, R. W. (1978) Interaction of barbiturates with dihydropicrotoxinin binding sites related to the GABA receptor-ionophore system. *Life Sci.* **22**, 1643–1652.

References

Warner, R. S. (1965) Management of the office patient with anxiety and depression. *Psychomatics* **6**, 347–351.

Westek, G. J., Speth, R. C., Reisine, T. D. and Yamamura, H. I. (1978) The effect of γ-aminobutyric acid on [^{3}H]-flunitrazepam binding in rat brain. *Eur. J. Pharmacol.* **50**, 445–447.

Wilkin, G., Wilson, J. E., Balazs, R., Schon, F. and Kelly, J. S. (1974) How selective is high affinity uptake of GABA into inhibitory nerve terminals. *Nature* **252**, 397–399.

Williams, M. and Risley, E. A. (1979) Enhancement of the binding of [^{3}H]-diazepam to rat brain membranes in vitro by SQ 20009, a novel anxiolytic agent, gamma-aminobuyric acid and muscimol. *Life Sci.* **24**, 833–841.

Williamson, M. J., Paul, S. M. and Skolnick, P. (1978) Labelling of benzodiazepine receptors in vivo. *Nature* **275**, 551–553.

Willow, M. and Johnson, G. A. R. (1980) Enhancement of GABA binding by pentobarbitone. *Neurosci. Lett.* **18**, 323–327.

Woolf, J. H. and Nixon, J. C. (1981) Endogenous effector of the benzodiazepine binding site: purification and characterisation. *Biochemistry,* **20**, 4263–4269.

Young, W. S. III and Kuhar, M. J. (1979) Autoradiographic localisation of benzodiazepine receptors in the brains of humans and animals. *Nature* **280**, 393–395.

Young, W. S. III and Kuhar, M. J. (1980) Radiohistochemical localisation of benzodiazepine receptors in rat brain. *J. Pharmacol. Exp. Ther.* **212**, 337–346.

4

Brain dopamine receptors: characterisation and isolation

Philip G. Strange

Department of Biochemistry, The Medical School, Queen's Medical Centre, Nottingham NG7 2UH, U.K.

Abbreviations used

CHAPS	–	3′-[(3-cholamidopropyl) dimethylammonio]-1-propane sulphonate
ED_{50}	–	concentration of substance giving 50% of maximal effect
IC_{50}	–	concentration of substance giving 50% inhibition of response or binding
LPC	–	lysophosphatidylcholine

1. INTRODUCTION

Dopamine is an important neurotransmitter in the brain and periphery. It is associated with motor and emotional function in the brain, in controlling pituitary prolactin release and in certain cardiovascular functions. Dopamine is also found in some ganglia and in amacrine cells in the retina. (For a review of the background see Cooper *et al.*, 1978.) These actions of dopamine are presumably mediated by binding to receptors and much evidence has accumulated recently from *in vivo* and *in vitro* studies suggesting the existence of multiple subclasses of dopamine binding sites and possibly receptors.

The discovery of the dopamine stimulated adenylate cyclase (see, for example, Iversen, 1975) and the advent of the ligand-binding technique has enabled much information to be obtained about dopaminergic receptors from *in vitro* tests. If, however, these *in vitro* methods are to be used in a valid way for defining dopamine receptors and possibly delineating receptor subpopulations it is important to state the properties expected of a dopamine receptor. Classically dopamine receptors have been defined using physiological or pharmacological tests and their sensitivity to dopaminergic antagonists such as neuroleptic drugs. Thus, for example, the behavioural effects of apomorphine, a dopaminergic agonist, and amphetamine, which causes dopamine release, are antagonised by neuroleptic drugs (Leysen et al., 1978b; Niemegeers and Janssen, 1979), the electrophysiological actions of dopamine may be blocked by neuroleptic drugs (Gallagher et al., 1980) and the inhibitory effect of dopamine on pituitary prolactin release is blocked by neuroleptic drugs (Ojeda et al., 1974; Caron et al., 1978). It seems reasonable, therefore, that in studies on dopamine receptors using *in vitro* methods such as binding assays or modulation of adenylate cyclase it should be possible to demonstrate antagonism of the actions or binding of dopamine by neuroleptic drugs and correlations between the potency of neuroleptic drugs in the *in vitro* system and in recognised *in vivo* tests for dopaminergic receptor activity.

1.1 Multiple dopamine binding sites

The results which have been obtained up to the present using *in vitro* binding and cyclase assays may be accounted for in terms of three dopaminergic sites (Withy et al., 1981a) which will be termed D_1, D_2 and D_3 according to the nomenclature of Creese (1982) (see Table 1). The D_1 binding site is associated

Table 1 — Dopaminergic binding sites.

	D_1	D_2	D_3
Agonist affinity	μM	$nM(R_H)$ $\mu M(R_L)$	nM
Antagonist affinity	nM for some neuroleptics; low affinity for benzamides and butyrophenones	nM	μM
Linkage to adenylate cyclase	stimulatory	inhibitory	unknown (possibly stimulatory)
Physiological function	parathyroid hormone release; poor correlation with most dopaminergic functions	behavioural effects of dopamine and inhibition by neuroleptics; inhibition of pituitary prolactin release	unknown

with dopamine stimulation of adenylate cyclase. The action of dopamine at this site is antagonised only by certain neuroleptic drugs and it has been suggested that functionally in rat striatum the D_1 sites are unimportant (Scatton, 1982) although *in vitro*, interaction between D_1 sites and D_2 sites (which will be shown to be receptors below) has also been reported (Stoof and Kebabian, 1981). In bovine parathyroid cells, however, stimulation of D_1 sites and hence adenylate cyclase by dopamine is correlated with stimulation of parathyroid hormone release (Brown *et al.*, 1977). Therefore, although in the central nervous system the function of D_1 sites is far from clear D_1 sites may play a physiological role in some systems and it may therefore be possible to draw an analogy between the D_1 site and the β-adrenergic receptor-stimulated adenylate cyclase, although there is still much work to be done in this field.

The D_3 site detected by ligand-binding studies with tritiated dopaminergic agonists shows a high affinity for dopaminergic agonists and a low affinity for neuroleptic drugs and there is little correlation between the affinities of various substances in behavioural tests and their affinities for binding to D_3 sites (Seeman, 1980). Therefore at present these sites do not fulfil the criteria for dopamine receptors and may be binding sites with no physiological relevance. It has been suggested recently that a part of the D_3 binding of [^{3}H] dopamine in rat brain represents a high affinity agonist state of the D_1 site (Hamblin and Creese, 1982).

The D_2 sites discovered by ligand-binding studies with tritiated butyrophenone neuroleptics and labelled in part by tritiated dopamine agonists show high affinities for neuroleptics and moderate affinities for agonists (see, for example, Seeman, 1980) and methods for investigating the properties of these sites will be considered in some detail in the next section. A number of correlations have been demonstrated between binding of substances to the D_2 sites and *in vivo* tests for dopaminergic activity (Leysen *et al.*, 1978b, Seeman, 1980; and see below) so it seems that the D_2 sites fulfil the criteria for dopaminergic receptors. Indeed, Laduron (1980) has argued for a single dopamine receptor complex with the properties of the D_2 sites. Agonist binding to the D_2 sites is heterogenous and may be accounted for in terms of two classes of sites with different agonist affinities but equal antagonist affinity (Withy *et al.*, 1981, Creese, 1982). On the basis of their affinities for neuroleptics these sub-sites may be considered as subpopulations of D_2 sites and Creese (1982) has termed them R_H and R_L sites (high and low affinity sites). In other classifications of dopamine receptors these high and low affinity agonist binding sites have been termed D_2 and D_4 (Sokoloff *et al.*, 1980) or D_4 and D_2 sites (Seeman, 1980). As it is not clear whether these are separate receptor subpopulations or different conformational states of the same receptor it is preferable to refer to both subclasses as D_2 receptors. There is evidence in several systems that D_2 receptors are coupled to inhibition of adenylate cyclase (Onali *et al.*, 1981; Stoof and Kebabian, 1981; Cote *et al.*, 1982) and this inhibition of cyclic AMP production may be involved in some of the observed physiological effects of D_2 receptors.

2. USE OF LIGAND-BINDING STUDIES WITH [^{3}H]SPIPERONE FOR STUDYING D$_2$ RECEPTORS IN THE BRAIN

Spiperone (spiroperidol) is a potent neuroleptic agent that was introduced in a tritiated form as the ligand of choice for assaying dopamine receptors (Fields *et al.*, 1977; Leysen *et al.*, 1978a). It is a useful ligand in that it has a very high affinity for D$_2$ receptors and shows relatively low non-specific binding in ligand-binding assays. It does, however, have a significant affinity for serotonergic receptors (Leysen *et al.*, 1978b) as do many classical neuroleptics. Many groups have used [^{3}H]spiperone for studying receptors (see Seeman, 1980) and we have used this ligand for studying D$_2$ receptors in bovine caudate nucleus, (Withy *et al.*, 1980, 1981a, 1982). The caudate nucleus receives a dopaminergic innervation from the substantia nigra (Anden *et al.*, 1966) and a serotonergic innervation from the dorsal raphe (Van der Kooy and Hattori, 1980) so that dopaminergic and serotonergic components of [^{3}H]spiperone binding might be expected. When we examined the binding of [^{3}H]spiperone to bovine caudate nucleus membranes in detail it was found that the binding was displaced in a homogeneous manner (slope factor close to unity) by potent neuroleptics, e.g. butaclamol, flupenthixol, whereas atypical neuroleptics, e.g. sulpiride and the selective serotonergic antagonist mianserin showed heterogeneous behaviour (slope factors less than unity, Fig. 1) (Withy *et al.*, 1980, 1981a). By using computer

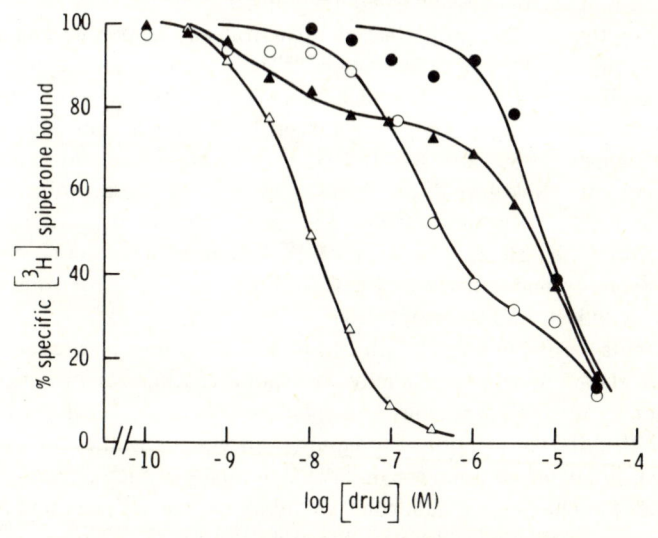

Fig. 1 – Displacement of [^{3}H]spiperone binding from bovine caudate nucleus microsomal preparation by various drugs. Displacement data for (+)-butaclamol (△), (−)butaclamol (●), sulpiride (○) and mianserin (▲) were obtained as described by Withy *et al.* (1981a). The lines are theoretical fits to models consisting of a single class of sites ((+), (−)-butaclamol) or two sets of sites with different affinities (sulpiride 68.6%, 0.16 μM. 31.4% 26.6 μM; mianserin 76.6%, 8.9 μM; 23.4%, 3.2 nM).

non-linear least squares fitting, the data could be shown to consist of contributions from two sets of sites with different pharmacological profiles (dopaminergic D_2 and serotonergic S_2) for [^{3}H]spiperone. From saturation analyses it was found that the binding parameters for the two classes of site in bovine striatum were: dopaminergic, B_{max} 453 fmol/mg protein, K_d [^{3}H]spiperone 0.072 nM; serotonergic B_{max} 703 fmol/mg protein, K_d [^{3}H]spiperone 1.66 nM (Withy et al., 1981a). Similar findings have been reported for rat striatum (Howlett and Nahorski, 1980; List and Seeman, 1981) although the proportion of serotonergic sites is lower and the affinities of the two sets of sites different. The labelling of serotonergic sites by [^{3}H]spiperone in the striatum may complicate the interpretation of earlier studies on D_2 receptors with this radioligand where it was assumed that striatal binding of [^{3}H]spiperone was to dopaminergic sites alone.

In these studies on bovine caudate nucleus it was found that sulpiride was a particularly selective dopaminergic agent showing a potency ratio (ratio of IC_{50} values at the two sets of sites) of 320 for dopaminergic sites versus serotonergic, whereas mianserin was found to be a particularly selective serotonergic agent showing a potency ratio of 270 for serotonergic versus dopaminergic sites. It was possible to take advantage of the selectivity of mianserin in defining conditions for using [^{3}H]spiperone for assaying D_2 receptors alone. Inclusion of 0.3 μM mianserin in all assays eliminated any binding of [^{3}H]spiperone to serotonergic receptors enabling the properties of the dopaminergic D_2 sites alone to be studied. It is to be emphasised that in other tissues and particularly in other species the binding properties of mianserin and spiperone and the proportions of the two sets of sites may be different and so the conditions for selective labelling of D_2 sites would have to be carefully worked out. Other selective serotonergic agents that have been used in place of mianserin are R43448 (Seeman, 1980) and R41468 (Leysen et al., 1981). Also the use of [^{3}H]domperidone (Sokoloff et al., 1980) may obviate these problems as it is reported to label D_2 receptors only. The data obtained using the approach described above for selective labelling of D_2 receptors show good correlations with anti-apomorphine activity and antipsychotic activity supporting the idea that these are indeed true D_2 receptors (Fig. 2).

When antagonist binding was studied to the dopaminergic sites alone it was found that the binding was generally to a homogeneous set of sites whereas agonist binding was heterogeneous and could be accounted for in terms of two sub-sites with different affinities for agonists but equal affinities for antagonists (see above). In bovine striatum roughly equal numbers of each class of site could be detected. Similar binding data can be determined for D_2 receptors in bovine anterior pituitary (Creese, 1982; S. H. Simmonds and P. G. Strange, unpublished observations) a tissue uncomplicated by serotonergic interactions. Treatment with guanine nucleotides lowers agonist affinities by causing an apparent conversion of the higher affinity agonist binding sites to the lower affinity agonist binding sites in brain and pituitary (Sokoloff et al., 1980; Creese, 1982) and

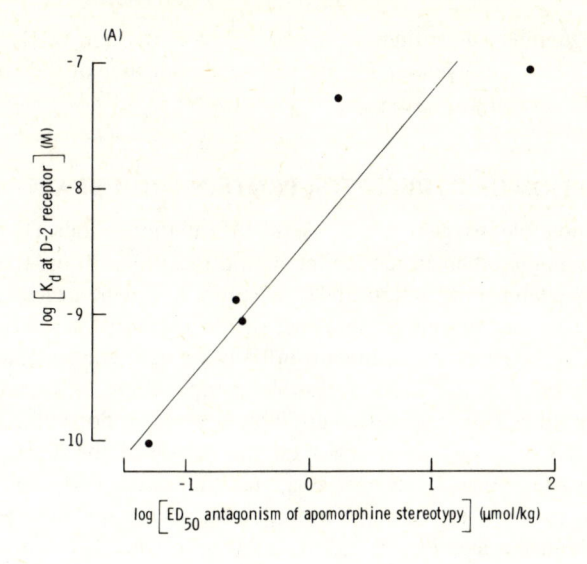

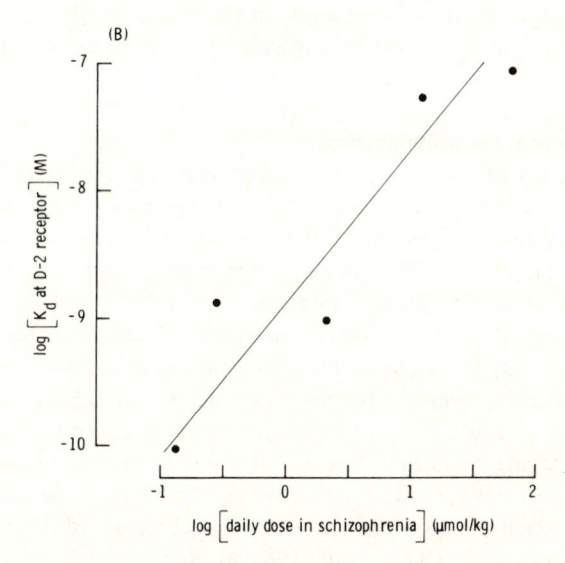

Fig. 2 — Correlation between ligand-binding data at D_2 receptors and physiological dopaminergic functions. Data for binding to D_2 receptors obtained as described in the text (Withy et al., 1981a) are plotted against (A) ED_{50} (concentration giving 50% antagonism) values for antagonism of apomorphine stereotypy taken from Leysen et al. (1978b) (B) average daily clinical dose of antipsychotic drugs for treating schizophrenia taken from Martindale (1977) except for metoclopramide which is from Lin et al. (1980). The drugs used and the log K_d at D_2 receptor are spiperone (−10.02), (+)-butaclamol (−9.02), α-flupenthixol (−8.89), metoclopramide (−7.27) and sulpiride (−7.07). Correlation coefficients: A 0.90, B 0.93.

these effects of guanine nucleotides are consistent with coupling of D_2 receptors to inhibition of adenylate cyclase. The functional significance of the two classes of agonist site is not clear at present.

3. SOLUBILISATION OF D_2 RECEPTOR PROTEIN FROM BRAIN

Although it is possible to gain much useful information about D_2 receptors from *in vitro* binding studies on membrane-bound receptors, in order to understand fully the mechanism of action of D_2 receptors it will be necessary to isolate and characterise the various components of the D_2 receptor system and reconstitute them in a defined environment. Initially we have concentrated on the receptor protein as it is the most accessible component of the system. The results of these studies may help define differences between dopamine receptors in different species and may throw light on the nature of the different dopaminergic binding sites. Also, it has been suggested that either Parkinson's disease or schizophrenia may be autoimmune in origin with the dopamine receptor as auto-antigen (Abramsky and Litvin, 1978). Isolation of the D_2 receptor protein will enable clarification of these suggestions in the same way that isolation of nicotinic acetylcholine receptors has for myasthenia gravis (see Fuchs and Bartfeld in chapter 7).

3.1 Choice of detergent for solubilisation

As the D_2 receptor is likely to be a membrane protein of the integral class it is necessary first to solubilise the protein from membranes using detergents. Reports have appeared of successful solubilisation of D_2 receptors from brain using 3'-[(3-cholamidopropyl) dimethyl-ammonio]-1-propane-sulphonate (CHAPS) (Lew *et al.*, 1981), cholate (Varmuza and Mishra, 1981), digitonin (Gorissen *et al.*, 1980; Davis *et al.*, 1981b) and lysophosphatidylcholine (Lerner *et al.*, 1981). High salt (potassium chloride) concentrations have also been reported to solubilise D_2 receptors (Clement-Cormier and Kendrick, 1981) but in the case of the muscarinic receptor a similar procedure gave a preparation of small membrane fragments rather than a true solubilisation (Gorissen *et al.*, 1981). In our own studies we have tested a series of detergents on bovine caudate. Solubilised binding sites were assayed by a charcoal adsorption assay (using [^{3}H] spiperone binding (Withy *et al.*, 1981b)). Receptor-associated binding was defined as the difference in binding in parallel assays containing 1 μM(+)- and (−)-butaclamol. These conditions have been demonstrated to detect specific receptor binding in experiments with membrane-bound species (see above) but may include dopaminergic and serotonergic contributions. Therefore the solubilised sites will be referred to as [^{3}H] spiperone binding sites until their pharmacological profile has been established. From the results (Withy *et al.*, 1981b; J. M. Hall, M. Wheatley and P. G. Strange, unpublished observations) it is clear that several detergents can solubilise [^{3}H] spiperone binding although CHAPS, cholate (in the presence of sodium chloride: cholate-salt) (see Carson, 1982), digitonin and lysophosphatidylcholine (LPC) give the best yields. It is noteworthy that two zwitterionic detergents are useful for this purpose and these may mimic the natural phospholipid associated with the receptor. In studies reported below we have concentrated mainly on the use of LPC.

3.2 Pharmacological characterisation of solubilised [^{3}H] spiperone binding sites

It is important to determine the nature of the solubilised [^{3}H] spiperone binding sites to show they are true solubilised D_2 receptors. This has been achieved by several approaches including displacement experiments with specific ligands (Fig. 3) (Wheatley and Strange, 1982).

For the LPC-solubilised preparation non-radioactive spiperone displaced 84% of the solubilised [^{3}H] spiperone binding indicating that 16% was non-specifically bound. The drug butaclamol, however, provides a more discriminating test for the nature of the solubilised binding. It exists in (+)- and (−)-isomeric forms, the (+)-isomer being clinically active the (−)-isomer being clinically inactive. For membrane-bound D_2 receptors the butaclamol potency ratio (IC_{50} (−)-butaclamol/IC_{50} (+)-butaclamol) is 800 (Withy et al., 1981b). In the LPC-solubilised preparation the butaclamol potency ratio is only 12 indicating that the solubilised binding sites have partly lost the ability to discriminate between the two isomers of butaclamol or that the solubilised binding sites consist of a mixture of solubilised receptors and sites showing little or no ability to discriminate (non-stereospecific sites). Careful examination of the displacement curves reveals a region of the binding (roughly 30%) where (+)-butaclamol displaces binding but (−)-butaclamol does not (butaclamol concentration (10–300 nM). The remainder of the displacement shows similar affinities for the two isomers. The selective dopaminergic antagonist sulpiride also displaces about 30% of the binding whereas the selective serotonergic antagonist mianserin shows little high affinity displacement. These data are consistent with a division of the total solubilised binding into 16% non-specific binding, 30% solubilised D_2 receptors (with a high butaclamol potency ratio) and 54% of sites that bind [^{3}H] spiperone but show a butaclamol potency ratio close to unity (non-stereospecific sites). The latter sites have been termed spirodecanone sites (Gorissen et al., 1980; Howlett et al., 1979) although their nature is not clear.

This analysis also validates the use of 1 μM (+)- and (−)-butaclamol for defining specific D_2 receptor binding of [^{3}H] spiperone to the LPC-solubilised preparation. The pharmacological profile of the digitonin-solubilised preparation from bovine caudate nucleus is broadly similar (P. A. Frankham, R. M. Withy and P. G. Strange, unpublished observations) and similar profiles have been reported for digitonin solubilisation of rat striatum (Gorissen et al., 1980) whereas for dog striatum (Gorissen et al., 1980) and human striatum (Davis et al., 1981b) there is less binding of [^{3}H] spiperone to spirodecanone sites and good agreement between the binding properties of solubilised and membrane-bound species.

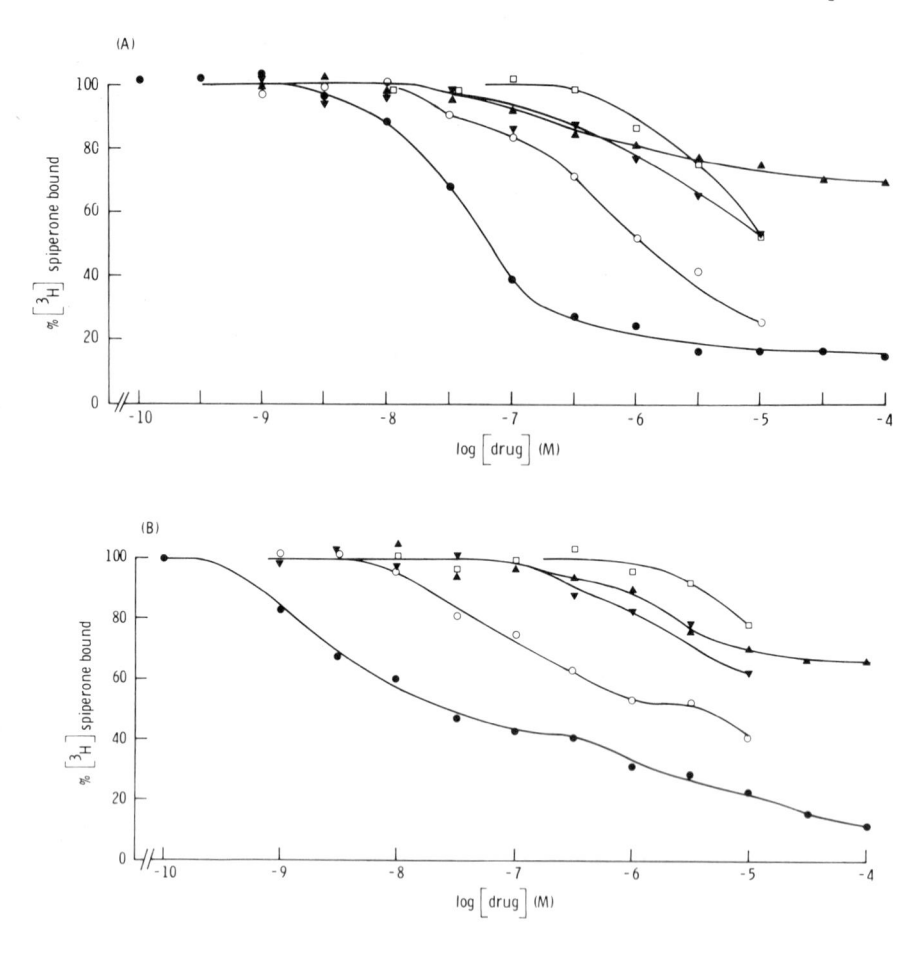

Fig. 3 — Pharmacological profile of [³H]spiperone binding to LPC-solubilised preparations from bovine caudate nucleus. LPC solubilisation was carried out as described (Withy et al., 1981b) in the presence of 0.1 mM EDTA (A) or 1 mM EDTA, 1 mM EGTA, 0.1 mM phenylmethane-sulphonylfluoride, 0.1 mM benzethonium chloride (B) and [³H]spiperone (1 nM approx.) binding determined in the presence of (+)-butaclamol (○), (−)-butaclamol (□), mianserin (▼), spiperone (●), and sulpiride (▲) at the concentrations indicated.

The CHAPS-solubilised preparation from bovine caudate nucleus (M. Wheatley and P. G. Strange, unpublished observations) shows evidence of a significant serotonergic component of [³H]spiperone binding in addition to dopaminergic and spirodecanone components. In agreement with this, Lew et al., (1981) have reported two affinities for specific [³H]spiperone binding to a CHAPS-solubilised preparation of rat striatum. The pharmacological profile of the cholate-salt solubilised preparation indicates that it contains a high proportion of solubilised D_2 sites with very low non-specific and spirodecanone contributions (J. M. Hall and P. G. Strange, in preparation).

In some receptor systems proteolysis of receptors during solubilisation has been shown to be a problem (see, for example, Lang et al., 1979). Therefore we have investigated the effect of including inhibitors of the four main classes of proteinase (Barrett, 1980; Wheatley et al., 1982) during solubilisation by LPC on the butaclamol potency ratio (see Table 2). Inclusion of proteinase inhibitors

Table 2 — Effect of proteinase inhibitors on butaclamol potency ratio.

Additions during solubilisation and assay	$\dfrac{IC_{50}\,(-)\text{-butaclamol}}{IC_{50}\,(+)\text{-butacamol}}$
–	12
Pepstatin A (18 μM)	7
Benzethonium chloride (0.1 mM)	170
Iodoacetic acid (5 mM)	11
Leupeptin (99 μM)	11

Solubilisation with 0.1% LPC was carried out as described by Withy et al. (1981b) in the presence of 1 mM EDTA, 1 mM EGTA, 0.1 mM phenylmethane-sulphonylfluoride and 0.1 mM dithiothreitol (omitted for iodoacetic acid and leupeptin) and further substances as indicated. The IC_{50} values for (+) and (−) butaclamol for displacement of specific [^{3}H]spiperone binding (defined as the difference in binding in the presence of 1 μM (+) and (−) butaclamol) were determined. EDTA, EGTA and phenylmethane-sulphonylfluoride themselves were without effect

changes the potency ratio considerably and from the data it seems that benzethonium chloride is the major contributor to this change. Benzethonium chloride has been reported to be an inhibitor of cysteine proteinases (Otto, 1971) and arylaminopeptidases (Makinen, 1968) but other inhibitors of cysteine proteinases, e.g. iodoacetic acid, leupeptin, were ineffective. Therefore benzethonium may be inhibiting a proteinase that is insensitive to these other agents or alternatively these effects may not be due to proteinase inhibition. Benzethonium, for example, shows some detergent properties (Ansel and Cadwallader, 1964) and interacts with other detergents (Moore and Tennant, 1966) so it may be that the combination of benzethonium chloride and LPC solubilises a different combination of proteins or the same proteins in a different conformational state.

When the pharmacological profile of the sites solubilised in the presence of benzethonium was determined by displacement experiments (Fig. 3) it was found that (−)-butaclamol was virtually inactive whereas (+)-butaclamol displaced 46% of the binding at moderate concentrations and little further displacement occurred. Displacement by non-radioactive spiperone was complex indicating the presence of high affinity sites and low affinity sites. Mianserin showed no high affinity displacement indicating the absence of serotonergic sites and sulpiride displaced almost the same amount of binding as (+)-butaclamol. These data show that about 50% of the [^{3}H]spiperone binding is to solubilised D_2 receptors. It is interesting that these manipulations have eliminated the sites that showed an affinity for spiperone and butaclamol but showed little or no ability to discriminate between the butaclamol isomers (spirodecanone sites). If the effects of benzethonium can be attributed to inhibition of proteolysis it is tempting to speculate that these sites are partially proteolysed receptors (D_2 dopamine or serotonergic receptors) which might show such properties. Alternatively the combination of LPC and benzethonium may solubilise a different mixture of proteins or a different conformational state of the same proteins or benzethonium may inhibit [^{3}H]spiperone binding to these spirodecanone sites. Whatever the explanation, the inclusion of benzethonium allows LPC to be used to obtain a preparation of solubilised D_2 sites with qualitatively similar properties to the membrane-bound species (see below) and considerably reduces interference from non-specific and non-stereospecific binding sites.

This analysis shows that the specific binding defined using 1 μM (+)- and (−)-butaclamol is to solubilised D_2 receptors. Therefore it is possible to carry out saturation binding experiments to the LPC-solubilised (+ benzethonium) D_2 receptors using these conditions. The data indicate a major class of sites with a dissociation constant of 0.20 ± 0.08 nM (mean ± S.D. 3 observations). There is some evidence for low affinity sites as well. From the capacity of the LPC-solubilised D_2 sites it is possible to estimate the yield of solubilisation of D_2 receptors by LPC from brain membranes as 10%. The IC_{50} values for binding to LPC-solubilised D_2 receptors have been corrected for occupancy of the receptors by [^{3}H]spiperone and a comparison with data obtained on the membrane bound D_2 receptor is given in Fig. 4. A good correlation is observed between the two sets of data although the affinities for binding to solubilised receptors are lower overall. Hence the LPC-solubilised (+ benzethonium) and membrane-bound species are quite similar pharmacologically although removal from the membrane has lowered ligand-binding affinities. It should be noted that a contributory factor to any differences observed may be the differences in assay conditions for membrane bound and solubilised sites (Withy *et al.*, 1981a, 1981b). For example, membrane-bound receptors are assayed at 25 °C whereas solubilised sites are assayed at 4 °C. In preliminary studies at 25 °C with LPC-solubilised receptors the binding properties were similar to those at 4 °C although much less non-specific binding was observed at 25 °C (M. Wheatley and P. G. Strange, unpublished observations).

In conclusion, therefore, the combination of LPC and benzethonium pro-

Sec. 3] Solubilisation of D_2 Receptor Protein from Brain

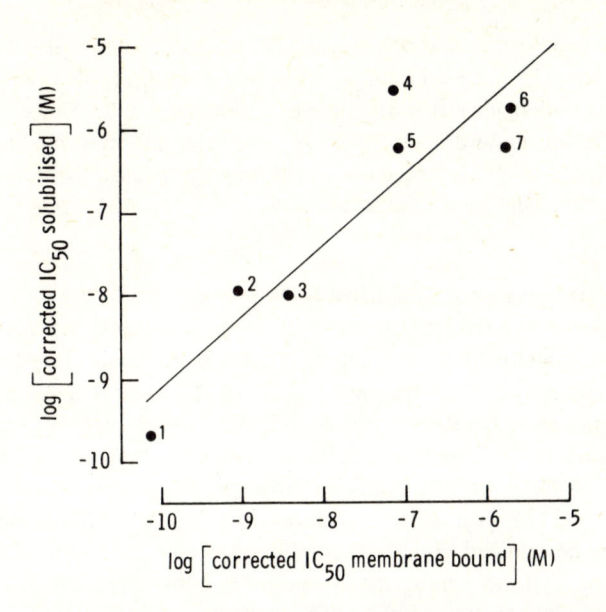

Fig. 4 — Comparison of drug potencies at membrane-bound and solubilised D_2 receptors. IC_{50} values for LPC-solubilised D_2 receptors obtained in the presence of 1 mM EDTA, 1 mM EGTA, 0.1 mM phenylmethane-sulphonyfluoride, 0.1 mM benzethonium chloride (M. Wheatley and P. G. Strange, 1982) were corrected for [^{3}H] spiperone concentration as described in Withy et al. (1981a). Values for membrane-bound receptors are from Withy et al. (1981a) except for (−)-butaclamol (Withy, 1980) and domperidone (calf caudate, Seeman, 1980) 1, spiperone; 2, (+)-butaclamol; 3, domperidone; 4, apomorphine; 5, sulpiride; 6, (−)-butaclamol; 7, mianserin; correlation coefficient 0.91.

duces a preparation of solubilised D_2 receptors showing quite similar pharmacological properties to membrane-bound D_2 receptors and suitable for further studies.

3.3 Effect of guanine nucleotides on solubilised D_2 receptors

As a further test of the integrity of the solubilised D_2 receptors we have investigated the effects of guanine nucleotides on agonist binding. As discussed in Section 2 guanine nucleotides lower agonist affinities at membrane-bound D_2 receptors by binding to a guanine nucleotide regulatory protein (N-protein) (see Chapter 12). For the LPC-solubilised D_2 receptors no similar effect of GTP could be found on apomorphine binding (IC_{50} − 0.1 mM GTP 0.38 μM; IC_{50} + 0.1 mM GTP 0.19 μM, (−benzethonium) P. A. Frankham and P. G. Strange, unpublished experiments). This could be because the receptors were uncoupled from the N-protein prior to solubilisation although GTP did not affect apomorphine binding to solubilised receptors if membranes were pretreated with apomorphine before solubilisation in order to couple receptor and N-protein (IC_{50} − 0.1 mM GTP 0.63 μM; IC_{50} + 0.1 mM GTP 0.56 μM). At present, therefore, it seems that no effect of guanine nucleotides can be demonstrated at solubilised D_2 receptors and a similar finding has been reported by Lew et al., (1981) in a CHAPS-solubilised preparation. It may be that the N-protein is lost or irreversibly altered during solubilisation.

3.4 Physical characteristics of solubilised D_2 receptors

We have also begun to characterise the solubilised D_2 receptors using sucrose density gradient centrifugation and gel filtration (P. A. Frankham, J. M. Hall and P. G. Strange, unpublished observations). The LPC-solubilised and digitonin-solubilised preparations have been run on 15–30% linear sucrose density gradients (Fig. 5) and the two preparations behave similarly. Typical soluble marker enzymes were included on the gradients and this allowed sedimentation coefficients of 11.9 S (LPC-solubilised receptors) and 11.4 S (digitonin-solubilised receptors) to be determined. These data are in agreement with those of Gorissen et al., (1979) for digitonin-solubilised D_2 receptors from dog striatum where the receptors gave a sedimentation coefficient between 11.3 S and 16.0 S.

Both the LPC and digitonin solubilised preparations have been chromatographed on Sepharose 4B or 6B. In preliminary experiments with the digitonin solubilised preparation it was found that recoveries from Sepharose 6B were poor unless some phospholipid was included in the column buffers, e.g. crude soyabean phospholipid. This may be because during chromatography, phospholipid attached to the receptor is lost and that the receptor requires phospholipid for stability as has been demonstrated for the sodium channel protein (Agnew and Raftery, 1979). When soluble marker enzymes were included with the sample during the chromatography on Sepharose 4B the digitonin-solubilised D_2 receptor showed a Stokes radius of 8 nm.

These findings show that the preparations obtained with LPC and digitonin are true solubilised preparations but because the solubilised receptor will be present in a complex of detergent and phospholipid it is difficult to use these data to determine a molecular weight for the receptor. In order to estimate the molecular weight of the receptor it will be necessary to carry out further work with parallel centrifugation runs in sucrose density gradients in water and deuterium oxide.

As a further test of the nature of the solubilised receptors the LPC-solubilised (+ benzethonium) preparation has been examined by electron microscopy and subjected to filtration through 200 nm filters. Electron microscopy (after glutaraldehyde fixation) showed no membranous structures. The level of specific binding of [^{3}H] spiperone was unaffected upon passage through the 200 nm filter. These findings are consistent with the truly solubilised nature of the preparation.

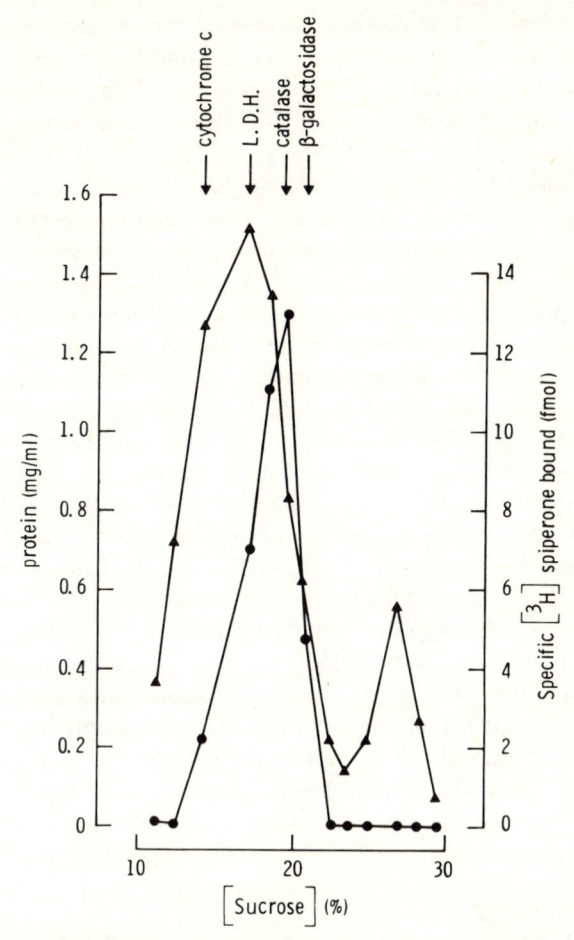

Fig. 5 — Sucrose density gradient fractionation of LPC-solubilised D_2 receptors. LPC-solubilised (+ benzethonium) D_2 receptor preparation together with marker enzymes were layered on to a 15–30% sucrose gradient containing 1 mM EDTA, 1 mM EGTA, 0.1 mM benzethonium chloride, 0.1 mM dithiothreitol, , 20 mM HEPES (pH 7.4) and 0.03% LPC and centrifuged for 20 hours at 95,000g. The positions of marker enzymes are indicated by arrows: protein solubilised (▲); [^{3}H] spiperone bound (●).

3.5 Purification of solubilised D_2 receptor protein

By using the techniques outlined in the previous section limited purification of the D_2 receptor protein (about threefold with sucrose density gradient centrifugation and about fourfold with gel filtration) may be obtained. Lilly *et al.* (1981) have used a combination of isoelectric focusing and gel filtration to obtain a 40-fold purification of digitonin-solubilised D_2 receptors from dog brain.

For more substantial purification, however, it will be necessary to employ some form of affinity chromatography. One possibility is to attach a specific dopaminergic ligand to sepharose and Ramwami and Mishra (1982) have reported a 153-fold purification of solubilised D_2 receptors from bovine striatum by using a haloperidol affinity column. The recent isolation of vipoxin from the venom of Russell's Viper (Freedman and Snyder, 1981) and the report that it binds specifically to biogenic amine receptors including D_2 receptors may offer an alternative substance for attachment to sepharose for affinity chromatography.

Some receptors have been shown to be glycoproteins (see, for example, Shorr *et al.*, 1978) so that if the D_2 receptor is also a glycoprotein, lectin affinity chromatography may offer a possible purification procedure. Accordingly, we have investigated the interaction of LPC-solubilised D_2 receptors with concanavalin A-sepharose. Solubilised receptors bind to concanavalin A-sepharose and the binding can be prevented by α-methyl mannoside showing that the receptors are indeed glycoproteins. Receptors adsorbed to concanavalin A-sepharose may be eluted by treatment with α-methyl mannoside resulting in a fivefold purification of receptors (J. M. Hall and P. G. Strange, unpublished observations). These are preliminary findings and this procedure should eventually be capable of effecting a substantial purification of solubilised receptors.

A powerful technique that should be applicable to purification of D_2 receptors in the near future will be the use of monoclonal antibodies attached to sepharose. The great specificity inherent in a monoclonal antibody should lead to a substantial purification of D_2 receptors that may not be attainable using the other methods described above. A preliminary report of the preparation of a monoclonal antibody has appeared (Davis *et al.*, 1981a) and much progress is to be expected in this area.

4. CONCLUSION

Considerable progress has been made in understanding and defining the various dopaminergic binding sites. The D_2 subclass is particularly well understood and appears to satisfy the criteria for being a true receptor. Methods for solubilising this dopamine receptor have been devised and the solubilised products characterised. In the future it is expected that the preliminary purification studies reported above will be greatly elaborated and that the solubilised receptors will be reconstituted into defined systems in order to study the mechanism of action.

ACKNOWLEDGEMENTS

The experimental work described above and carried out in the author's laboratory was supported by grants from the S.E.R.C./Pfizer Central Research, The Wellcome Trust, The M.R.C. and the S.E.R.C./Fisons Pharmaceuticals.

REFERENCES

Abramsky, O. and Litvin, Y. (1978) Autoimmune response to dopamine-receptor as a possible mechanism in the pathogenesis of Parkinson's Disease and Schizophrenia. *Perspectives in Biology and Medicine,* **22,** 104–114.

Agnew, W. S. and Raftery, M. A. (1979) Solubilised tetrodotoxin binding component from the electroplax of Electrophorus electricus. Stability as a function of mixed lipid-detergent micelle composition. *Biochemistry,* **18,** 1912–1919.

Anden, N., Fuxe, K., Hamberger, B. and Hokfelt, T. (1966) A quantitative study on the nigro-neostriatal dopamine neuron system in the rat. *Acta. Physiol. Scand.* **67,** 306–312.

Ansel, H. C. and Cadwallader, D. E. (1964) Hemolysis of erythrocytes by antibacterial preservatives. *J. Pharm. Sci.* **53,** 169–172.

Barrett, A. (1980) Introduction: the classification of proteinases. Ciba Foundation Symposia **75,** 1–13.

Brown, E. M., Carroll, R. J. and Aurbach, G. D. (1977) Dopaminergic stimulation of cyclic AMP accumulation and parathyroid hormone release from dispersed bovine parathyroid cells. *Proc. Natl. Acad. Sci. USA* **74,** 4210–4213.

Caron, M. G., Beaulieu, M., Raymond, V., Gagne, B., Brouin, J., Lefkowitz, R. J. and Labrie, F. (1978) Dopaminergic receptors in the anterior pituitary gland. *J. Biol. Chem.* **253,** 2244–2253.

Carson, S. (1982) Cholate-salt solubilisation of bovine brain muscarinic receptors. *Biochem. Pharmacol.* **31,** 1806–1809.

Clement-Cormier, Y. and Kendrick, P. E. (1981) Solubilisation of the dopamine receptor: a status report. *Biochem. Pharmacol.* **30,** 2197–2202.

Cooper, J. R., Bloom, F. E. and Roth, R. H. (1978) *The Biochemical Basis of Neuropharmacology.* Oxford University Press, New York.

Cote, T. E., Grewe, C. W., Tsuruta, K., Stoof, J. C., Eskay, R. L. and Kebabian, J. W. (1982) D-2 dopamine receptor-mediated inhibition of adenylate cyclase activity in the intermediate lobe of the rat pituitary gland requires GTP. *Endocrinology,* **110,** 811–819.

Creese, I. (1982) Dopamine receptors explained. *Trends in Neurosciences* **5,** 40–43.

Davis, A., Fraser, C. M., Lilly, L., Madras, B. K., Venter, J. C. and Seeman, P. (1981a) Monoclonal antibodies to partially purified D_2 dopamine receptors. *Soc. Neurosci. Abs.* 10.

Davis, A., Madras, B. and Seeman, P. (1981b). Solubilisation of neuroleptic/dopamine receptors of human brain striatum. *Eur. J. Pharmacol.* **70,** 321–329.

Fields, J. Z., Reisine, T. D. and Yamamura, H. I. (1977) Biochemical demonstration of dopaminergic receptors in rat and human brain using [^{3}H]-spiroperidol. *Brain Research,* **136,** 578–584.

Freedman, J. E. and Snyder, S. H. (1981) Vipoxin. *J. Biol. Chem.* **256**, 13172–13179.

Gallagher, J. P., Inokuchi, H. and Shinnick-Gallagher, P. (1980) Dopamine depolarisation of mammalian primary afferent neurones. *Nature* **283**, 770–772.

Gorissen, H., Aerts, G. and Laduron, P. (1979) Characterisation of solubilised dopamine receptors from dog striatum. *FEBS Letts.* **100**, 281–285.

Gorissen, H., Ilien, B., Aerts, G. and Laduron, P. (1980) Differentiation of solubilised dopamine receptors from spirodecanone binding sites in rat striatum. *FEBS Letts.* **121**, 133–138.

Gorissen, H., Aerts, G., Ilien, B. and Laduron, P. (1981) Solubilisation of muscarinic acetylcholine receptors from mammalian brain: an analytical approach. *Anal. Biochem.* **111**, 33–41.

Hamblin, M. W. and Creese, I. (1982) [^{3}H] Dopamine binding to rat striatal D-2 and D-3 sites; enhancement by magnesium and inhibition by guanine nucleotides and sodium. *Life Sci.* **30**, 1587–1595.

Howlett, D. R., Morris, H. and Nahorski, S. R. (1979) Anomalous properties of [^{3}H] spiperone binding sites in various areas of the rat limbic system. *Mol. Pharmacol.* **15**, 506–514.

Howlett, D. R. and Nahorski, S. R. (1980) Quantitative assessment of heterogeneous [^{3}H] spiperone binding to rat neostriatum and frontal cortex. *Life Sciences* **26**, 511–517.

Iversen, L. L. (1975) Dopamine receptors in the brain. *Science* **188**, 1084–1089.

Laduron, P. (1980) Dopamine receptor: from an *in vivo* concept towards a molecular characterisation. *Trends in Pharmacological Sciences* **1**, 471–474.

Lang, B., Barnard, E. A., Chang, L. and Dolly, J. O. (1979) Putative benzodiazepine receptor: a protein solubilised from brain. *FEBS Letts.* **104**, 149–153.

Lerner, M. H., Rosengarten, H. and Friedhoff, A. J. (1981) Solubilisation and characterisation of [^{3}H] spiroperidol binding sites in calf caudate. *Life Sci.* **29**, 2367–2374.

Leysen, J. E., Gommeren, W. and Laduron, P. M. (1978a) Spiperone: a ligand of choice for neuroleptic receptors. *Biochem. Pharmacol.* **27**, 307–316.

Leysen, J. E., Niemegeers, C. J. E., Tollenaere, J. P. and Laduron, P. M. (1978b) Serotonergic component of neuroleptic receptors. *Nature*, **272**, 168–171.

Leysen, J. E., Awouters, F., Kennis, L., Laduron, P., Vanderberk, J. and Janssen, P. A. J. (1981) Receptor binding profile of R41468, a novel antagonist at 5HT$_2$ receptors. *Life Sci.* **28**, 1015–1022.

Lew, J. Y., Fong, J. C. and Goldstein, M. (1981) Solubilisation of the neuroleptic binding receptor from rat striatum. *Eur J. Pharmacol.* **72**, 403–405.

Lilly, L., Davis, A., Madras, B. K., Fraser, C. M., Venter, J. C. and Seeman, P. (1981) Brain D$_2$ dopamine receptors: partial purification. *Soc. Neurosci. Abs.* 10.

References

Lin, C. W., Maayani, S. and Wilk, S. (1980) The effect of typical and atypical neuroleptics on binding of [^{3}H] spiroperidol in calf caudate. *J. Pharmacol. Exp. Ther.* **212**, 462–468.

List, S. J. and Seeman, P. (1981) Resolution of dopamine and serotonin receptor components of [^{3}H] spiperone binding to rat brain regions. *Proc. Nat. Acad. Sci. USA* **78**, 2620–2624.

Makinen, K. K. (1968) Inhibition of arylaminopeptidases and proline iminopeptidases of human whole saliva by benzethonium chloride. *FEBS Letts.* **2**, 101–104.

Martindale (1977) *The Extra Pharmacopoeia* (A Wade, ed.). The Pharmaceutical Press, London.

Moore, C. D. and Tennant, D. J. (1966) Germicides based on surface active agents. *Manufacturing Chemist* **37** (Oct.) 56–57.

Niemegeers, C. J. E. and Janssen, P. A. J. (1979) A systematic study of the pharmacological activities of dopamine antagonists. *Life Sciences* **24**, 2201–2216.

Ojeda, S. R., Harris, P. G. and McCann, S. M. (1974) Effect of blockade of dopaminergic receptors on prolactin and LH release: median eminence and pituitary sites of action. *Endocrinology* **94**, 1650–1657.

Onali, P., Schwartz, J. P. and Costa, E. (1981) Dopaminergic modulation of adenylate cyclase stimulation by vasoactive intestinal peptide in anterior pituitary. *Proc. Natl. Acad. Sci. USA* **78**, 6531–6534.

Otto, K. (1971) Cathepsins B1 and B2, in *Tissue Proteinases.* (Barrett, A. J. and Dingle, J. T. (eds.). North-Holland, Amsterdam

Ramwami, J. and Mishra, R. K. (1982) Purification of dopamine (D2) receptor by affinity chromatography. *Fred. Proc.* **41**, 1325.

Seeman, P. (1980) Brain dopamine receptors. *Pharmacol. Rev.* **32**, 229–313.

Scatton, B. (1982) Effect of dopamine agonists and neuroleptic agents on striatal acetylcholine transmission in the rat: evidence against dopamine receptor multiplicity. *J. Pharmacol. Exp. Ther.* **220**, 197–202.

Shorr, R. G., Dolly, J. O. and Barnard, E. A. (1978) Composition of acetylcholine receptor protein from skeletal muscle. *Nature* **274**, 283–284.

Sokoloff, P., Martres, M. P. and Schwartz, J. C. (1980) Three classes of dopamine receptor (D-2, D-3, D-4) identified by binding studies with [^{3}H]-apomorphine and [^{3}H] domperidone. *Naunyn-Schmiedeberg's Arch. Pharmacol.* **315**, 89–102.

Stoof, J. C. and Kebabian, J. W. (1981) Opposing roles for D-1 and D-2 dopamine receptors in efflux of cyclic AMP from rat neostriatum. *Nature* **294**, 366–368.

Van der Kooy, D. and Hattori, T. (1980) Dorsal raphe cells with collateral projections to the caudate-putamen and substantia nigra: a fluorescent retrograde double labelling study in the rat. *Brain Res.* **186**, 1–7.

Varmuza, S. and Mishra, R. K. (1981) A rapid and simple method for assaying [^{3}H] spiroperidol binding to solubilised dopamine receptors. *Pharmacol. Res. Commun.* **13**, 587–605.

Wheatley, M., Frankham, P. A., Hall, J. M. and Strange, P. G. (1982) Characterisation of solubilised dopamine receptors. *Biochem. Soc. Trans.* **10**, 373.

Wheatley, M. and Strange, P. G. (1982) Characterisation of brain D_2 dopamine receptors solubilised by lysophosphatidycholine. *FEBS Letts.*, in press.

Withy, R. M. (1980) The characterisation of dopamine receptors in bovine caudate nucleus. Ph.D. Thesis, University of Nottingham, U.K.

Withy, R. M., Mayer, R. J. and Strange, P. G. (1980) [^{3}H] Spiroperidol binding to brain neurotransmitter receptors. *FEBS Letts.* **112**, 293–295.

Withy, R. M., Mayer, R. J. and Strange, P. G. (1981a) Use of [^{3}H] spiperone for labelling dopaminergic and serotonergic receptors in bovine caudate nucleus. *J. Neurochem.* **37**, 1144–1154.

Withy, R. M., Wheatley, M., Frankham, P. A. and Strange, P. G. (1981b) Solubilisation of brain dopamine receptors. *Biochem. Soc. Trans.* **9**, 416.

Withy, R. M., Mayer, R. J. and Strange, P. G. (1982) The distribution of radioligand binding and 5'-nucleotidase activities in bovine caudate nucleus subcellular fractions. *J. Neurochem.* **38**, 1348–1355.

5

Receptors for dihydropyridine calcium channel antagonists

Peter Bellemann and **Andreas Schade**

Department of Pharmacology, Bayer AG, D-5600 Wuppertal, West Germany

---

Abbreviations used
    DHP   —   1,4-dihydropyridine
    DMSO   —   dimethylsulphoxide.

1. INTRODUCTION

The calcium entry blockers, or 'calcium antagonists' (Fleckenstein, 1977), are now widely applied in the therapy of many cardiovascular disorders (Stone et al., 1980). They belong to a pharmacologically potent group of drugs whose mechanism of action is postulated to be inhibition of the slow inward current of calcium ions in several tissues (Fleckenstein, 1977; Rosenberg and Triggle, 1978), particularly in the heart (Kohlhardt and Fleckenstein, 1977) and cerebral vascular smooth muscle (Shimizu et al., 1980; Towart, 1981). The structurally heterogeneous compounds have been typified by verapamil, diltiazem, and 1,4-dihydropyridines (DHP) (Vater et al., 1972), of which nifedipine (Fig. 1) has been the most studied. In 1981, we identified a high-affinity binding site for DHP in cardiac membranes (Bellemann et al., 1981) that has been confirmed also in other tissues (Bolger et al., 1982; Ehlert et al., 1982). The present report demonstrates receptor binding sites in brain membranes for DHPs with tritiated nimodipine, a potent analogue of nifedipine with centrally active properties (Towart et al., 1982; Hoffmeister et al., 1982).

2. EXPERIMENTAL PROCEDURES

2.1 Materials

Tritiated isopropyl-(2-methoxy-ethyl)-1,4-dihydro-2,6-dimethyl-4-(3-nitrophenyl)-3,5-pyridinedicarboxylate ([^{3}H]nimodipine), radiolabelled by New England

Dihydropyridines

	R_1	R_2	R_3
Nifedipine	CH_3-	$-CH_3$	$2\text{-}NO_2$
Nimodipine	$\begin{array}{c}CH_3\\CH_3\end{array}\!\!\!>\!\!CH-$	$-C_2H_4\text{-}O\text{-}CH_3$	$3\text{-}NO_2$
Niludipine	$C_3H_7\text{-}O\text{-}C_2H_4-$	$-C_2H_4\text{-}O\text{-}C_3H_7$	$3\text{-}NO_2$
Nisoldipine	CH_3-	$-CH_2\text{-}CH\!\!<\!\!\begin{array}{c}CH_3\\CH_3\end{array}$	$2\text{-}NO_2$
Nitrendipine	CH_3-	$-C_2H_5$	$3\text{-}NO_2$

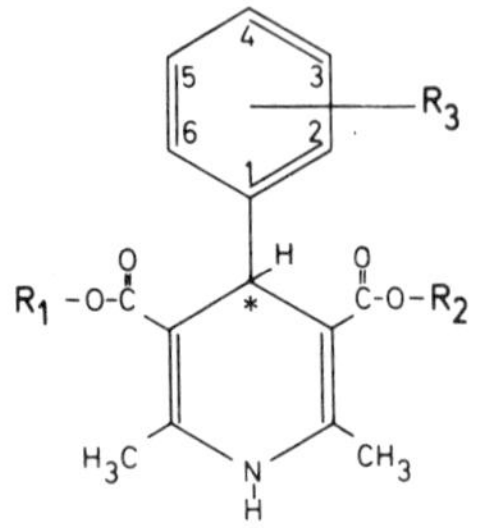

Fig. 1 — The structures of 1,4-dihydropyridine (DHP) derivatives.

Nuclear, Boston, Mass., U.S.A., had a specific activity of 160–180 Ci/mmol and was stored light-protected at −30 °C.

2.2 Methods

Adult male Wistar rats (250–280 g) were killed by cervical dislocation. The brain was rapidly removed, gently homogenised (Potter–Elvehjem homogeniser) in 10 volumes of ice-cold 0.32 M sucrose, centrifuged at 1,000 g for 10 min, and recentrifuged at 40,000 g (20 min) to yield a crude membrane fraction that was stored under liquid nitrogen at −190 °C.

Binding assays were performed under strict sodium light to prevent breakdown of DHPs which can occur at shorter wavelengths. The membrane protein (60–90 µg of protein per assay) was incubated at 37 °C in 50 mM Tris-HCl-buffer, pH 7.4 containing 150 mM NaCl, 1 mM $CaCl_2$, the indicated concentration of [^{3}H]nimodipine, and various additions, e.g. 1,4-dihydropyridine derivatives as presented in Table 2, in a final assay volume of 0.25 ml. DHP derivatives were dissolved in DMSO as the solvent, and a constant ratio of DHP/DMSO was used in the assay. After the time indicated particle-bound and free radioligand were separated after dilution by rapid vacuum filtration through Whatman GF/C filters, and two additional washes (3.5 ml) with ice-cold Tris-HCl-buffer, pH 7.4. Bound [^{3}H]nimodipine was measured by conventional scintillation counting (counting efficiency: 35–40%). Data were plotted according to Scatchard, and displacement curves were analysed with computer programmes employing a Commodore computer (model 4032).

3. RESULTS AND CONCLUSIONS

[^{3}H]Nimodipine has been shown by thin layer chromatography to remain stable under all experimental conditions reported here. Protein dependent binding of [^{3}H]nimodipine was linear up to about 120 µg/0.1 ml. The non-specific binding,

Table 1 — Regional distribution of DHP-receptor in various rat brain areas. Dissociation constants and maximal binding (capacity) of [^{3}H]nimodipine binding in seven rat brain regions. The values are the means of 4 to 8 experiments (r: 0.933–0.982) using at least 3 different protein preparations.

	K_D (nM)	B_{max} (pmol/mg protein)
Cortex	1.11	0.50
Cerebellum	1.17	0.18
Mesencephalon	1.08	0.16
Hypothalamus	0.69	0.26
Hippocampus	0.60	0.54
Septum/basal ganglia	0.76	0.36
Pons/medulla	1.86	0.16

Table 2 — Inhibition of specific [^{3}H]nimodipine binding (1.5 nM) to rat cortical membranes by various non-radioactive DHP derivatives.

Compounds	K_i (nM)[a]
Bay k 5552 (nisoldipine)	<0.01
(±) Bay e 6927	<0.05
Bay a 7168 (niludipine)	0.09
(±) Bay e 5009 (nitrendipine)	0.93
(±) Bay e 9736 (nimodipine)	1.44
Bay a 1040 (nifedipine)	7.0
(−) Bay e 9736	1.04
(+) Bay e 9736	2.4
(−) Bay e 5009	0.43
(+) Bay e 5009	8.8
(−) Bay e 6927	0.003
(+) Bay e 6927	27.5

Values are the means of 4 to 8 experiments, each in triplicate, using at least 3 different protein preparations. The chemical structures are given in the references cited in the text.

[a] Calculation: $K_i = IC_{50}/(1 + LC/K_D)$, LC = ligand concentration; IC_{50}, concentration causing 50% inhibition of specific [^{3}H]nimodipine binding.

defined as binding in the presence of excess unlabelled ligand was reasonably low. Addition of nimodipine at 10 μM to the binding assay displaced 85–90% of the total [^{3}H] nimodipine binding.

Specific [^{3}H] nimodipine binding was maximal after 8 min, and reversible (Fig. 2). The kinetically determined dissociation constant (K_D value) using the association and dissociation rates (Fig. 2) was 0.4 nM. The specific [^{3}H]-nimodipine binding was saturable and a plateau level was achieved at about 3–4 nM of radioligand (Fig. 3). Scatchard analyses of the saturation isotherm (Fig. 3, insert) revealed a single straight line in the concentration range below 6 nM, indicating the presence of a single binding site. The equilibrium dissociation constant K_D was 1.11 ± 0.15 nM ($r > 0.95; n = 8$), and the total number of binding sites (density) B_{max} amounted to 0.50 ± 0.12 pmol/mg of protein ($r > 0.95; n = 8$). Hill plots of the [^{3}H] nimodipine saturation isotherm (Fig. 4) had slopes of 0.91–1.02 ($r > 0.958–0.996; n = 8$) indicating absence of co-operativity.

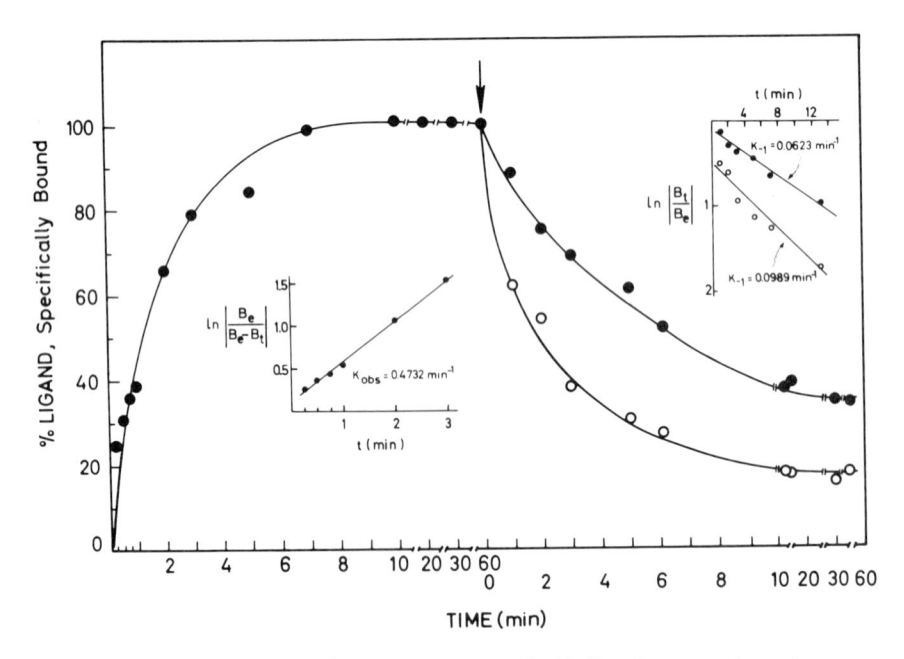

Fig.2 – Time course of [^{3}H] nimodipine specific binding demonstrating saturability and reversibility by pulse/chase-experimentation. The figure shows a representative experiment performed with partially purified membranes from rat cortex at 37 °C, pH 7.4 in 50 mM Tris-HCl-buffer (150 mM NaCl, 1 mM CaCl$_2$). The association reaction has a half-life time of approximately 1.6 min, and the dissociation reaction (arrow) a half-life time of 1.7 min by addition of excess (10 μM) unlabelled nimodipine (o), and 6 min when the chase was performed by dilution (•). The inserts show the linear transformation of the data and the kinetic constants (correlation coefficient $r > 0.95$).

Specific [³H] nimodipine binding in seven rat brain regions (Table 1) demonstrated only one population of binding sites in each brain area tested, with no major regional differences in the dissociation constant and density: cortex: K_D 1.11 nM, B_{max} 0.50 pmol/mg of protein; cerebellum: 1.17, 0.18; mesencephalon: 1.08, 0.16; hypothalamus 0.69, 0.26; hippocampus: 0.60, 0.54; septum/basal ganglia: 0.76, 0.36; pons medulla: 1.86, 0.16 (n = 6–8 experiments, $r > 0.95$).

The DHP-receptor is highly selective: interaction of nimodipine with ten different receptors, e.g. the muscarinic cholinergic-, α_1- and α_2-, β-adrenergic-, benzodiazepine-, dopamine-, GABA-, histamine-, opiate-, and serotonin receptor, revealed very low displacement potency.

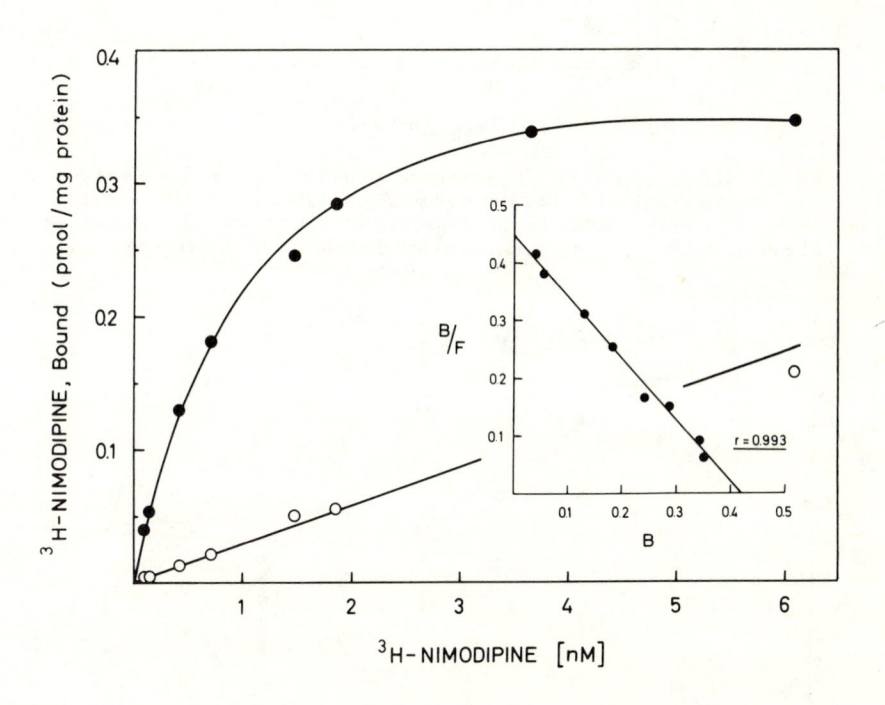

Fig. 3 – Saturation isotherm of [³H] nimodipine to partially purified membranes of rat cortex. [³H] nimodipine, 0.11–6.25 nM, was incubated in triplicate with or without excess of unlabelled nimodipine. Specific binding (●), defined as conditions where non-specific binding remains linear, was calculated as the difference between total binding and that not displaced by excess (10 μM) nimodipine (○). *Insert:* Scatchard analysis of the specific binding data (ratio of Bound/Free versus Bound ligand, B/F vs. B) revealed linearity, and indicates one binding site with an equilibrium dissociation constant K_D = 0.94 nM (reciprocal of the slope value) and maximal binding (B_{max} value) of 0.42 pmol/mg of protein (r = 0.993, intercept on abscissa). The experiment was repeated several times using different protein preparations and yielded closely similar results (r: 0.95–0.993).

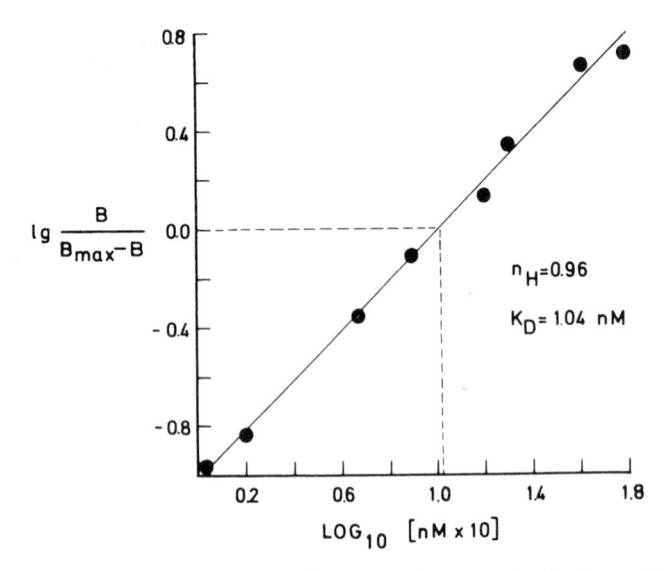

Fig. 4 — Hill analysis of the [³H]nimodipine saturation isotherm. The dissociation constant calculated by the Hill analysis, $K_D = 1.04$ nM ($r = 0.98$), corresponds well with that elicited by the Scatchard plot (Fig. 3). The Hill coefficient (slope) $n_H = 0.96 \pm 0.06$ ($r > 0.96; n = 8$) indicates absence of cooperativity.

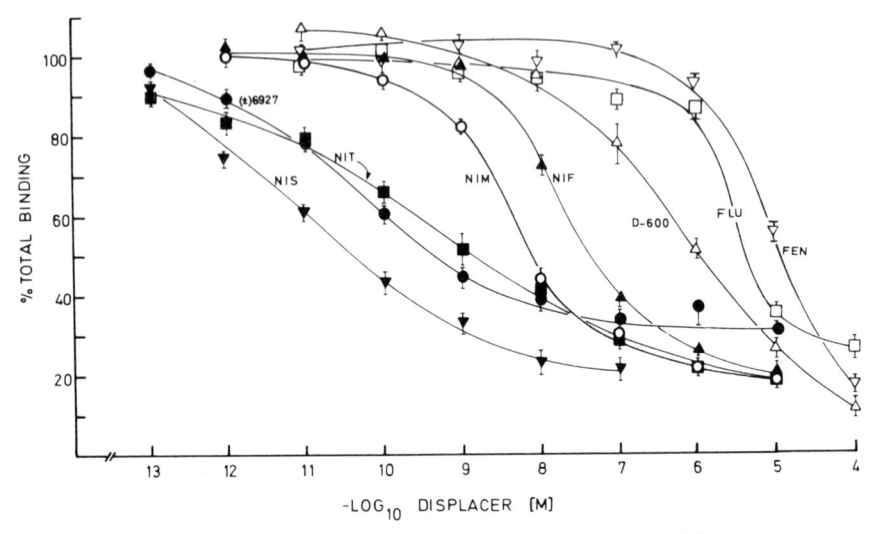

Fig. 5 — Displacement experiments with [³H]nimodipine (<1.5 nM) and various potent 'calcium antagonists' demonstrating high specificity for DHP derivatives. Total binding is plotted against $-\log_{10}$ (M) of displacer: (▼), NIS, nisoldipine; (●), (±) Bay e 6927; (■), NIT, nitrendipine; (○), NIM, nimodipine; (▲), NIF, nifedipine. (△), D-600, gallopamil; (□), FLU, flunarizine. (▽), FEN, fendiline. The results are the means ± S.E. of 3 to 8 experiments using at least three different protein preparations.

The [³H]nimodipine binding site proved to be highly specific for various 1,4-dihydropyridine derivatives, and discriminated between their optical isomers. Several DHP derivatives potently displaced [³H]nimodipine (Fig. 5) with inhibition constants (K_i values) in the nano- or subnanomolar range (Table 2). Structurally different calcium antagonists, on the other hand, e.g. verapamil, its methoxy derivative gallopamil (D-600), flunarizine, or fendiline were much less potent (Fig. 5) and exhibited K_i values greater than 10^{-6} M. Diltiazem did not displace at all, but interestingly seemed to increase the amount of [³H]-nimodipine bound to the receptor.

The stereoselectivity of the receptor, an additional criterion of specificity of action, has been demonstrated with the enantiomers of nimodipine, nitrendipine, and Bay e 6927 (Table 2, lower panel). In each case, e.g. nitrendipine (Fig. 6), the (−)-stereoisomer displayed much higher affinity for [³H]nimodipine binding sites than its (+)-stereoisomer. Interestingly, the displacement potency of the racemic compounds was always closer to that of the (−)-stereoisomer indicating that the (−)-enantiomer of the racemic compounds

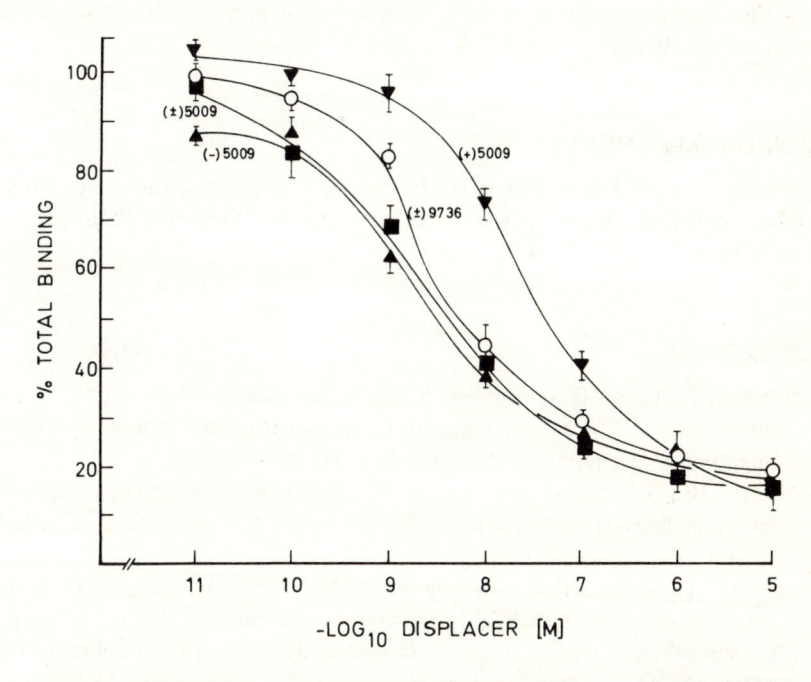

Fig. 6 − Displacement experiments with [³H]nimodipine (< 1.5 nM) and the enantiomers of nitrendipine (Bay e 5009). Total binding is plotted against −log₁₀ (M) of displacer. (▲), (−) Bay e 5009; (▼), (+) Bay e 5009; (■), nitrendipine, ((±) Bay e 5009); (○), nimodipine ((±) Bay e 9736). Data demonstrating stereoselectivity are the means ± S.E. of 3 to 5 experiments using different protein preparations.

actually contributes to the DHP specific action. The differences in the K_i values of the binding data of the three enantiomers correspond well with the concomitant functional response (Towart et al., 1981).

Saturability, reversibility and markedly selective displacement activities do not necessarily imply a physiologically meaningful recognition site. However, the DHP-receptor reported here seems to be physiologically significant, since its binding characteristics can be well correlated to pharmacological activities. Excellent correlation over five orders of magnitude (Bellemann et al., 1983) was achieved between [^{3}H] nimodipine displacement potency and the inhibition of K$^+$-stimulated contraction of rabbit aortic strips, or Ba^{2+}-induced contraction of guinea-pig ileum (Towart et al., 1981).

In summary, receptor binding sites for DHP with high affinity and stereospecificity have been demonstrated with the radiolabelled calcium antagonist nimodipine, a potent inhibitor of depolarisation-induced contractions of peripheral or cerebral vascular smooth muscle. [^{3}H] nimodipine binding, its inhibition by structurally different classes of calcium entry blockers, and the receptor interaction with mono-, di-, and trivalent cations should facilitate the localisation and characterisation of these binding sites that may be associated with the putative Ca^{2+}-channel.

ACKNOWLEDGEMENTS

The authors thank Drs H. Meyer and E. Wehinger, Bayer AG, Wuppertal, for the enantiomers, and Drs R. Towart and S. Kazda for providing their data for correlation.

REFERENCES

Bellemann, P., Ferry, D., Lübbecke, F. and Glossmann, H. (1981) [^{3}H]-Nitrendipine, a potent calcium antagonist, binds with high affinity to cardiac membranes. *Arzneim. Forsch./Drug Res.* **31**, 2064–2067.

Bellemann, P., Schade, A. and Towart, R. (1983) Dihydropyridine-receptor in rat brain labelled with [^{3}H] nimodipine. *Proc. Natl. Acad. Sci., U.S.A.* in press.

Bolger, G. T., Gengo, P. J., Luchowski, E. M., Siegel, H., Triggle, D. J. and Janis, R. A. (1982) High affinity binding of a calcium channel antagonist to smooth and cardiac muscle. *Biochem. Biophys. Res. Commun.* **104**, 1604–1609.

Ehlert, F. J., Itoga, E., Roeske, W. R. and Yamamura, H. I. (1982) The interaction of [^{3}H]-nitrendipine with receptors for calcium antagonists in the cerebral cortex and heart of rats. *Biochem. Biophys. Res. Commun.* **104**, 937–943.

References

Fleckenstein, A. (1977) Specific pharmacology of calcium in myocardium, cardiac pacemakers, and vascular smooth muscle. *Ann. Rev. Pharmacol. Toxicol.* **17**, 149–166.

Hayashi, S. and Toda, N. (1977) Inhibition by Cd^{2+}, verapamil and papaverine of Ca^{2+}-induced contractions in isolated cerebral and peripheral arteries of the dog. *Br. J. Pharmacol.* **60**, 35–43.

Hoffmeister, F., Benz, U., Heise, A., Krause, H. P. and Neuser, V. (1982) Behavioral effects of nimodipine in animals. *Arzneim. Forsch./Drug Res.* **32**, 347–360.

Kohlhardt, M. and Fleckenstein (1977) Inhibition of the slow inward current by nifedipine in mammalian ventricular myocardium. *Naunyn-Schmiedeberg's Arch. Pharmacol.* **293**, 267–272.

Rosenberg, L. and Triggle, D. J. (1978) Calcium, calcium translocation and specific calcium antagonists, in *Calcium and Drug Action* (Weiss G. B. and Goodman, F. R., eds.) pp. 3–31. Plenum Press, New York.

Shimizu, K., Onta, T. and Toda, N. (1980) Evidence for greater susceptibility of isolated dog cerebral arteries to Ca antagonists than peripheral arteries. *Stroke* **11**, 261–266.

Stone, P. H., Antman, E. M., Muller, J. E. and Braunwald, E. (1980) Calcium channel blocking agents in the treatment of cardiovascular disorders. Part II: Hemodynamic effects and clinical applications. *Ann. Intern. Med.* **93**, 886–904.

Towart, R. (1981) The selective inhibition of serotonin-induced contractions of rabbit cerebral vascular smooth muscle by calcium-antagonistic dihydropyridines. An investigation of the mechanism of action of nimodipine. *Circ. Res.* **48**, 650–657.

Towart, R., Wehinger, E. and Meyer, H. (1981) Effects of unsymmetrical ester substituted 1,4-dihydropyridine derivatives and their optical isomers on contraction of smooth muscle. *Naunyn-Schmiedeberg's Arch. Pharmacol.* **317**, 183–185.

Towart, R., Wehinger, E., Meyer, H. and Kazda, S. (1982) The effects of nimodipine, its optical isomers and metabolites on isolated vascular smooth muscle. *Arzneim. Forsch./Drug Res.* **32**, 338–346.

Vater, W., Kroneberg, G., Hoffmeister, F., Kaller, H., Meng, K., Oberdorf, A., Puls, W., Schlossmann, K. and Stoepel, K. (1972) Zur Phamakologie von 4-(2-nitrophenyl)-2,6-dimethyl-1,4-dihydropyridin-3,5-dicarbonsäuredimethylester (nifedipine, BAY a 1040). *Arzneim. Forsch./Drug Res.* **22**, 1–14.

6

Coexistence of neuropeptides and monoamines: possible implications for receptors

P. C. Emson, J. M. Lundberg[†] and R. F. T. Gilbert
MRC Neurochemical Pharmacology Unit, MRC Centre, Medical School, Hills Road, Cambridge, CB2 2QH, U.K.

Abbreviations used

ACh	—	acetylcholine
AChE	—	acetylcholinesterase
ChAT	—	choline acetyltransferase
CCK	—	cholecystokinin
HPLC	—	high performance liquid chromatography
5-HT	—	5-hydroxytryptamine
NMPB	—	N-methyl-4-piperidinyl benzilate
SP	—	substance P
TRH	—	thyrotropin releasing hormone
VIP	—	vasoactive intestinal polypeptide

It is becoming an increasingly common finding that neuropeptides are found in mammalian neurones which contain other putative neurotransmitters such as the biogenic amines (5-hydroxytryptamine, dopamine, noradrenaline and acetylcholine) or amino acids (γ-aminobutyric acid) (for review see Hökfelt *et al.*, 1980a and Table 1). In the majority of these cases, the peptide-like immunoreactivity has not been characterised and there is no evidence that the peptide material can be released to influence a post-synaptic cell. However, the demonstration of coexistence raises the possibility of a variety of functional interactions between the coexisting "transmitters". Some of the possible roles of

[†]Department of Pharmacology, Karolinska Institute, Stockholm, Sweden

coexisting peptides have been discussed by Hökfelt et al., (1980a) and include possible pre- or post-synaptic sites of action as well as non-transmitter trophic or hormonal roles for coexisting peptides. In this short chapter the three currently most studied examples of coexistence between biogenic amines and peptides will be discussed. In these cases there is some preliminary evidence available that the neuropeptide (if released) can act at both pre-or post-synaptic sites and in binding to its receptor can inflence the receptor of a biogenic amine to modify its potency as an agonist.

Table 1 – Examples of neurones and endocrine cells containing both a classical transmitter and a peptide.

Classical transmitter	Peptide	Area (species)
Serotonin	Substance P	Medulla oblongata (rat)
Serotonin	TRH	Medulla oblongata (rat)
Serotonin	Substance P and TRH	Medulla oblongata (rat)
Serotonin	β-endorphin	Medulla oblongata (cat)
Serotonin	Enkephalin	Medulla oblongata (cat)
Noradrenaline	Pancreatic polypeptide	Locus coeruleus (rat) Adrenal medulla (rat)
Noradrenaline	Enkephalin	Adrenal medulla (cat)
Adrenaline	Enkephalin	Adrenal medulla (cat)
Noradrenaline	Neurotensin	Adrenal medulla (cat)
Dopamine	CCK	Ventral tegmental area (rat, human)
Acetylcholine	VIP	Autonomic ganglia (cat)
GABA	Motilin	Cerebellum (rat)

Modified from Hökfelt et al., (1980).

1. COEXISTENCE OF 5-HYDROXYTRYPTAMINE, SUBSTANCE P AND TRH IN THE BULBO-SPINAL NEURONES OF THE RAT SPINAL CORD

The original observations that 5-hydroxytryptamine (5-HT) and substance P might coexist in medullary raphe neurones (Hökfelt *et al.*, 1978) were followed by several other studies confirming the immunohistochemical observations and extending them to include the possibility that certain medullary raphe neurones might contain another peptide, thyrotropin releasing hormone (TRH) as well as substance P and 5-HT (Chan-Palay *et al.*, 1978; Gilbert *et al.*, 1982; Johansson *et al.*, 1981). In the ventral lumbar spinal cord of the rat (Fig. 1) the coincidence between 5-HT, SP and TRH terminal immunoreactivity is very striking. This TRH-like and SP-like immunoreactivity is lost from the ventral spinal cord following the intracerebroventricular injection of the serotonin directed toxins 5,6 or 5,7-dihydroxytryptamine which deplete the spinal cord of its 5-HT content (Gilbert *et al.*, 1982; Johansson *et al.*, 1981). Peptide characterisation studies have shown that the TRH-like and SP-like immunoreactivity in the rat ventral cord is indistinguishable from the authentic tri- or undecapeptides by gel chromatography or HPLC (Gilbert *et al.*, 1982).

Pharmacological studies have also indicated that the spinal cord SP and TRH contents can be influenced by the monomine depleting drugs (reserpine or tetrabenazine) (Gilbert *et al.*, 1981). However, the time course of these changes in content are much slower than the depletion of 5-HT produced by reserpine or tetrabenazine and it is suggested they may reflect changes in impulse flow (see for example Viveros *et al.*, 1980). Similarly although there is one report of co-storage of SP and 5-HT immunoreactivity within the same vesicle, (Pelletier *et al.*, 1981) pharmacological studies suggest that the 5-HT may be released independently of TRH release (Marsden *et al.*, 1982).

1.1 Possible functional interactions

Immunohistochemical studies (Johansson *et al.*, 1981; Gilbert *et al.*, 1982) have shown that 5-HT, SP and TRH-like immunoreactivity coexists in the intermediolateral columns and around the lumbar ventral cord motor neurones. 5-HT has previously been implicated in the modulation of activity of preganglionic sympathetic neurones in the thoracic intermediolateral columns (Wing and Chalmers, 1974; Cabot *et al.*, 1979) and ventral horn motor neurones (Anderson and Shibuya, 1966; McCall and Aghajanian, 1979; White and Neuman, 1980). Interestingly, SP and TRH have very similar effects to 5-HT on motor neurone excitability (Fig. 2) in that they increase neuronal excitability without usually causing the cell to fire (Nicoll *et al.*, 1980). These observations prompted Gilbert (1981) to suggest that the coexisting substances may be acting as modulators to set the 'gain' and to influence the response of the motor neurones to a fast-acting transmitter, such as glutamate.

Although it seems reasonable to assume that 5-HT and the two putative

Sec. 1] Coexistence of 5-hydroxytryptamine 113

peptide transmitters (SP and TRH) which coexist in the ventral cord are actually released upon nervous activation, this has not yet been demonstrated. There are, however, two preliminary sets of experiments which indicate that the coexisting peptides may act at different sites, In one series of experiments Barbeau and Bedard (1981) showed in the rat that chronic spinal cord transection produced

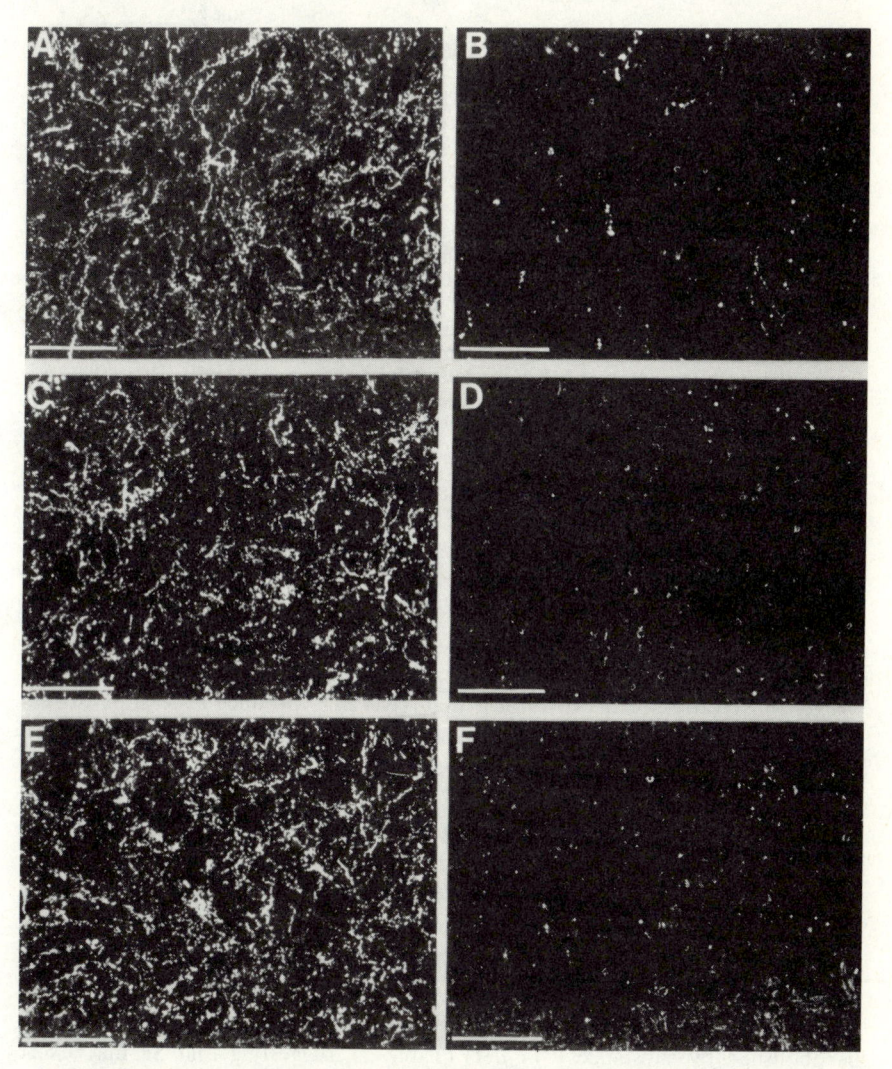

Fig. 1 – Horizontal sections of sacral ventral horn (S2) viewed under dark-field illumination stained for 5-hydroxytryptamine (A,B) thyrotropin-releasing hormone (C,D) and substance P (E,F). A,C,E are from a control animal, whilst B,D,F are from an animal pretreated with 200 µg 5,7-DHT (free base) i.c.v three weeks previously. Scale bars indicate 100 µm.

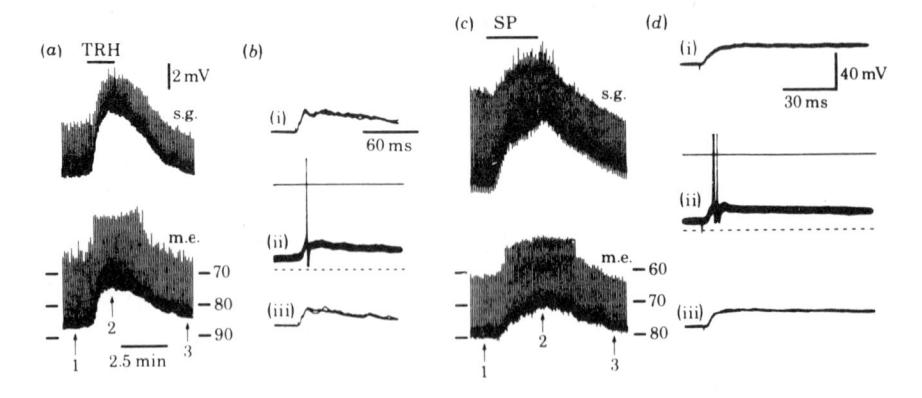

Fig. 2 — TRH and substance P bring subthreshold e.p.s.p's. to threshold. The top record (s.g.) in (a) monitors the effect of TRH with sucrose gap recording from the ventral root. The bottom record (m.e.) is a microelectrode intracellular record from a motoneuron. The upward deflexions are e.p.s.p's. generated by stimulation of the dorsal root. (b) Sample film records of the responses obtained at the indicated numbers below the intracellular chart record in (a). The action potential during the depolarisation is severely attenuated in the pen record. (c), (d) A similar set of records obtained with substance P from another motoneurone.

an increase in electromyograph activity (EMG) in the fast extensor muscles of the rat thigh. This background EMG activity could be substantially increased by treatment with 5-HTP or 5-HT agonists and was only blocked by 5-HT antagonists (such as cyproheptadine). It was particularly interesting to note that TRH could produce a similar effect which was also blocked by 5-HT antagonists (Fig. 3). The observation seems to suggest that TRH may produce at least part of its effects on motor neurones via a mechanism which somehow involves a 5-HT receptor (defined as inhibited by cyproheptadine, Fig. 3). In contrast to the possible post-synaptic interactions of TRH and 5-HT, SP may act pre-synaptically to enhance the release of 5-HT from the bulbo-spinal terminals. Experiments by Mitchell and Fleetwood-Walker (1981) have shown that 5-HT (1µM) will reduce the release of [^{3}H] 5-HT from ventral spinal cord slices *in vitro*. This type of effect is commonly seen in a number of monoamine-containing neuronal systems and is believed to reflect the presence on the 5-HT neurone of pre-synaptic 5-HT autoreceptors which can inhibit the release of 5-HT. Interestingly this autoreceptor effect was unchanged in the presence of TRH (10 µM) but could be abolished by substance P (4 µM) (Table 2), suggesting that SP may exert part of its effect via a mechanism which results in blockade of the 5-HT autoreceptor. SP may thus exert a mainly pre-synaptic action to block the 5-HT autoreceptor whilst TRH interacts in some ill-defined manner with a post-synaptic 5-HT receptor. It is important to note that although in the ventral

Sec. 1]   Coexistence of 5-hydroxytryptamine   115

cord SP and TRH may influence either pre- or post-synaptically localised 5-HT receptors there is no evidence for any direct interaction of TRH or SP with the relevant 5-HT receptor and we assume that both SP and TRH produce their effects via pre- or post-synaptic SP or TRH receptors.

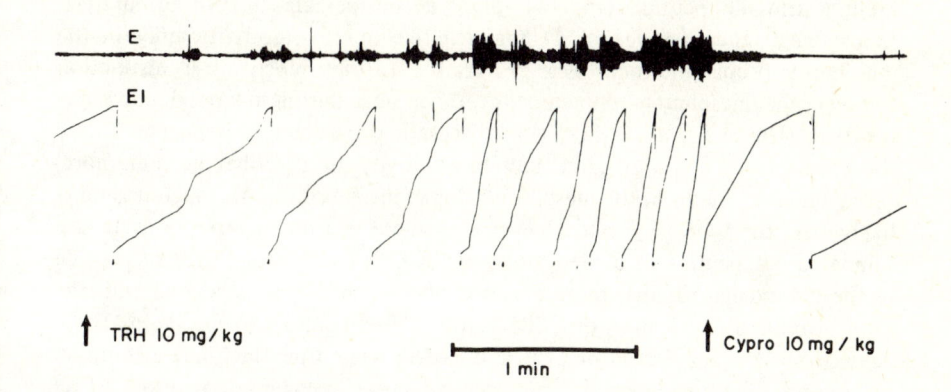

Fig. 3 — Electromyographic (EMG) tracing of extensor (E) (right quadriceps femoris) muscle of a chronic spinal rat. The lower tracing represents the integrated EMG of the extensor muscle (E1) muscle. Note the rapid onset of the effect of TRH and the rapid return to the baseline level of activity after cyproheptadine (10 mg/kg).

Table 2 — Antagonism by substance P of the autoreceptor effect of 5-HT on [^{3}H] 5-HT release from the rat ventral spinal cord *in vitro*.

[^{3}H] 5-HT release (% increase over basal efflux)	48 mM K$^+$	Autoreceptor effect of 5-HT on [^{3}H] 5-HT release (% increase over basal efflux)	48 mM K$^+$
K$^+$ alone	39.0 ± 2.6	K$^+$ alone	35.4 ± 2.9
K$^+$ + SP (4μM)	38.9 ± 3.5	K$^+$ + 5-HT (1μM)	14.4 ± 2.4
K$^+$ + TRH (10μM)	36.3 ± 3.3	+ 5-HT (1μM) + SP (4μM)	33.8 ± 1.8
		+ 5-HT (1μM) TRH (10μM)	15.4 ± 3.7

From Mitchell and Fleetwood-Walker (1981).

2. CHOLECYSTOKININ AND DOPAMINE

The demonstration by Hokfelt et al. (1980b) and Vanderhaeghen et al. (1980) that a number of mesencephalic dopamine neurones projecting to mesolimbic regions (such as the nucleus accumbens, bed nucleus of the stria terminalis, central amygdaloid nucleus and olfactory tubercle) also contained a cholecystokinin/gastrin-like peptide (Fig. 4) raised a considerable amount of interest. In particular the presence of CCK/gastrin-like immunoreactivity only in the mesolimbic dopamine neurones provided a further biochemical distinction between the mesolimbic dopamine neurones and the nigro-striatal dopamine neurones (Hökfelt et al., 1980c). In this case it is particularly important to use the expression CCK/gastrin-like immunoreactivity to describe the immunoreactive material found in the mesolimbic dopamine neurones. Although it is most likely that the CCK/gastrin-like peptide in these neurones corresponds to the sulphated octapeptide of cholecystokinin (CCK-8) (the major CCK-like peptide in the mammalian brain), there are a number of problems associated with the unambiguous identification of CCK/gastrin related peptides in the brain. These come from the fact that both CCK and gastrin related peptide share a common carboxyterminal pentapeptide sequence which is recognised by most CCK/gastrin directed antibodies. This means that careful gel chromatography is necessary to resolve CCK and gastrin peptides in the brain. A further complication comes from the observation that in some regions of the CNS, CCK/gastrin-directed antibodies will visualise 'CCK-immunoreactivity' which is not detected in CCK/gastrin directed radioimmunoassays (for detailed discussion see Schultzberg et al., 1982; Marley et al., 1982). The implication of these observations is that there may be another peptide in the central nervous system which has some cross-reactivity with the CCK/gastrin antibodies used in histochemistry but is not detected in the radioimmunoassay.

Despite these reservations over the exact nature of the CCK-like peptide in the dopamine neurones a number of workers have looked for pharmacological interactions between CCK-peptides and the dopamine systems. Initial observations (Fuxe et al., 1980) showed that intracerebral administration of CCK-8 increased dopamine turnover in the striatum, whilst decreasing dopamine turnover in the nucleus accumbens. Parallel electrophysiological studies by Skirboll et al. (1981) demonstrated that within the substantia nigra and ventral tegmental area CCK excited dopamine cells only in regions where CCK-immunoreactivity could be demonstrated. These observations suggest that the dopamine neurones carry presynaptic CCK receptors which may regulate the release of dopamine from their terminals. In agreement with this suggestion, initial work on the release of [^{3}H] dopamine from striatal slices indicated that CCK-8 (10μM) would enhance the release of [^{3}H] dopamine *in vitro* (consistent with a presynaptic action of CCK-8) (Kovacs et al., 1981). Attempts to replicate this finding by other workers (Markstein personal communication) have not provided any clear-cut results. Despite these confusing results on release, several

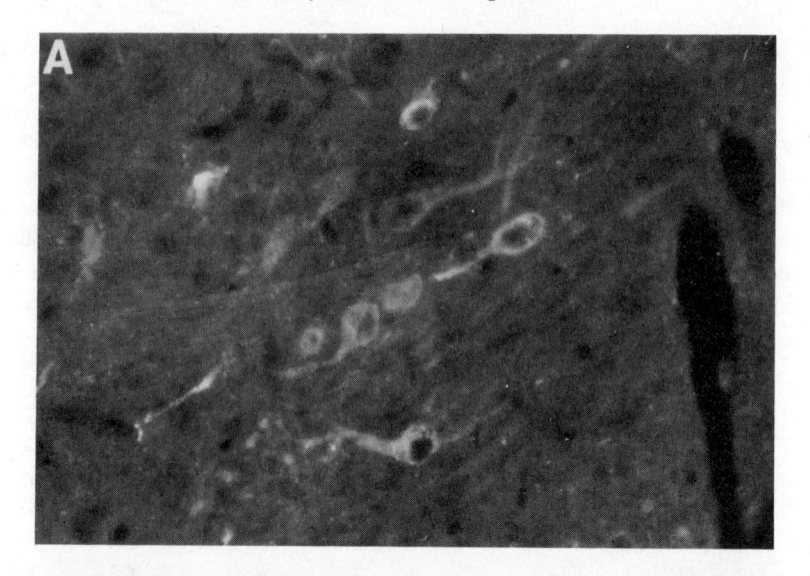

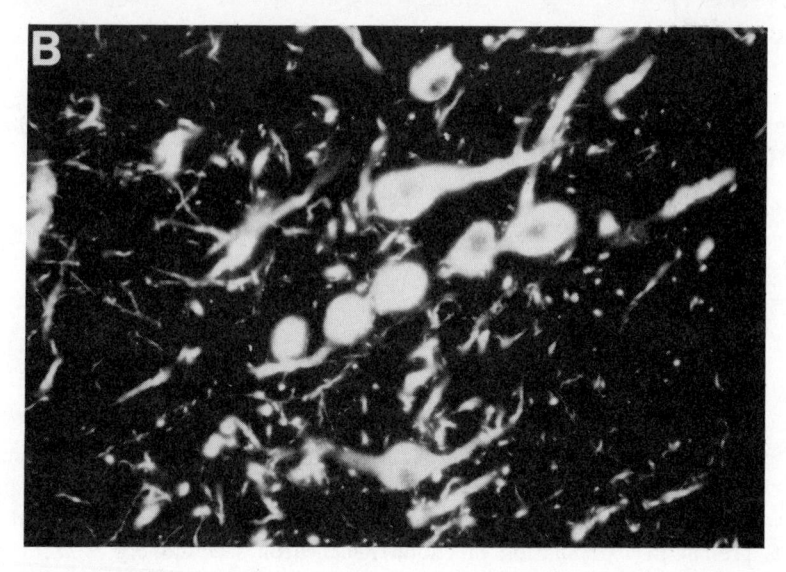

Fig. 4 – Immunofluorescence micrographs of the rat ventral tegmental area after incubation with an antiserum to gastrin/cholecystokinin (A) and tyrosine hydroxylase (B). A and B show the same section after elution of the gastrin/cholecystokinin antibody (A) and 'restaining' with antiserum to tyrosine hydroxylase. Note the coexistence of gastrin/cholecystokinin with tyrosine hydroxylase in some of the neurones (arrowheads) from Hökfelt et al. (1980c).

groups have reported effects of CCK-8 on the binding of dopamine agonists ([³H] dopamine) or dopamine antagonists ([³H] spiperone) to membrane preparations from the striatum or nucleus accumbens (putative D_2 receptors). These results show wide discrepancies (Table 3) so that the trends observed must be treated with some caution (Fuxe et al., 1981; Bhoola et al., 1982; Murphy and Schuster, 1982). Despite this confusion it does seem clear that CCK-8 does not affect the dopamine stimulated adenylate cyclase (Fuxe et al., 1981; Bhoola et al., 1982).

Table 3 — Effects of CCK on dopamine receptor binding.

	Control		CCK		CCK
	K_D (pM)	B_{max}(fmol/ mg protein)	K_D (pM)	B_{max}(fmol/ mg protein)	concentration
Striatum					
Fuxe et al. (1981) Ligand [³H] Spiperone	76	496	50	466	10 nM
Bhoola et al. (1982) Ligand [³H] Spiperone	579	319	385	279	1000 nM
Murphy and Schuster (1982) [³H] Dopamine	24,000	153	6000	67	1000 nM
Nucleus accumbens					
Bhoola et al. (1982) Ligand [³H] Spiperone	390	152	1800	466	1000 nM

3. VASOACTIVE INTESTINAL POLYPEPTIDE AND ACETYLCHOLINE

The discovery that about 10–15% of the neurones in the seventh lumbar sympathetic ganglion of the cat were vasoactive intestinal polypeptide (VIP) immunoreactive and acetylcholinesterase positive (AChE) lead to the suggestion that VIP was present in a population of cholinergic neurones (Lundberg et al., 1980). This seemed particularly likely since 10–15% of the neurones in this ganglion are non-adrenergic and the ganglion was known to contain choline acetyltransferase (ChAT) (Buckley et al., 1967). VIP containing AChE-rich neurones were also found in the submandibular and sphenopalatine ganglion which innervate exocrine elements and blood vessels (Fig. 5). At least some of the ganglion cells of the sphenopalatine ganglion seem to contain both VIP and ChAT (Lundberg et al., 1981a). It should be emphasised that most peripheral cholinergic neurones do not seem to contain VIP (see Lundberg, 1981). VIP

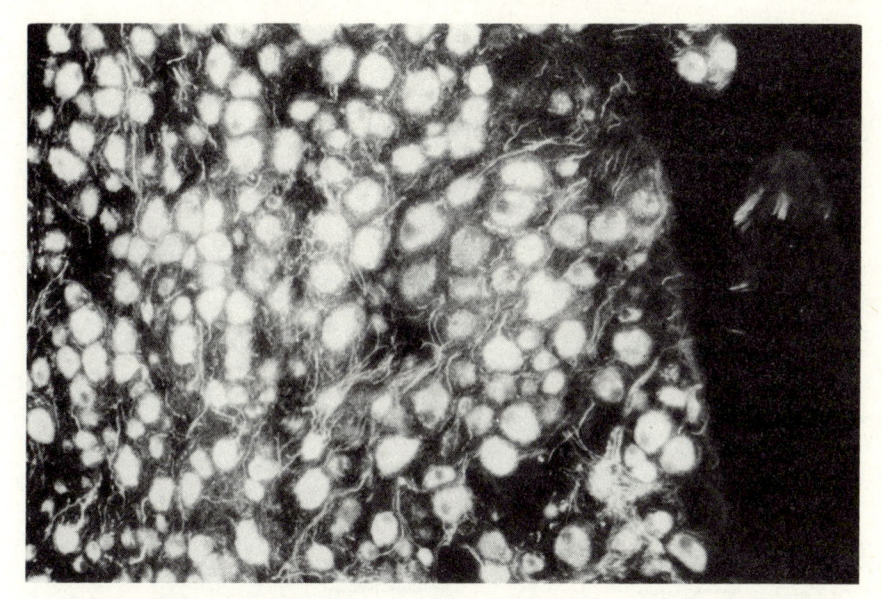

Fig. 5 – Immunofluorescence micrograph from the cat sphenopalatine ganglion after incubation with antiserum to VIP. Note that virtually all the ganglion cells are VIP immunoreactive.

seems to be present in large dense-cored vesicles in nerve endings with a 'cholinergic appearance' (i.e. dominance of small clear vesicles which probably store acetylcholine) (Johansson and Lundberg, 1981; Lundberg et al., 1981b).

The cat submandibular gland has been extensively used to study functional interactions of VIP and acetylcholine in causing vasodilation and exocrine secretion (Lundberg, 1981). The salivary secretion induced by parasympathetic nerve stimulation is completely blocked by atropine suggesting a cholinergic mechanism. On the other hand the concomitant vasodilation is reduced by atropine at low frequencies and is atropine-resistant at high stimulation frequencies. Recently it has been possible to detect simultaneous release of VIP acetylcholine upon stimulation of the parasympathetic nerve innervating the submandibular gland. The differential atropine sensitivity of the vasodilatory response may be explained by a preferential release of acetylcholine at low frequencies while the VIP release is most pronounced at high frequencies (Lundberg et al., 1982a). Atropine treatment seems to increase the VIP content of the venous effluent from the submandibular gland during nerve stimulation. This may suggest the presence of presynaptic muscarinic receptors on VIP nerves which inhibit VIP release (Lundberg et al., 1981a, 1982a). An increased VIP release may also account for the vasodilatory response which follows atropine treatment at high stimulation frequencies (Fig. 6) (Lundberg et al., 1981a, 1982a).

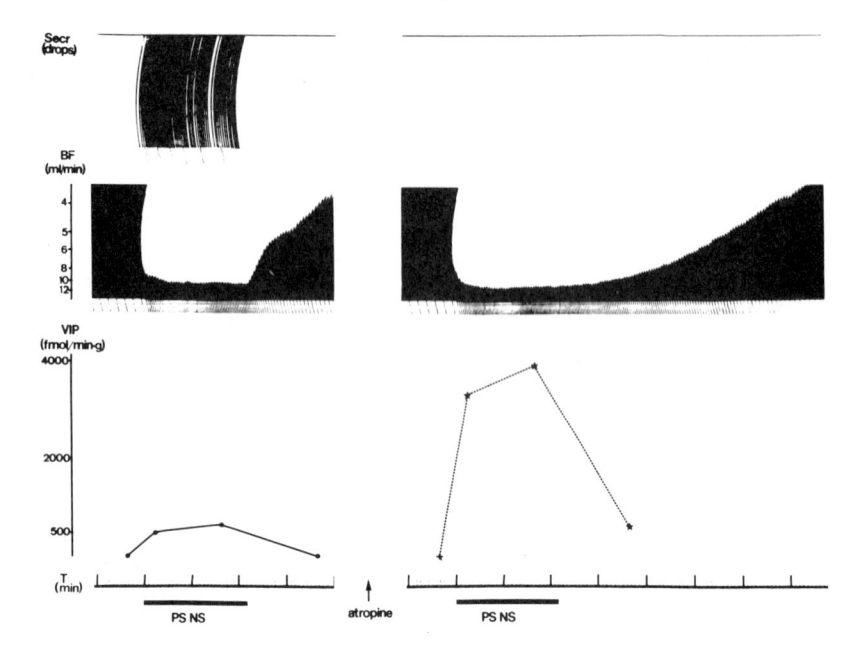

Fig. 6 — Effect of parasympathetic nerve stimulation (15 Hz) on cat submandibular gland blood flow (BF ml/min), salivary secretion (secr/drops) and VIP output (fmol/min/g) in the local venous effluent. Time scale is one min between large bars. Atropine 0.5 mg/kg i.v. was given at arrow (from Lundberg, 1981).

Given the biochemical and physiological evidence for coexistence and co-release of VIP and acetylcholine it was of particular interest to note that there were signs of synergistic interactions between VIP and acetylcholine in that VIP potentiated the effect of acetylcholine on secretion (Fig. 7) (Lundberg et al., 1981a). These observations suggested that VIP and acetylcholine might also be interacting at their post-synaptic receptor sites. In order to investigate this possibility Lundberg et. al. (1982b) used cat submandibular salivary gland membranes and looked at the effects of purified porcine VIP on binding of the muscarinic antagonist ($[^3H]$ N-methyl-4-piperidinyl benzilate ($[^3H]$ 4-NMPB). These results showed that porcine VIP enhanced the rate of association of the labelled ligand with the muscarinic binding site. There was, however, no change in the B_{max} or dissociation rate in the presence of VIP. The effect on association rate was present only for a short period (5–10 minutes) and was maximal between 10^{-9}–10^{-8} M VIP, being absent at higher concentrations of VIP. This result suggests that in binding to its receptor VIP may be producing an allosteric effect on the muscarinic receptor to enhance its affinity for ACh. In further support of this suggestion Lundberg et al. (1982b) also found that VIP could markedly enhance the agonist potency of carbamylcholine or acetylcholine in

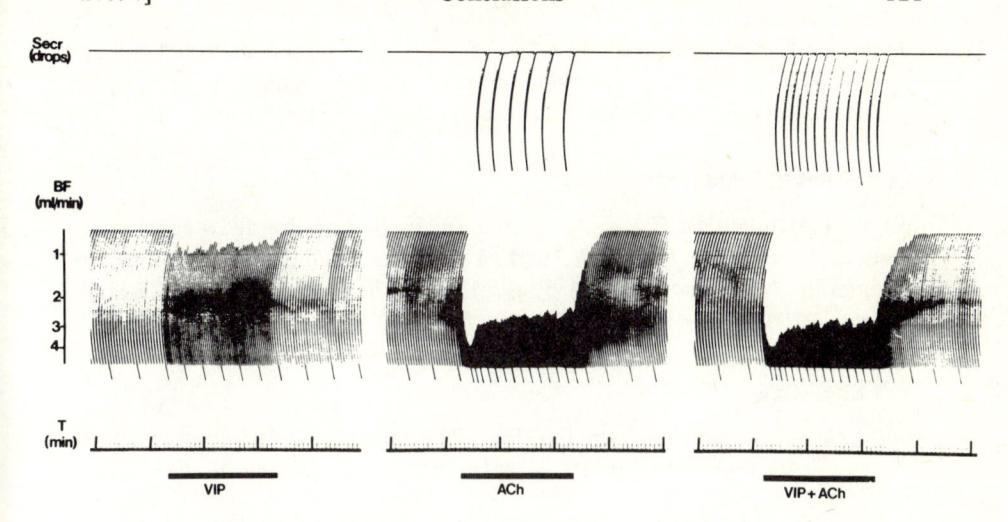

Fig. 7 — Effect of local intra-arterial infusion of VIP and acetylcholine (ACh) on cat submandibular gland blood flow and salivary secretion (legend as Fig. 6). Combination of VIP and ACh causes a potentiated salivary response (from Lundberg, 1981).

displacing [^{3}H] 4-NMPB from the salivary gland binding sites. These results are particularly interesting as they support the physiological observations. The effect was again only short-lived (maximum 10 minutes) and it is possible that this reflects degradation of VIP. Alternatively, this short-term effect may be a characteristic of the response perhaps due to the desensitisation of the VIP receptor.

The interactions between VIP and acetylcholine also seem to be reflected in the intracellular production of second messengers such as cyclic nucleotides. Thus it was found that acetylcholine and VIP have a synergistic effect on cyclic AMP production in slices or isolated acini from the cat submandibular gland (Fredholm and Lundberg, 1982; Enyedi et al., 1982).

4. CONCLUSIONS

The data presented here provide some initial hints that coexisting peptides may interact at both pre- and post-synaptic receptors and that in binding to their receptors they may also influence the receptors of the more established transmitter candidate (such as 5-HT, dopamine and ACh) with which they may coexist. Apart from these possible receptor interactions the coexisting peptides may also act as growth factors or trophic factors ensuring the correct organisation and continuing connectivity of the neurones with the appropriate post-synaptic cell. It will be fascinating to see how this topic develops further and it should perhaps be pointed out that if there is an endogenous ligand for the benzodiazepine receptor this would provide another example of a receptor interaction involving the GABA-benzodiazepine receptor (see Chapter 3).

ACKNOWLEDGEMENTS

We are grateful to Dr. R. A. Nicoll and Dr. H. Barbeau for allowing us to reproduce Figs. 2 and 3 and Dr. L. L. Iversen for reading and criticising this manuscript. JML was supported by grants from the Swedish Medical Research Council. RFTG was an MRC Scholar.

REFERENCES

Anderson, E. G. and Shibuya, T. (1966) The effects of 5-hydroxytryptophan and L-tryptophan on spinal synaptic activity. *J. Pharmacol. Exp. Ther.* **153**, 352–360.

Barbeau, H. and Bedard, P. (1981) Similar motor effects of 5-HT and TRH in rats following chronic spinal transection and 5,7-dihydroxytryptamine injection. *Neuropharmacology* **20**, 471–481.

Bhoola, K. D., Dawbarn, D., O'Shaughnessy, C. and Pycock, C. (1982) Modulation of dopamine receptor activation by the neuropeptides VIP and CCK. *Brit. J. Pharmacol.*, **77**, 334P.

Buckley, G., Consolo, S., Giacobini, E. and Sjoquist, F. (1967) Choline acetyltransferase in innervated and denervated sympathetic ganglia and ganglion cells of the cat. *Acta Physiol. Scand.* **71**, 348–356.

Cabot, J. C., Wild, J. M. and Cohen, D. H. (1979) Raphe inhibition of sympathetic preganglionic neurons. *Science* **203**, 184–186.

Chan-Palay, V., Johansson, G. and Palay, S. (1978) On the coexistence of serotonin and substance P in neurons of the rat's central nervous system. *Proc. Natl. Acad. Sci. USA* **75**, 1582–1586.

Enyedi, P., Fredholm, B., Lundberg, J. and Anggard, A. (1982) Carbachol potentiates the cyclic AMP stimulating effect of VIP in cat submandibular gland. *Eur. J. Pharmacol.* **79**, 139–143.

Fredholm, B. and Lundberg, J. M. (1982) VIP induced cyclic AMP formation in cat submandibular gland. Potentiation by charbacholine. *Acta Physiol. Scand.* **114**, 157–159.

Fuxe, K., Andersson, K., Locatelli, V., Agnati, L. F., Hökfelt, T., Skirboll, L. and Mutt, V. (1980) Cholecystokinin peptides produce marked reductions of dopamine turnover in discrete areas in the rat brain following intraventricular injection. *Eur J. Pharmacol.* **67**, 329–332.

Fuxe, K., Agnati, L. F., Benferati, F., Cimmino, M., Algeri, S., Hökfelt, T. and Mutt, V. (1981) Modulation by cholecystokinin of ^{3}H-spiroperidol binding in rat striatum: evidence for increased affinity and reduction in the number of binding sites. *Acta Physiol. Scand.* **113**, 567–569.

References

Gilbert, R. F. T. (1981) Neuronal coexistence of peptides and amines. Ph.D. thesis, University of Cambridge.

Gilbert, R. F. T., Bennett, G. W., Marsden, C. A. and Emson, P. C. (1981) The effects of 5-hydroxytryptamine-depleting drugs on peptides in the ventral spinal cord. *Eur. J. Pharmacol.* **76**, 203–210.

Gilbert, R. F. T., Emson, P. C., Hunt, S. P., Bennett, G. W., Marsden, C. A., Sandberg, B. E. B., Steinbusch, H. W. M. and Verhofstad, A. A. J. (1982) The effect of monamine neurotoxins on peptides in the rat spinal cord. *Neuroscience* **7**, 69–87.

Hökfelt, T., Ljungdahl, A., Steinsbusch, H., Verhofstad, A., Nilsson, G., Brodin, E., Pernow, B. and Goldstein, M. (1978) Immunohistochemical evidence of substance P-like immunoreactivity in some 5-hydroxytryptamine-containing neurons in the rat central nervous system. *Neuroscience* **3**, 517–538.

Hökfelt, T., Johansson, O., Ljungdahl, A., Lundberg, J. and Schultzberg, M. (1980a) Peptidergic neurones. *Nature (Lond)* **284**, 515–521.

Hökfelt, T., Rehfeld, J. F., Skirboll, L., Ivemark, B., Goldstein, M. and Markey, K. (1980b) Evidence for coexistence of dopamine and CCK in mesolimbic neurones. *Nature* **285**, 476–478.

Hökfelt, T., Skirboll, L., Rehfeld, J. F., Goldstein, M., Markey, K. and Dann, O. 1980c) A subpopulation of mesencepahlic dopamine neurons projecting to limbic areas contain cholecystokinin-like peptide. Evidence from immunohistochemistry combined with retrograde tracing. *Neuroscience* **5**, 2093–2124.

Johansson, O., Hökfelt, T., Pernow, B., Jeffcoate, S. L., White, N., Steinsbusch, H. W. M., Verhofstad, A. A. J., Emson, P. C. and Spindel, E. (1981) Immunohistochemical support for three putative transmitters in one neuron: coexistence of 5-hydroxytryptamine, substance P and thyrotropin releasing hormone-like immunoreactivity in medullary neurons projecting to the spinal cord. *Neuroscience* **6**, 1857–1881.

Kovacs, G. A., Szabo, G., Penke, B. and Tleegdy. G. (1981) Effects of cholecystokinin on striatal dopamine metabolism and apomorphine induced stereotyped cage climbing in mouse. *Eur. J. Pharmacol.* **69**, 313–319.

Lundberg, J. M. (1981) Evidence for coexistence of vasoactive intestinal polypeptide (VIP) and acetylcholine in neurones of cat exocrine glands. Morphological, biochemical and functional studies. *Acta Physiol. Scand.* **112**, 469, 1–57.

Lundberg, J. M., Änggard, A., Fahrenkrug, J., Hökfelt, T. and Mutt, V. (1980) Vasoactive intestinal polypeptide (VIP) in cholinergic neurons of exocrine glands. Possible functional significance of coexisting transmitters for vasodilation and secretion. *Proc. Natl. Acad. Sci. USA,* **77**, 1651–1655.

Lundberg, J. M., Änggard, A., Emson, P. C., Fahrenkrug, J., and Hökfelt, T. (1981a) Co-existence of vasoactive intestinal polypeptide (VIP) and acetylcholine: possible significance for neural control, vasodilation and secretion. *Proc. Natl. Acad. Sci. USA.,* **78**, 5255–5295.

Lundberg, J. M. Fried, G., Fahrenkrug, J., Holmstedt, B., Lundgren, G., Lagercrantz, H., Anggard, A. and Hökfelt, T. (1981b) Subcellular fractionation of cat submandibular gland. Comparative studies on the distribution of acetylcholine and vasoactive intestinal polypeptide (VIP). *Neuroscience* 6, 1001–1010.

Lundberg, J. M., Anggard, A., Fahrenkrug, J., Lundgren, G. and Holmstedt, B. (1982a) Corelease of VIP and acetylcholine in relation to blood flow and salivary secretion in cat submandibular salivary gland. *Acta Physiol. Scand.,* 115, 525–528.

Lundberg, J. M., Hedlund, B. and Bartfai, T. (1982b) Vasoactive intestinal polypeptide enhances muscarinic ligand binding in cat submandibular gland. *Nature* 295, 147–149.

McCall, R. B. and Aghajanian, G. K. (1979) Serotonergic facilitation of facial motorneuron excitation. *Brain Res.* 169, 11–27.

Marley, P. D., Nagy, J. I., Emson, P. C. and Rehfeld, J. F. (1982) Cholecystokinin in the rat spinal cord: distribution and lack of effect of neonatal capsaicin and rhizotomy. *Brain Res.* 238, 494–498.

Marsden, C. A., Bennett, G. W., Irons, J., Gilbert, F. R. T. and Emson, P. C. (1982) Localisation and release of 5-hydroxytryptamine, thyrotrophin releasing hormone and substance P in rat ventral spinal cord. *Comp. Biochem. Physiol.,* 72C, 263–270.

Mitchell, P. and Fleetwood-Walker, S. (1981) Substance P, but not TRH modulates the 5-HT autoreceptor in ventral lumbar spinal cord. *Eur. J. Pharmacol.* 76, 119–120.

Murphy, R. B. and Schuster, D. I. (1982) Modulation of [^{3}H]-dopamine binding by cholecystokinin octapeptide (CCK-8). *Peptides.* In press.

Nicoll, R. A., Alger, B. E. and Jahr, C. E. (1980) Peptides as putative excitatory neurotransmitters, carnosine, enkephalin, substance P and TRH. *Proc. R. Soc. Lond. B.* 210, 133–149.

Pelletier, G., Steinsbusch, H. W. M. and Verhofstad, A. A. J. (1981) Immunoreactive substance P and serotonin are contained in the same dense core vesicles. *Nature* 293, 71–72.

Schultzberg, M., Dockray, G. J. and Williams, R. G. (1982) Capsaicin depletes CCK-like immunoreactivity detected by immunohistochemistry but not that measured by radioimmunoassay in rat dorsal spinal cord. *Brain Res.* 235, 205–211.

Skirboll, L. R., Grace, A. A., Hommer, D. W., Rehfeld, J. F., Goldstein, M., Hökfelt, T. and Bunney, B. S. (1981) Peptide-monamine coexistence: studies of the actions of cholecystokinin-like peptides on the electrical activity of midbrain dopamine neurons. *Neuroscience* 6, 2111–2124.

Vanderhaeghen, J. J., Lotstra, F., DeMey, J. and Gilles, C. (1980) Immunohistochemical localization of cholecystokinin and gastrin-like peptides in the brain and hypophysis of the rat. *Proc. Natl. Acad. Sci. USA* **77**, 1190–1194.

Viveros, O. H., Diliberto, E. J., Hazum, E. and Chang, K. J. (1980) Enkephalins as possible adrenomedullary hormones: storage, secretion and regulation of synthesis., in *Neural Peptides and Neuronal Communication* (Costa E. and Trabucchi, M., eds.) pp. 191–204. Raven Press, New York.

White, S. R. and Neuman, R. S. (1980) Facilitation of spinal motor neurone excitability by 5-hydroxytryptamine and noradrenaline. *Brain Res.* **188**, 119–127.

Wing, L. M. H. and Chalmers, J. P. (1974) Participation of central serotoninergic neurons in the control of the circulation of the unanaethetised rabbit. *Circ. Res.* **35**, 504–513.

7

Structure and function of the nicotinic acetylcholine receptor: proteolytic digestion and subunit specificity

Sara Fuchs and Daniel Bartfeld

Department of Chemical Immunology, The Weizmann Institute of Science, Rehovot 76100, Israel

---

Abbreviations used

AChR	—	acetylcholine receptor (nicotinic)
anti-nAChR	—	antibody against conformational determinants of AChR
anti-dAChR	—	antibody against non-structural determinants of AChR
anti-d$_{TR}$ AChR	—	antibody against trypsin resistant non-structural determinants of AChR
anti-d$_{TS}$ AChR	—	antibody against trypsin sensitive non-structural determinants of AChR
BAC	—	bromoacetylcholine
α-Bgt	—	α-bungarotoxin
EAMG	—	Experimental autoimmune myasthenia gravis
Pap-AChR	—	affinity purified, papain treated AChR
Pap-T-AChR	—	affinity purified, papain and trypsin treated AChR
RCM-AChR	—	reduced carboxymethylated AChR
SDS PAGE	—	sodium dodecyl sulphate polyacrylamide gel electrophoresis
T-AChR	—	affinity purified trypsin treated AChR

1. INTRODUCTION

The nicotinic acetylcholine receptor (AChR) is the first neurotransmitter receptor to be isolated and purified, and the most studied receptor on the molecular level. AChR is one of the key proteins governing the function of the neuromuscular junction. This receptor is a high molecular weight glycoprotein present in the membrane of the post-synaptic cell. Like other neuroreceptors it exists in very minute amounts in mammalian tissues. The advantage of this system over other receptors is the availability of electric organs of electric fish (electric eel or electric rays) which provide a rich source for AChR, where it exists in amounts three to four orders of magnitude higher than its amount in muscles. AChR can be isolated and purified in substantial amounts from this source. For most of our studies we have been using AChR purified from the electric organ of *Torpedo californica*. The main purification step and also the pharmacological assay for AChR, is based on the fact that α-neurotoxins from certain snake venoms bind specifically and with a high affinity for the nicotinic receptor. The receptor is extracted from the membranes of the electric organ by non-ionic detergents and purified essentially by one step of affinity chromatography on a toxin-resin, followed by elution of the purified receptor from the resin by the cholinergic ligand carbamylcholine (Aharonov *et al.*, 1977). The detergent-purified AChR from *Torpedo californica* is a multi-subunit glycoprotein with a molecular weight around 250,000 daltons. It is composed of four different kinds of subunits (α, β, γ and δ), ranging in molecular weights from 40,000 daltons (α-subunit) to 65,000 daltons (δ-subunit). The unit of the receptor molecule is a five polypeptide oligomer with the composition $\alpha_2\beta\gamma\delta$. The cholinergic binding site of the receptor is within the α subunit. The detailed biochemical and pharmacological properties of AChR have been summarized in several recent review articles (Karlin, 1980; Changeux, 1981; Conti-Tronconi and Raftery 1982, see also Chapter 14).

Special interest in the immunological properties of AChR arises from the involvement of an autoimmune response against this receptor in myasthenia gravis and its animal model, experimental autoimmune myasthenia gravis (EAMG), for reviews see Fuchs (1979a), Lindstrom (1979) and Vincent (1980). In our laboratory we are studying several topics related to the immunology and immunochemistry of AChR, with an emphasis on the involvement of AChR in myasthenia gravis. These topics include structure and function analysis of AChR, the nature and regulation of the immune response to AChR, monoclonal antibodies to AChR, anti-idiotypes and their possible function in regulating myasthenia, AChR biosynthesis, the role of the thymus in myasthenia, etc. In this report we will describe in more detail our recent results obtained from studies on proteolytic digestion of AChR and on the immunological specificity of the various receptor subunits.

2. MOLECULAR DISSECTION OF AChR: DENATURATION AND PROTEOLYTIC DIGESTION

A molecular dissection of AChR, followed by pharmacological and immunological analysis may lead to the isolation and identification of fragments or sites in the molecule which are responsible for its specific cholinergic and/or pathologic myasthenic activity. By following such an analysis it was possible to design derivatives of AChR with a potential to regulate specifically the immune response to AChR, and to immunosuppress EAMG, and to identify small fragments of AChR carrying the biological functions of the intact molecule.

The first derivative of AChR that we have studied in detail was a denatured preparation of the receptor (Bartfeld and Fuchs, 1977). AChR was irreversibly denatured by reduction and carboxymethylation in 6 M guanidine hydrochloride. The resulting reduced-carboxymethylated AChR (RCM-AChR) did not bind cholinergic ligands and did not induce EAMG in rabbits. Nevertheless, RCM-AChR is a good immunogen and elicits antibodies which cross-react with the native receptor. Anti-AChR sera show only a partial cross-reactivity with RCM-AChR, whereas anti-RCM-AChR sera react equally and identically with both RCM-AChR and AChR. Thus, the immune response to RCM-AChR is directed only against non-structural antigenic determinants of AChR which are shared by both the native and denatured receptor and which are not involved in the myasthenic activity of AChR. The difference in specificity between anti-AChR and anti-RCM-AChR is reflected also by their effects on the binding of α-bungarotoxin (α-Bgt) to AChR. Whereas anti-AChR antibodies block the *in vitro* binding of α-Bgt to AChR very effectively, anti-RCM-AChR antibodies block this binding only to a very limited extent (Bartfeld and Fuchs, 1977).

Denatured AChR preparation was shown to be an appropriate derivative for specific regulation of EAMG (Bartfeld and Fuchs, 1978; Fuchs, 1979a, 1979b; Fuchs *et al.*, 1980). RCM-AChR, which by itself is devoid of any myasthenic activity was shown to have an immunosuppressive effect on EAMG, either by preventing the onset of the disease or by reversing the clinical symptoms in myasthenic rabbits (Bartfeld and Fuchs, 1978). For the prevention experiments rabbits were first immunised several times with RCM-AChR and were then injected with the intact AChR. Under these conditions there was either a delay in the onset of EAMG or a complete prevention of the disease. Moreover, in addition to the protective potential of RCM-AChR, its therapeutic effect on myasthenic rabbits was observed as well. A single administration of RCM-AChR in complete Freund's adjuvant to myasthenic rabbits has resulted in many instances in a gradual suppression of the disease and complete reversal of symptoms. In all cases the therapeutic effect of RCM-AChR on EAMG was accompanied by a change in the specificity of the humoral immune response, from that typical of AChR-injected rabbits towards that of non-myasthenic, RCM-AChR injected rabbits (Bartfeld and Fuchs, 1978). The cross-reactivity between the intact and denatured AChR and the non-pathogenicity of the latter appear to be crucial in governing the immunosuppressive effect of RCM-AChR on EAMG.

A different strategy for structure and function analysis of AChR was applied by proteolytic degradations of the receptor molecule. This approach proved

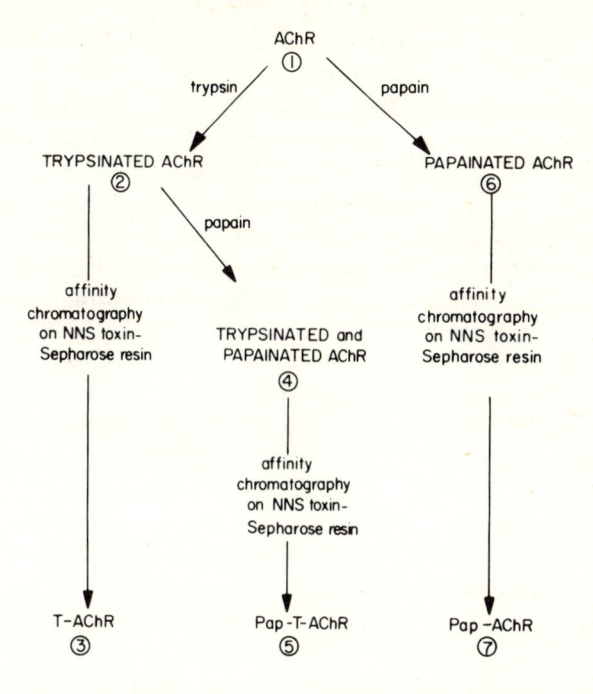

Fig. 1 — Schematic description of the various proteolytic digestions performed on AChR and the products obtained following their purification by affinity chromatography.

Table 1 — Specific binding activity of AChR derivatives following proteolytic digestions.

	Sample	Protein concentration (mg/ml)	Toxin binding (pmole α-Bgt/ml)	Specific activity (pmole α-Bgt/mg)
1.	AChR	1.50	16,250	10,830
2.	Trypsinated AChR	1.50	15,860	10,500
3.	T-AChR	0.33	9,600	29,000
4.	Trypsinated and papainated AChR	1.0	9,900	9,900
5.	Pap-T-AChR	0.24	6,380	26,580
6.	Papainated AChR	1.0	9,900	9,900
7.	Pap-AChR	0.3	7,370	24,500

useful for attempting the isolation and analysis of fragments of the receptor molecule which retain the original specific cholinergic binding site, or the pathologic myasthenic activity, or both. Studies with trypsin (Bartfeld and Fuchs, 1979a) and more recently with papain (Bartfeld et al., 1982) were performed. The various proteolytic digestions applied to AChR with trypsin, papain and their combination are described schematically in Fig. 1. The specific binding activity of the various products is summarised in Table 1.

We have previously shown that a controlled tryptic digestion of AChR did not change either the pharmacological specificity or the pathological myasthenic activity of the receptor molecule (Bartfeld and Fuchs, 1979a). The product obtained after tryptic digestion and repurification by affinity chromatography on a toxin-Sepharose resin was designated T-AChR (see Fig. 1). T-AChR is a simpler and smaller molecule than AChR; sodium dodecyl sulphate-polyacrylamide gel electrophoresis (SDS-PAGE) of T-AChR indicates that it is composed mainly of one type of subunit, with a molecular weight of about 27,000 daltons, which is smaller than any of the four subunits present in the intact receptor (Fig. 2a). Both the cholinergic and the myasthenic sites of AChR reside within the trypsin resistant region of AChR. Immunologically, T-AChR binds mainly to that fraction of anti-AChR antibodies which is directed against conformational antigenic determinants. The pharmacological specificity of T-AChR is identical to that of the intact AChR molecule (Bartfeld and Fuchs, 1979a). However, the specific binding activity of T-AChR, following the repurification step, is about three times higher than that of AChR (Table 1).

Affinity labelling experiments and immunological analysis of T-AChR support the fact that the 27K subunit originates from the 40K subunit (α-subunit) of the receptor. Affinity labelling of trypsinated AChR or of T-AChR with tritiated bromoacetylcholine ([^{3}H]BAC), which labels specifically the 40K subunit in AChR (Damle et al., 1978; Moore and Raftery, 1979; Lyddiatt et al., 1979), results in an exclusive labelling of the 27K subunit (Fig. 2b). Similarly, trypsinisation of [^{3}H]BAC-labelled AChR results in a radioactive band only at the 27K region (Fig. 3). The 27K subunit originates probably from the n-terminal end of the α subunit. This is supported by previous reports of Wennogle and Changeux (1980) who have demonstrated that tryptic digestion of AChR-rich membrane preparation starts at the C-terminal end of the 40K (α) chain, and that the NH$_2$-terminal sequence of the tryptic fragment is identical to that of 40K chain.

Rabbit antibodies elicited against the 27K subunit eluted from SDS-preparative gels of T-AChR, bind to the α subunit to the same extent as they bind to the immunising 27K subunit (Fuchs et al., 1981). Only very low binding is obtained with the β, γ or δ subunits, though a significant cross-inhibition between the various subunits was observed from cross-inhibition experiments using the anti-27K antibodies. These results suggest again that the 27K subunit originates exclusively from the α-subunit and that the cross-reactivity observed

Sec. 2] **Molecular Dissection of AChR** 131

with the other subunits is due to structural homology between the subunits as has been demonstrated by Raftery *et al.* (1980). However, our experiments cannot rule out completely the possibility that the β, γ and δ subunits, or part of them, might be also degraded by trypsin to a 27K subunit. The possibility

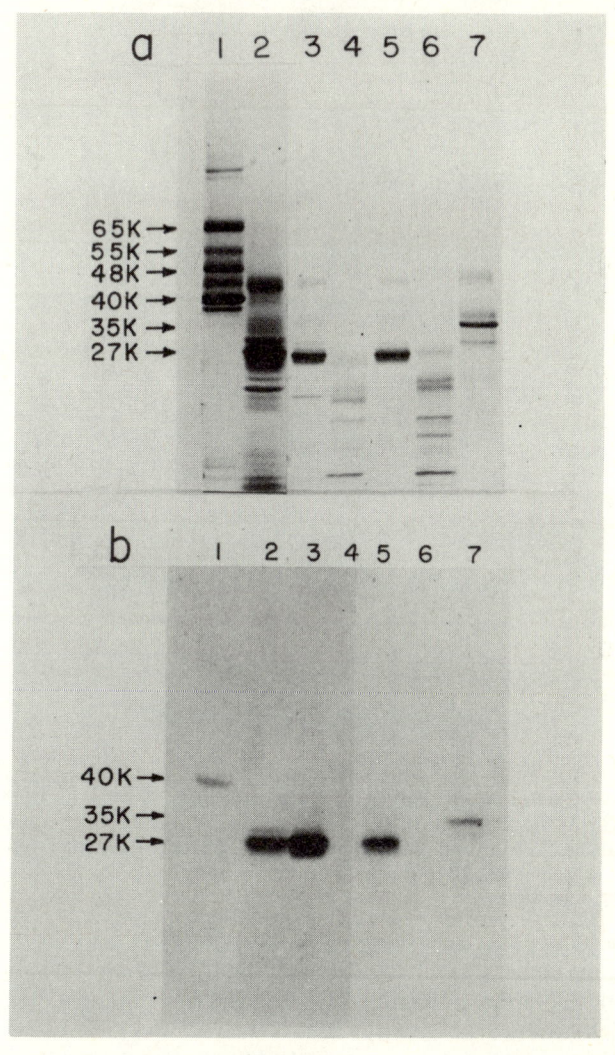

Fig. 2 — SDS-PAGE (concentration gradient of 7.5% to 15% polyacrylamide) of AChR and the products obtained following proteolytic digestion and then affinity labelling with [^{3}H]BAC. (1) AChR; (2) trypsinated AChR; (3) T-AChR; (4) trypsinated and papainated AChR (5) Pap-T-AChR (6) papainated AChR; (7) Pap-AChR. (See Fig. 1 for designation of samples.) A: Protein staining. B: Fluorogram.

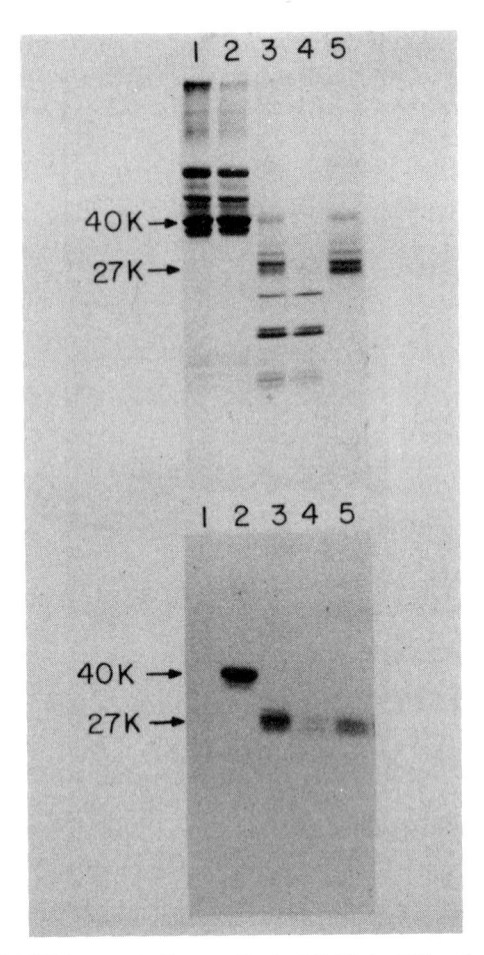

132 **Structure and Function of the Nicotinic Acetylcholine Receptor** [Ch. 7

Fig. 3 — SDS-PAGE (concentration gradient of 7.5% to 15% polyacrylamide) of [³H] BAC-labelled AChR and the products obtained following its tryptic digestion. (1) AChR (unlabelled); (2) [³H] BAC-labelled AChR; (3) trypsinated [³H]-BAC-labelled AChR, and the effluent (4), and eluate (5), obtained following its affinity chromatography on a toxin-Sepharose adsorbent. Top: Protein staining. Bottom: Fluorogram.

of a common domain to all the various subunits may be compatible with the idea that all the receptor subunits have diverged from one ancestral gene, as was proposed by Raftery et al. (1980).

In addition to tryptic digestion of AChR we have also applied papain digestion to the receptor (Bartfeld et al., 1982). Papain digestion of AChR or of trypsinated AChR were performed for 1 h with 0.5% papain (w/w). The reaction was stopped by the addition of excess iodoacetamide. In either case such a digestion resulted in the disappearance of the original high molecular weight subunits

(Fig. 2a, slots 4 and 6). However, no significant loss of binding properties to α-Bgt were observed following papain digestion (Table 1). This is similar to our earlier observations with trypsinated AChR and the observations made by Lindstrom et al. (1980) and Huganir and Racker (1980) following papain and pronase digestion of AChR, respectively.

In an attempt to isolate a small active AChR fragment obtained following papain digestion, both papainated AChR as well as trypsinated and papainated AChR (Fig. 1, products 6 and 4, respectively) were applied to an affinity chromatography resin of Naja naja siamensis toxin bound to Sepharose. The adsorbed components were eluted with 0.7 M carbamylcholine and were designated Pap-AChR and Pap-T-AChR, respectively. The specific binding activity of Pap-AChR and Pap-T-AChR to α-bungarotoxin is more than twice that of AChR or of the respective digests before the purification step by affinity chromatography (Table 1). This increased value in specific activity is lower than the binding activity of T-AChR, suggesting that smaller amounts of digested material were removed by the purification step.

On SDS-PAGE Pap-AChR appears as one major band with a molecular weight of 35K (Fig. 2a, slot 7). Some additional weak bands at molecular weights range of 43K, 37K and 31K are observed occasionally. Pap-T-AChR appears as one major band with a molecular weight of 27K, similar to the one observed with T-AChR (Fig. 2a, slot 5). The lack of the 35K and 27K band in the papain digests of AChR and trypsinated AChR, respectively, may be attributed to a more extended digestion with papain before the addition of iodoacetamide.

In order to test whether the major polypeptide chains obtained following papain digestion and repurification bear the binding site of AChR, affinity labelling experiments with [^{3}H] BAC were performed. As can be seen in Fig. 2b, the 35K band is exclusively labelled in Pap-AChR (slot 7) whereas the 27K band is the only one labelled in trypsinated AChR, T-AChR or Pap-T-AChr (slots 2, 3 and 5). Intact AChR is labelled only on the 40K subunit (α) as reported earlier (Damle et al., 1978; Lyddiatt et al., 1979).

The results of the papain digestion experiments suggest that under the conditions applied papain might cleave less of the α-subunit than trypsin does. The trypsin resistant 27K domain is probably resistant also to additional papain proteolysis.

3. ON THE ANTIGENIC SPECIFICITY OF THE SUBUNITS OF AChR

Torpedo AChR is a multi-subunit molecule, the unit structure of which is built of five polypeptide chains with the composition $\alpha_2\beta\gamma\delta$. The cholinergic binding site is on the α subunit. The biological function of the other subunits as well as the possible structural relationship between them are still being investigated. Antigenic characterisation of the individual subunits should be most helpful in elucidating these points. For functional studies we are interested mainly in characterising the immunogenic determinants of AChR which are expressed following immunisation or autosensitisation with the intact molecule. Our approach was therefore to study the reactivity of the various subunits with antibodies elicited against *intact* AChR or with antibody subpopulations fractionated from these antibodies.

The various antibody fractions used in this study are described schematically in Fig. 4. By specific immuno-adsorptions it is possible to fractionate anti-AChR antibodies according to their antigenic specificity (Bartfeld and

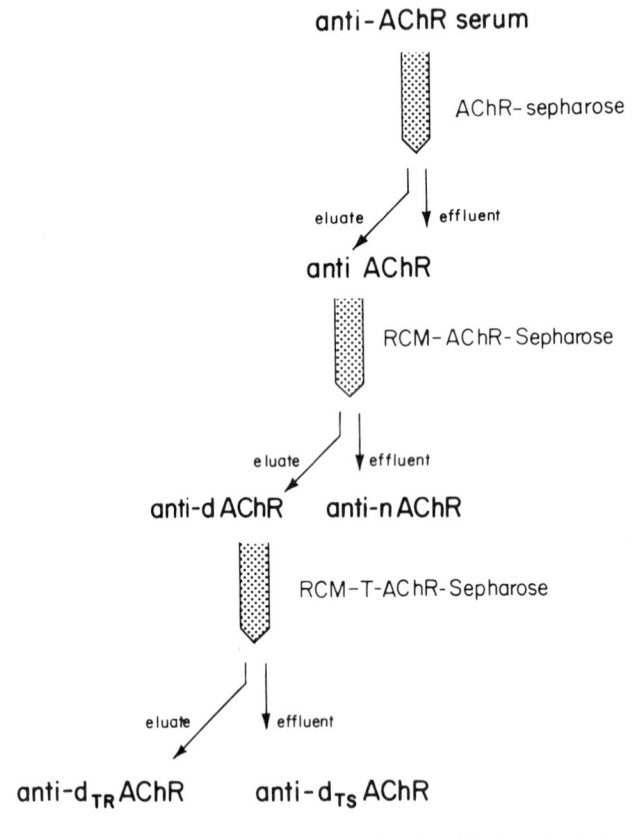

Fig. 4 — Fractionation of anti-AChR antibodies. Specifically adsorbed antibodies were eluted from the Sepharose adsorbents with 0.15 M NH$_4$OH and dialysed against phosphate-buffered saline. Anti-nAChR represent the antibody fraction directed against conformational determinants of AChR present only on the native intact molecule. Anti-dAChR are antibodies directed against non-structural determinants present also on denatured AChR. Anti-d$_{TS}$AChR and anti-d$_{TR}$AChR are antibodies directed against trypsin-sensitive and trypsin-resistant non-structural determinants.

Fuchs, 1979b; Fuchs et al., 1981). By fractionation of anti-AChR antibodies on a resin of RCM-AChR-Sepharose two antibody subpopulations were obtained: antibodies directed against conformational antigenic determinants of AChR, designated anti-nAChR, and antibodies against non-structural (or sequential) determinants of AChR which are present also on the denatured receptor, and designated anti-dAChR. Further fractionation of anti-dAChR on RCM-T-AChR-Sepharose or on T-AChR-Sepharose (Fig. 4) resulted in two additional antibody subpopulations: Antibodies directed against non-structural determinants in AChR which are resistant to tryptic digestion (anti-d_{TR} AChR) and others directed against trypsin sensitive non-structural determinants of AChR (anti-d_{TS} AChR). Anti-d_{TR} AChR antibodies are probably those responsible for the partial binding of T-AChR to anti-dAChR antibodies (Bartfeld and Fuchs, 1979a), and anti-d_{TS} AChR represent antibodies against determinants which are present in unfolded regions in the intact receptor molecule and are available to tryptic digestion.

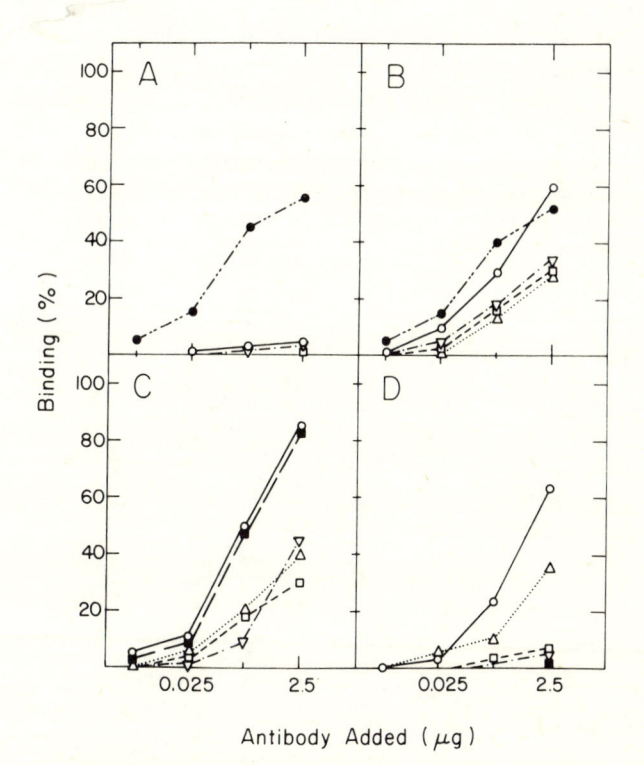

Fig. 5 — Antigenic specificity of the receptor polypeptide chains. Binding of [^{125}I] labelled AChR (●), α (○), β (∇), γ (□), δ (△) and 27K subunit (■) to: (a) anti-nAChR antibodies. (B) anti-dAChR antibodies; (C) anti-d_{TR} AChR antibodies; (D) anti-d_{TS} AChR antibodies.

The individual subunits of AChR, like the whole denatured receptor, bind to both anti-AChR and to anti-RCM-AChR sera (Fuchs et al., 1981). However, in anti-AChR antibodies, the subunits bind only to that fraction of antibodies which is directed against non-structural antigenic determinants (anti-dAChR) and do not bind to anti-nAChR (Fig. 5, A and B). Following fractionation of anti-dAChR on RCM-T-AChR-Sepharose adsorbent, anti-d$_{TR}$AChR antibodies bind all the four receptor subunits (Fig. 5C), whereas a weaker binding of the subunits to anti-d$_{TS}$AChR is observed, particularly of the γ and δ subunits (Fig. 5D). The 27K subunit isolated from T-AChR does not bind at all to anti-d$_{TS}$AChR antibodies (Fig. 5D). In all cases the binding of the α subunit to the antibodies was higher than that of all the other subunits, suggesting that the most immunopotent determinants of AChR are on the α subunit, as has been previously reported (Tzartos and Lindstrom, 1980) from studies with monoclonal antibodies.

The antigenic specificity of the isolated subunits was further analysed by inhibition experiments. Each of the four radiolabelled subunits were reacted with anti-dAChR, anti-d$_{TR}$AChR or anti-d$_{TS}$AChR antibodies and the displacement of each particular binding by the homologous unlabelled subunits or by the other three unlabelled subunits was measured (Fig. 6). By this assay it is possible to select from a certain antibody population only antibodies which bind to one particular subunit, even if they represent a minor antibody fraction, and to analyse their specificity. Although in all cases the highest displacement

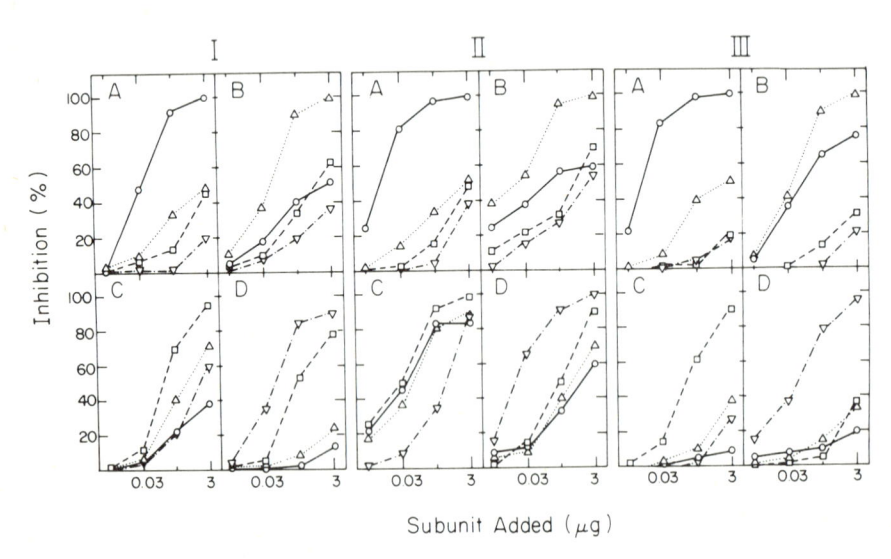

Fig. 6 — Antigenic specificity of the receptor subunits. Inhibition of the binding of A, [^{125}I]α; B, [^{125}I]β C [^{125}I]γ and D, [^{125}I]δ to anti-dAChR(I), anti-d$_{TR}$AChR (II) and anti-d$_{TS}$AChR (III) by unlabelled subunits: α (○); β (△); γ (□) and δ (▽).

was achieved by the homologous unlabelled subunit, indicating the presence of some distinct antigenic determinants specific for the individual subunits, a significant cross-reactivity between the various subunits was observed (Fig. 6). The cross-inhibition among the various subunits may be interpreted either by some shared immunogenic determinants between the various subunits expressed at the level of antibody production, or by antigenic cross-reactivity between common or similar determinants which can bind to a certain extent to antibodies raised by a particular subunit. The α-subunit is most likely the most immunogenic of all the receptor subunits when immunisation with intact receptor is performed. Also, the cross-inhibition of the β, γ and δ subunits with antibodies that bind to the [^{125}I] α subunit is weaker than the cross-inhibition of the non-homologous subunits with antibodies which bind to radiolabelled β, γ or δ subunits (Fig. 6). This may indicate a higher degree of subunit specificity of the α-subunit.

By analysing the reactivity of the isolated subunits with anti-d$_{TR}$ AChR and anti-d$_{TS}$ AChR (Fig. 6, II and III) we could test which of the two antibody fractions contributes to the cross reactivity between the subunits, as observed with anti-dAChR (Fig. 6, I). These cross-inhibition experiments have demonstrated that a higher degree of subunit cross-reactivity is expressed by anti-d$_{TR}$ AChR antibodies than by anti-d$_{TS}$ AChR antibodies. This may suggest that most of the antigenic determinants which are distinct for the various receptor chains are present in the exposed regions of the molecule which are sensitive to trypsin, and that the undigested portion of the molecule contains the determinants which are more cross-reactive either on the level of immunogenicity or antigenic specificity. If the trypsin resistant portion of AChR originates from the N-terminal part of the α-subunit, as can be implicated by sequence experiments by Wennogle and Changeux (1980), the high cross-reactivity of the subunits with antibodies directed against this portion (anti-d$_{TR}$ AChR) is compatible with the sequence homology between the N-termini of the four receptor subunits (Raftery *et al.*, 1980).

Our experiments using antibodies raised against *intact* AChR and cross-inhibition assays demonstrate that there are antigenic similarities between the various polypeptide chains of *Torpedo* AChR. Such a similarity has been observed by Mehraban *et al.* (1982) using antibodies raised against 'renatured' AChR polypeptide chains. Cross-reactivity between various AChR subunits was also shown by using monoclonal antibodies selected from immunisation with AChR (Souroujon *et al.*, 1982; Tzartos and Lindstrom, 1980). In earlier studies of subunit specificity, Lindstrom *et al.* (1978) have elicited antibodies against the isolated subunits. By using this approach they could not detect any cross-reactivity between the subunits and concluded that the various receptor chains are immunologically distinct from each other. However, it should be borne in mind that the individual receptor subunits can be isolated only following denaturation. Under such conditions the subunits may express different antigenic determinants than those present in the intact molecule in the membrane, or even in the detergent purified form. Indeed, from analysis of the specificity of antibodies elicited in rats against the isolated denatured receptor chains, Lindstrom *et al.* (1978, 1979) have concluded that the most immunogenic determinants on the native receptor are lost upon denaturation and that the antibodies induced by the isolated chains are of different specificities than those induced by native receptor. Mehraban *et al.* (1982) could demonstrate antigenic cross-reactivity among the receptor chains, similar to that observed in our experiments, when they used antibodies elicited against the isolated chains following renaturation. Under these conditions some of the original determinants which were lost by the denaturation step might have been reformed. However, it is not known whether the renaturation procedure leads to an identical folding and spatial structure of the receptor polypeptide chains as in the intact receptor, in the presence of the neighbouring subunits. Therefore, the use of polyclonal or monoclonal antibodies elicited against the *intact* receptor may be advantageous.

4. CONCLUDING REMARKS

We have reported here on some recent studies related to structural and immunochemical aspects of the nicotinic AChR. This is just one part of our research programme on the immunology of this receptor and its involvement in myasthenia gravis. Myasthenia gravis is just one case and by now probably the most established one, of an anti-receptor autoimmune disease. The advantage of this particular system is in the availability of large quantities of the purified autoantigen and of an appropriate experimental model disease. As this book is devoted to basic and clinical aspects of cell surface receptors, it may be appropriate to mention that several other diseases were shown to have autoantibodies to cell surface receptors. Thus, antibodies to insulin receptor were demonstrated in cases of insulin-resistant diabetes (Flier *et al.*, 1975), antibodies to thyrotropin receptor in Graves's disease (Rees-Smith and Hall, 1974) and antibodies to β_2-adrenergic receptor in allergic rhinitis and asthma (Ventner *et al.*, 1980). The list may be longer and include additional diseases involving autoimmune response to other neuroreceptors and hormone receptors. Besides the great potential of anti-receptor antibodies in studying the respective receptors, the findings that some diseases as mentioned are of an autoimmune nature, give a new dimension to understanding the mechanism underlying such diseases and may contribute to their better control.

ACKNOWLEDGEMENTS

We wish to thank Mrs Dora Barchan for technical help. This work was supported by grants from the Muscular Dystrophy Association of America and the Los Angeles Chapter of the Myasthenia Gravis Foundation.

REFERENCES

Aharonov, A., Tarrab-Hazdai, R., Silman, I. and Fuchs, S. (1977) Immunochemical studies on acetylcholine receptor from Torpedo californica. *Immunochemistry* **14**, 129–137.

Bartfeld, D. and Fuchs, S. (1977) Immunological characterization of an irreversibly denatured acetylcholine receptor. *FEBS Lett.* **77**, 214–218.

Bartfeld, D. and Fuchs, S. (1978) Specific immunosuppression of experimental autoimmune myasthenia gravis by denatured acetylcholine receptor. *Proc. Natl. Acad. Sci. USA* **75**, 4006–4010.

Bartfeld, D. and Fuchs, S. (1979a) Active acetylcholine receptor fragment obtained by tryptic digestion of acetylcholine receptor from Torpedo californica. *Biochem. Biophys. Res. Commun.* **89**, 512–519.

Bartfeld, D. and Fuchs, S. (1979b) Fractionation of antibodies to acetylcholine receptor according to antigenic specificity. *FEBS Lett.* **105**, 303–306.

Bartfeld, D., Barchan, D. and Fuchs, S. (1982) Proteolytic cleavage of the nicotinic acetylcholine receptor: studies with trypsin and papain. *Israel J. Med. Sci.* **In press.**

Changeux, J. P. (1981) The acetylcholine receptor: An 'allosteric' membrane protein. *The Harvey Lectures, Series* 75, 85–254.

Conti-Tronconi, B. M. and Raftery, M. A. (1982) The nicotinic cholinergic receptor: Correlation of molecular structure with functional properties. *Ann. Rev. Biochem.* **51**, 491–530.

Damle, V. N., McLaughlin, M., and Karlin, A. (1978) Bromoacetylcholine as an affinity label of the acetylcholine receptor from Torpedo californica. *Biochem. Biophys. Res. Commun.* **84**, 845–851.

Flier, J. S., Kahn, C. R., Roth, J. and Bar, R. S. (1975) Antibodies that impair receptor binding in an unusual diabetic syndrome with severe insulin resistance. *Science* **190**, 63–65.

Fuchs, S. (1979a) Immunology of the nicotinic acetylcholine receptor. *Current Topics in Microbiology and Immunology* **85**, 1–29.

Fuchs, S. (1979b) Immunosuppression of experimental myasthenia, in *Plasmaphoresis and the Immunobiology of Myasthenia Gravis* pp. 20–31 (ed. P. C. Dau) Houghton Mifflin. pp. 20–31.

Fuchs, S., Bartfeld, D., Eshhar, Z., Feingold, C., Mochly-Rosen, D., Novick, D., Schwartz, M. and Tarrab-Hazdai, R. (1980) Immune regulation of experimental myasthenia. *J. Neurol. Neurosurg. and Psych.* **43**, 634–643.

Fuchs, S., Bartfeld, D., Mochly-Rosen, D., Souroujon, M. and Feingold, C. (1981) Acetylcholine receptor: Molecular dissection and monoclonal antibodies in the study of experimental maysthenia. *Ann. N.Y. Acad. Sci.* **377**, 110–124.

Huganir, R. L. and Racker, E. Endogenous and exogenous proteolysis of the acetylecholine receptor from Torpedo californica. *J. Supramol Structure* **14**, 13–19.

Karlin, A. (1980) Molecular properties of nicotinic acetylcholine receptors, in *The Cell Surface and Neuronal Function* (C. W. Cotman, G. Poste and G. L. Nicolson, eds.) pp. 191–260. Elsevier, North-Holland Biomedical Press.

Lindstrom, J. (1979) Autoimmune response to acetylcholine receptors in myasthenia gravis and its animal models. *Adv. Immunol.* **27**, 1–50.

Lindstrom, J., Einarson, B. and Merlie, J. (1978) Immunization of rats with polypeptide chains from Torpedo acetylcholine receptor causes an autoimmune response to receptors in rat muscle. *Proc. Natl. Acad. Sci.* **75**, 769–773.

Lindstrom, J., Walter, B. and Einarson, B. (1979) Immunochemical similarities between subunits of acetylcholine receptors from Torpedo, electrophorus, and mammalian muscle. *Biochemistry* **18**, 4470–4480.

Lindstrom, J., Gullick. W., Conti-Tronconi, B. and Ellisman, M. (1980) Proteolytic nicking of the acetylcholine receptor. *Biochemistry* **19**, 4791–4795.

Lyddlat, A., Sumikawa, K., Wolosin, J. M., Dolly, J. O. and Barnard, E. A. (1979) Affinity labelling by bromoacetylcholine of a characteristic subunit in the acetylcholine receptor from muscle and Torpedo electric organ. *FEBS Lett.* **108**, 20–24.

Mehraban, F., Dolly, O. and Barnard, E. A. (1982) Antigenic similarities between the subunits of acetylcholine receptor from Torpedo marmorata. *FEBS Lett.* **141**, 1–5.

Moore, H. H. and Raftery, M. A. (1979) Studies of reversible and irreversible interactions of an alkylating agonist with Torpedo californica acetylcholine receptor in membrane-bound and purified states. *Biochemistry* **18**, 1862–1867.

Raftery, M. A., Hunkapiller, M. W., Strader, C. D. and Hood, L. E. (1980) Acetylcholine receptor: Complex of homologous subunits. *Science* **208**, 1454–1457.

Rees-Smith, B. and Hall, R. (1974) Thyroid stimulating immunoglobulins in Graves' disease. *Lancet* ii, 427–430.

Souroujon, M. C., Mochly-Rosen, D., Bartfeld, D. and Fuchs, S. (1982) Experimental autoimmune myasthenia gravis: Specificity of antibodies to acetylcholine receptor, in *Molecular and Cellular Control of Muscle Development*. H. Epstein and M. Pearson, eds., Cold Spring Harbor Laboratory. In press.

Tzartos, S. J. and Lindstrom, J. M. (1980) Monoclonal antibodies used to probe acetylcholine receptor structure: Localization of the main immunogenic region and detection of similarities between subunits. *Proc. Natl. Acad. Sci. USA* **77**, 755–759.

Venter, J. C., Fraser, C. M. and Harrison, L. C. (1980). Autoantibodies to β_2-adrenergic receptors: A possible cause of adrenergic hyporesponsiveness in allergic rhinitis and asthma. *Science* **207**, 1361–1363.

Vincent, A. (1980) Immunology of acetylcholine receptors in relation to myasthenia gravis. *Physiol. Rev.* **60**, 756–824.

Wennogle, L. P. and Changeux, J. P. (1980) Transmembrane orientation of proteins present in acetylcholine receptor-rich membranes from Torpedo marmorata studied by selective proteolysis. *Eur. J. Biochem.*, **106**, 381–393.

8

Adaptive changes in dopamine neuronal function in response to chronic neuroleptic administration

P. Jenner and C. D. Marsden

University Department of Neurology, Institute of Psychiatry and King's College Hospital Medical School, Denmark Hill, London SE5, U.K.

1. INTRODUCTION

Neuroleptic drugs are commonly accepted as being of benefit in the treatment of schizophrenia owing to their ability to block cerebral dopamine receptors (Carlsson and Lindqvist, 1963). Such an action is borne out by the acute pharmacological profile of these compounds, all of which have a common mechanism of action to inhibit cerebral dopamine function (Anden et al., 1970; Janssen et al., 1965; Niemegeers and Janssen, 1979). However, rapid tolerance develops to the acute pharmacological actions of neuroleptic compounds. After a few days or few weeks of neuroleptic administration, the ability of the neuroleptic drugs to induce catalepsy wanes, the compensatory increase in dopamine turnover produced by acute neuroleptic administration disappears (Scatton, 1977), and animals show less inhibition of apomorphine-induced stereotyped behaviour (Asper et al., 1973; Sayers et al., 1975). Other studies have shown that repeated administration of neuroleptic drugs for a period of some weeks, followed by drug withdrawal, can lead to an exaggerated response to dopamine receptor activation. Animals treated in this manner show exaggerated stereotyped response to apomorphine (Tarsy and Baldessarini, 1973, 1974; von Voigtlander et al., 1975). This is accompanied by increased dopamine receptor numbers as shown by changes in ligand binding experiments using [^{3}H]spiperone or [^{3}H]-haloperidol (Burt et al., 1977; Muller and Seeman, 1977). This evidence is taken to indicate the development of dopamine receptor supersensitivity in response to continued neuroleptic administration. Dopamine receptor supersensitivity may even develop following the administration of a single dose of neuroleptic (Moller-Neilsen et al., 1976; Martres et al., 1977). Obviously adaptive changes in response to neuroleptic administration occur rapidly in response to neuroleptic therapy.

We were interested, therefore, to discover how neuroleptic drugs work on chronic administration, for it is common for these compounds to be administered for many years to disturbed psychiatric patients. We wished also to discover why, tardive dyskinesia, a persistent abnormal movement disorder, appears during the course of chronic neuroleptic therapy. Tardive dyskinesia is associated with over-activity of cerebral dopamine function, rather than with dopamine receptor blockade (Klawans, 1973). We embarked upon a series of chronic neuroleptic studies in which a variety of neuroleptic drugs were administered to rats in their drinking water for periods of up to 18 months.

2. TIME COURSE OF CHANGES OCCURRING IN STRIATAL DOPAMINE RECEPTOR FUNCTION IN RESPONSE TO CHRONIC ADMINISTRATION OF TRIFLUOPERAZINE

In an initial investigation, the time course of changes occurring in cerebral dopamine function in response to chronic neuroleptic treatment in rats was investigated. Trifluoperazine dihydrochloride (2.5–3.5 mg/kg/day) was administered for periods up to 12 months in the drinking water (Clow *et al.*, 1979a, 1979b; 1980a, 1980b, 1980c). At intervals during the course of drug administration we measured behavioural and biochemical parameters of cerebral dopaminergic activity (Fig. 1).

Apomorphine-induced stereotyped behaviour was inhibited 2 weeks and 1 month after starting drug intake, but by 3 months had returned to control levels. However, by 6 months and 12 months of trifluoperazine administration apomorphine produced an enhanced stereotyped response compared to that observed in age-matched control animals. So, despite continued intake of high doses of neuroleptic, adaptation to the presence of the drug occurred such that the stereotyped response to apomorphine became exaggerated. The number of dopamine receptor binding sites labelled by [^{3}H] spiperone in striatal preparations was decreased after 1 month's drug intake. However, by 6 and 12 months of drug administration, the number of binding sites identified by this ligand was increased compared to that found in tissue from control animals. Similarly, striatal dopamine sensitive adenylate cyclase was inhibited after one month of drug treatment, but by 3 months of trifluoperazine administration the response to dopamine was not different to that observed in control animals. Again, after 6 or 12 months' drug intake the dopamine stimulation of cyclic AMP formation was enhanced compared to that observed in age-matched control groups.

The conclusion from this study was that despite continued high dosage neuroleptic intake, dopamine receptors adapted to the presence of neuroleptic drug such that a normal or supersensitive response was elicited, associated with an increase in receptor numbers as measured by [^{3}H] spiperone binding. It was also clear, however, that such changes were not straightforward. When the effect of drug intake on stereotyped behaviour was examined in detail, it was apparent that only some components of the stereotyped response were enhanced (Fig. 2).

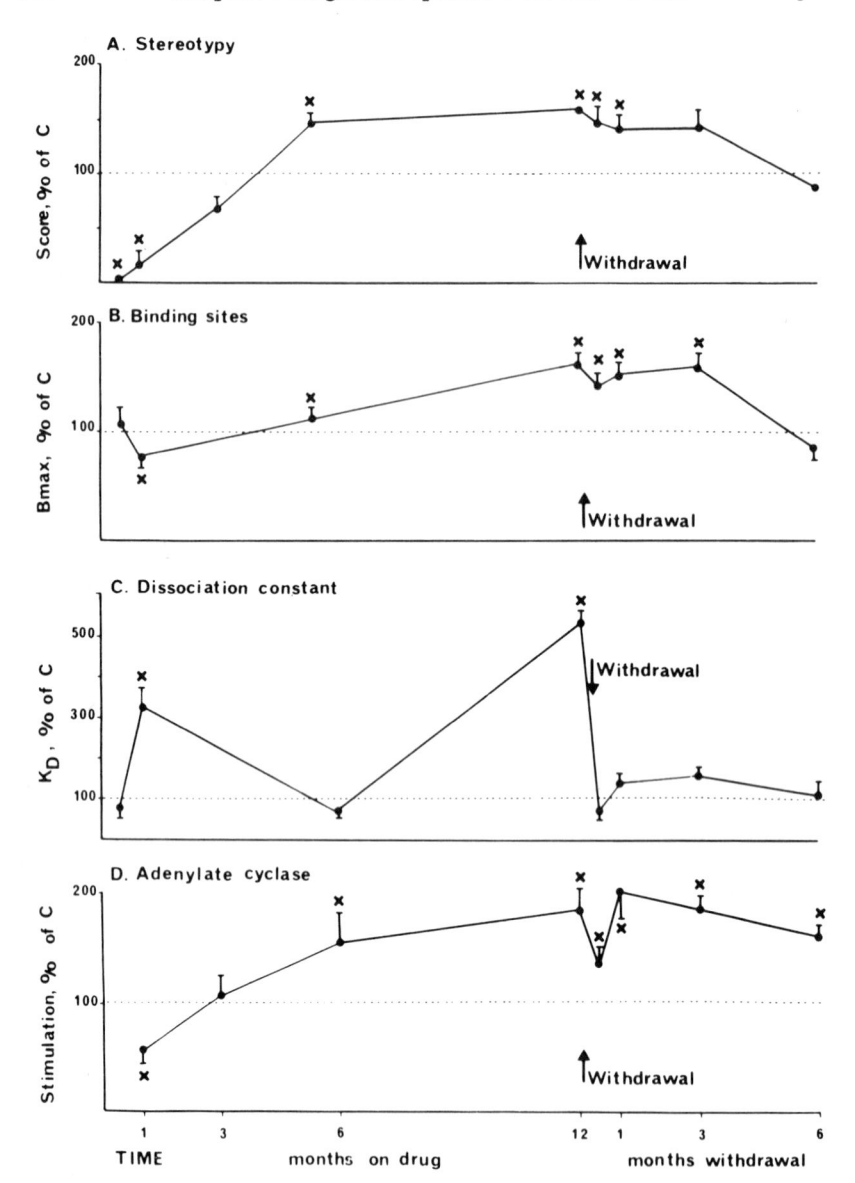

Fig. 1 — The effect of continuous administration of trifluoperazine dihydrochloride (2.5 mg/kg/day po) to rats for 12 months and its subsequent withdrawal for 6 months on (a) apomorphine (0.5 mg/kg sc)-induced stereotypy, (b) the number of striatal binding sites (B_{max}) for [^{3}H]spiperone, (c) the dissociation constant (K_D) for striatal [^{3}H]spiperone binding and (d) the dopamine (50 μM) stimulation of striatal adenylate cyclase. Values are expressed as a percentage of those obtained for age-matched control animals. Taken from Marsden and Jenner (1980). * $p < 0.05$ compared to age-matched control animals.

Sec. 2] Changes Occurring in Striatal Dopamine Receptor Function 145

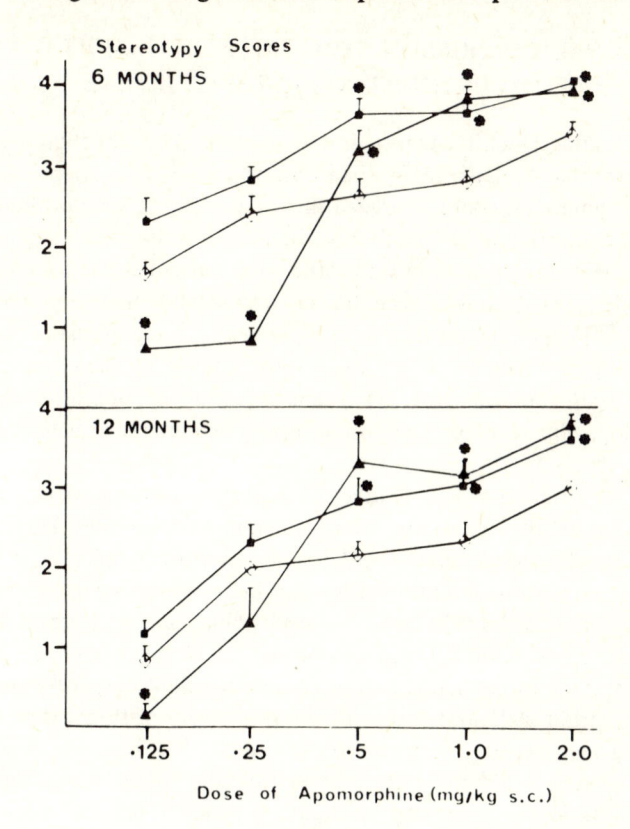

Fig. 2 — Apomorphine (0.125–2.0 mg/kg sc 15 min previously)-induced stereotypy in rats treated continuously with trifluoperazine dihydrochloride (2.5–3.5 mg/kg/day) (▲) or thioridazine dihydrochloride (30–40 mg/kg/day) (■) for 6 and 12 months, compared to age-matched control rats (◊). Taken from Clow et al. (1979b). *$p < 0.05$ compared to control rats.

After 6 or 12 months trifluoperazine intake, the stereotyped response to high doses of apomorphine (0.5–2.0 mg/kg sc) was enhanced with mainly an increase in peri-oral licking, gnawing or biting movements. In contrast, the effect of low doses of apomorphine (0.125 or 0.25 mg/kg sc) remained inhibited. At these doses the stereotyped response comprised mainly sniffing and locomotion. Perhaps different dopamine-containing areas of the brain, or different dopamine receptors within a single area of the brain, are involved in the genesis of the different components of stereotyped behaviour. We therefore investigated the changes occurring in dopamine receptor function within the striatum in more detail.

3. FUNCTIONAL CHANGES IN STRIATAL ACETYLCHOLINE CONTENT IN RESPONSE TO CONTINUED NEUROLEPTIC INTAKE

Prior to evaluating specific changes in dopamine receptor function it was necessary to ensure that the changes we had observed were of functional significance in the intact animal. It could be argued that an increase in apomorphine-induced stereotyped behaviour is observed only because this dopamine agonist in high doses displaces neuroleptic drug from otherwise blocked dopamine receptors, to unmask underlying dopamine receptor supersensitivity. Similarly, the increase in striatal [^{3}H] spiperone binding to washed tissue preparations may represent an increase in receptors which are adequately blocked by the concentration of neuroleptic drug found *in vivo*. So it was necessary to look at some index of enhanced dopamine receptor function occurring beyond post-synaptic dopamine receptors in striatum.

In the rat striatum a large number of post-synaptic dopamine receptors are thought to lie on the cell bodies of cholinergic interneurones. Dopamine normally inhibits cholinergic function. Administration of a single dose of neuroleptic drug removes this inhibitory tone and increases cholinergic action. This is reflected by increased utilisation of acetylcholine and a decrease in striatal acetylcholine content (Sethy and van Woert, 1974; Stadler *et al.*, 1973). As tolerance develops to the acute actions of neuroleptic drugs, acetylcholine concentrations within striatum return to control values within a few weeks of commencing neuroleptic administration (Sethy, 1976). However, if after some weeks of neuroleptic administration the neuroleptic drug is then withdrawn and a few days are allowed for drug washout, dopamine receptor supersensitivity is apparent. This is accompanied by an increase in the ability of apomorphine to elevate striatal acetylcholine content due to the increased inhibitory tone on cholinergic function (Choi and Roth, 1978; Consolo *et al.*, 1978). In our chronic neuroleptic studies, we predicted that if the changes we have observed in response to continuous chronic neuroleptic treatment are of functional significance in the intact animal, then they too should be accompanied by increased striatal acetylcholine content.

To examine this hypothesis, the effect of continuous chronic administration of *cis*- or *trans*-flupenthixol to rats for a period of 14 months was investigated (Murugiaiah *et al.*, 1982a, 1982b). At this time we looked at a number of parameters associated with cerebral dopamine function and also measured striatal acetylcholine content. After 18 months of neuroleptic treatment there was no residual compensatory increase in dopamine turnover (Table 1). Neither *cis*- nor *trans*-flupenthixol altered the striatal dopamine content, and administration of *cis*-flupenthixol was associated with a reduction of HVA and DOPAC levels below those observed in control animals; *trans*-flupenthixol did not cause any change in the levels of these dopamine metabolites. This suggests that a compensatory decrease in dopamine turnover had occurred perhaps as a response to en-

Table 1 — The effect of continuous administration of *cis*-flupenthixol (0.8–1.2 mg/kg/day) or *trans*-flupenthixol (0.9–1.2 mg/kg/day) for 14 months compared to age matched control animals on striatal dopamine (DA), HVA and DOPAC concentrations, apomorphine (0.5 mg/kg sc 15 min previously)-induced stereotyped behaviour and specific (dopamine 10^{-4}M) striatal [^{3}H] spiperone binding (B_{max} and K_D).

Drug treatment (dose mg/kg/day)	Striatal DA (µg/g)	Striatal HVA (ng/g)	Striatal DOPAC (ng/g)	Stereotypy score	Striatal [^{3}H] spiperone binding[a] B_{max} (pmoles/g tissue)	K_D (nM)
Controls (distilled water alone)	9.9 ± 0.6	878 ± 121	879 ± 60	2 ± 0	10.8 ± 1.2 10.1 ± 0.3	0.17 ± 0.05 0.09 ± 0.09
cis-flupenthixol (0.8–1.2)	11.0 ± 0.7	489 ± 121[c]	492 ± 147[c]	3.9 ± 0.1[c]	8.3 ± 0.5[b] 9.7 ± 0.7	0.25 ± 0.03 0.18 ± 0.04
trans-flupenthixol (0.9–1.2)	11.4 ± 0.9	632 ± 91	768 ± 59	2 ± 0	8.8 ± 1.0 10.3 ± 0.9	0.15 ± 0.05 0.11 ± 0.03

[a] The individual values for determination on two separate tissue pools are presented.
[b] Following a 2-week withdrawal period B_{max} was 14.2 ± 0.5 pmoles/g tissue ($p < 0.05$) compared to 10.2 ± 0.5 pmoles/g tissue for control animals.
[c] $p < 0.05$ compared to control values.

hancement of post-synaptic dopamine receptor supersensitivity. Indeed, apomorphine-induced stereotyped behaviour was markedly enhanced in animals receiving *cis*-flupenthixol, but *trans*-flupenthixol treated animals showed a response identical to that observed in the control group (Table 1). Interestingly, there was no increase in specific [^{3}H] spiperone binding to striatal preparations from animals receiving either *cis*- or *trans*-flupenthixol administration for 18 months (Table 1). However, following a 2-week withdrawal period a 40% increase in [^{3}H]-spiperone binding was observed in the *cis*-flupenthixol group, but not in tissue from animals receiving *trans*-flupenthixol. This suggests that the underlying increase in dopamine was adequately masked by continued administration of *cis*-flupenthixol. However, such data does not provide any index of the extent to which available receptors may be occupied by transmitter or by drug.

Acute administration of *cis*-flupenthixol caused a small decrease in striatal acetylcholine content; this was not observed when *trans*-flupenthixol was administered. After 14 months' continuous intake of *cis*-flupenthixol the striatal acetylcholine content was double that observed in control animals, while

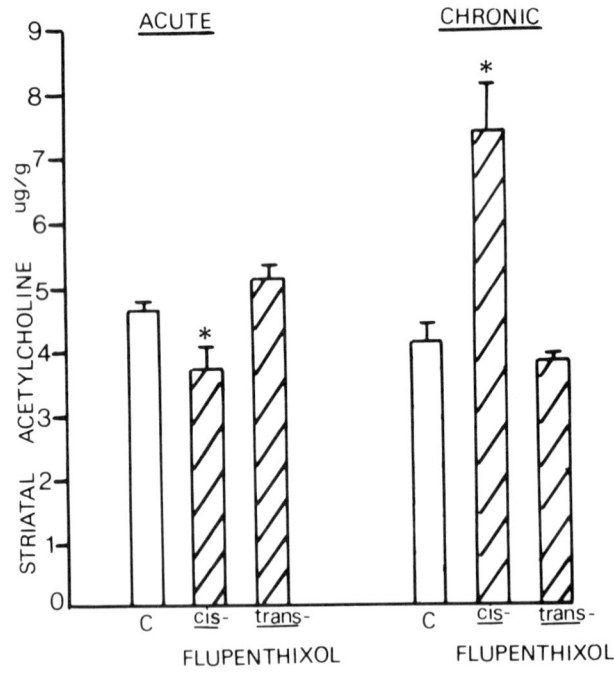

Fig. 3 — The effect on striatal acetylcholine concentrations of acute administration of *cis*- or *trans*-flupenthixol (1 mg/kg po 3 h previously) or continuous administration of *cis*-flupenthixol (0.8–1.2 mg/kg/day) or *trans*-flupenthixol (0.9–1.2 mg/kg/day) for 14 months compared to animals receiving distilled water alone. Taken from Murugaiah *et al.*, 1982b. *$p < 0.05$ compared to age-matched control animals.

[Sec. 4] **Alterations in the Characteristics of Striatal Dopamine Receptors** 149

trans-flupenthixol had no effect on this parameter (Fig. 3). This supports the hypothesis that the dopamine receptor supersensitivity we have observed is of functional significance in the intact animal, resulting in enhanced inhibitory dopamine tone on cholinergic function in the striatum, leading to an accumulation of acetylcholine.

4. ALTERATIONS IN THE CHARACTERISTICS OF STRIATAL DOPAMINE RECEPTORS

The studies so far suggest that striatal dopamine receptors can adapt to the presence of neuroleptic drug such that they exhibit functional dopamine receptor supersensitivity. If this is due simply to an increase in the number of available receptors then one might expect that these would be adequately blocked by the high concentration of neuroleptic drug found *in vivo*. The fact that they are not suggests that some change occurs in the nature of these dopamine receptors which are present in neuroleptic treated animals.

There is some evidence to support such a suggestion from behavioural observations made in animals treated with *cis*-flupenthixol for up to 18 months (Murugaiah *et al.*, 1982a). Two behaviours which are thought to be mediated by dopamine function in the striatum, namely catalepsy and apomorphine-

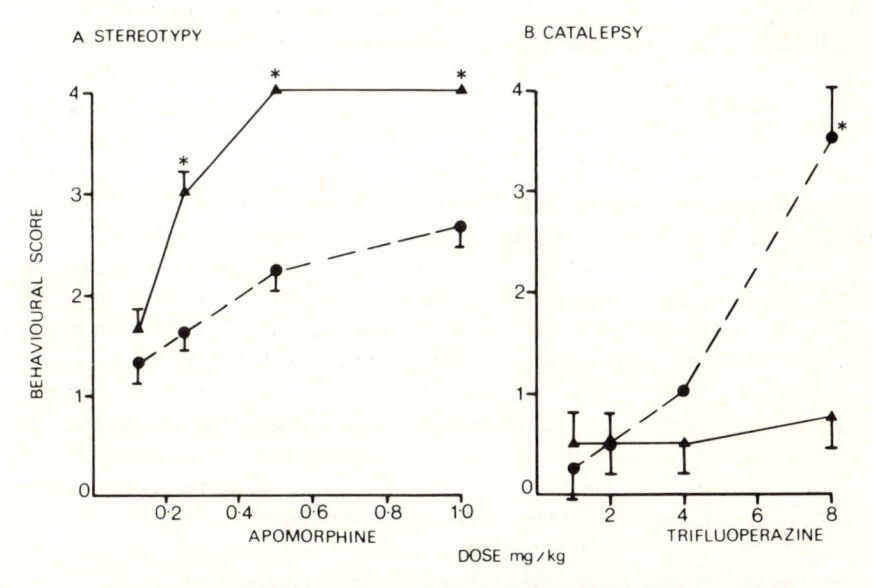

Fig. 4 – Dose response curves for (a) the induction of apomorphine (0.125–1.0 mg/kg sc 15 min previously)-induced stereotypy and (b) trifluoperazine (1–8 mg/kg/ip 3 h previously)-induced catalepsy in rats receiving *cis*-flupenthixol (0.8–1.2 mg/kg/day po) for 18 months (▲) compared to control animals receiving distilled water alone (●). Taken from Murugaiah *et al.*, 1982a. *$p < 0.05$ compared to age-matched control animals.

induced stereotypy, are altered differentially by chronic neuroleptic treatment (Fig. 4). Administration of a range of doses of apomorphine to animals treated with *cis*-flupenthixol for 18 months produced a far greater stereotyped response than is observed in age-matched control animals. This suggests that such animals are showing agonist supersensitivity. In contrast, when such animals were treated with a range of concentrations of trifluoperazine less cataleptic activity was observed compared to control animals. This suggests that the receptors present are showing antagonist subsensitivity. The characteristics of the dopamine receptor appear to have changed to that of agonist supersensitivity with antagonist subsensitivity.

The examination of [^{3}H] spiperone binding to striatal preparations also provides some evidence that may indicate altered receptor characteristics (Murugaiah *et al.*, 1982a). Thus in 4 long-term neuroleptic experiments we have carried out we have observed a reproducible pattern of change in both the number of binding sites (B_{max}) and in the dissociation constant (K_D) for [^{3}H] spiperone binding to striatal preparations (Fig. 5). In brief, in each of these experiments we observed a gradual increase in B_{max} over the period of the experiment such that by 6 or 12 months of drug administration there was a clear increase in the number of [^{3}H] spiperone binding sites. The pattern of change in the dissociation constant (K_D), however, was more complex. One month after the start of drug intake K_D was overall elevated, which is attributed to the presence of neuroleptic drug in the tissue preparations. However, by 3 or 6 months K_D returned to, or towards, normal but by 9 or 12 months of drug administration K_D again was elevated. It can be argued that this pattern of change is due to inherent variation in our assay procedure, but this is unlikely since we have repeated these experiments over a 5-year period using a range of neuroleptic compounds. The fall in K_D at 6 months might be due to a decrease in effective drug concentration, as a result of some phenomenon such as hepatic enzyme induction. However, our evidence is that drug levels are either stable or rise during the course of chronic administration (unpublished data). Circulating serum prolactin levels also remain elevated throughout the time course of drug administration (Dyer *et al.*, 1981). The alternative explanation is that the change in K_D which occurred at 6 months, at a time when the number of [^{3}H] spiperone binding sites was increasing, was due to the new population of receptors having different kinetic characteristics to the endogenous receptor sites. To examine this we looked carefully at Scatchard plots of the data, but could see no deviation from linearity to suggest the existence of two receptor populations. We also subjected the data to Hill plot analysis but, again, could see no deviation from linearity. There is no obvious explanation for this consistent pattern of change in K_D, but it could represent some kinetic change which occurs in receptor characteristics at a time when receptor numbers start to increase. The changes are reminiscent of the alterations in both B_{max} and K_D which occur in the striatum following 6-hydroxydopamine lesions (Feuerstein *et al.*, 1981a, 1981b).

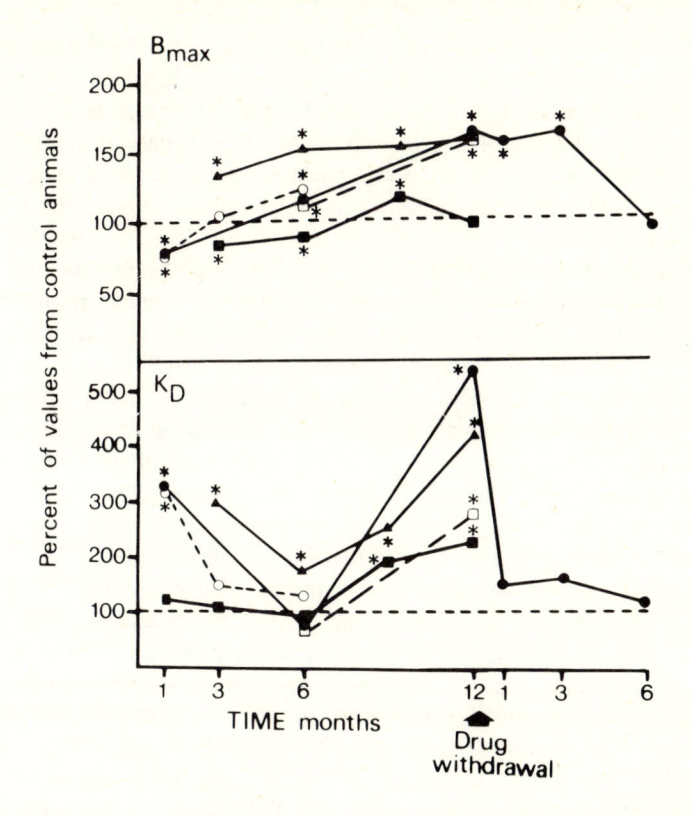

Sec. 4]   Alterations in the Characteristics of Striatal Dopamine Receptors   151

Fig. 5 – Comparison of the alteration in dopamine (10^{-4} M) specific [^{3}H] spiperone (0.125–4.0 nM) binding to rat striatal preparations produced by the administration of trifluoperazine, thioridazine or cis-flupenthixol continuously for up to 12 months in four distinct studies. (A) Receptor numbers B_{max} (B) Dissociation constant (K_D). Experiment 1. Trifluoperazine dihydrochloride 2.5–3.5 mg/kg/day, ●; thioridazine dihydrochloride 30–40 mg/kg/day, ▫; Experiment 2. Trifluoperazine dihydrochloride 2.8–4.0 mg/kg/day, ○. Experiment 3. Trifluoperazine dihydrochloride 4.5–5.6 mg/kg/day, ▲. Experiment 4. cis-Flupenthixol hydrochloride 0.8–1.2 mg/kg/day, ■. *$p < 0.05$ compared to appropriate age-matched control animals.

More than one population of dopamine receptors may exist within rat striatum. The most commonly quoted division of dopamine receptors is into those linked to the enzyme adenylate cyclase (D-1) and those which act independently of this enzyme (D-2) (Chapter 4 and Kebabian and Calne, 1979). Agonist ligands may label a site which is distinct from that labelled by antagonist ligands (Creese and Sibley, 1979). We have investigated the effect of chronic neuroleptic administration on D-1, D-2 and agonist binding sites within rat striatum. [^{3}H]-Spiperone has been used to label D-2 receptors, and [^{3}H] N,n-propylnorapomorphine to identify agonist binding sites within striatum. There is no specific ligand for the identification of D-1 receptors. However, [^{3}H] piflutixol labels both D-1 and D-2 receptors (Hyttel, 1981). By using the specific D-2 antagonist, sulpiride, to mask binding of [^{3}H] piflutixol to D-2 receptors, it is possible to use this ligand to identify D-1 receptors alone. The changes in the binding of these ligands to striatal receptor sites have been examined in animals treated with *cis*- or *trans*-flupenthixol for a period of 18 months, in comparison to age-matched control animals.

We firstly compared changes in the binding of [^{3}H] spiperone and [^{3}H]-N,*n*-propylnorapomorphine to adenylate cyclase independent receptor sites (Table 2). After 18 months drug intake there was a marked increase in B_{max} for [^{3}H] spiperone binding in animals receiving *cis*-, but not *trans*-flupenthixol; K_D increased in those animals receiving *cis*-flupenthixol, but no change was apparent in tissue from animals treated with *trans*-flupenthixol. In contrast, there was a marked decline in the number of [^{3}H] N,*n*-propylnorapomorphine binding sites in animals receiving *cis*-flupenthixol compared to animals treated with *trans*-flupenthixol; K_D of [^{3}H] N,*n*-propylnorapomorphine decreased in the *cis*-flupenthixol group compared to the *trans*-flupenthixol group.

This data suggests that there are reciprocal changes in the number and affinity to those binding sites labelled by [^{3}H] spiperone and [^{3}H] N,*n*-propylnorapomorphine in animals receiving chronic *cis*-flupenthixol. The results may explain the apparent agonist supersensitivity and antagonist subsensitivity exhibited in the behavioural experiments in animals treated in this manner. They have a larger number of neuroleptic binding sites of a low affinity, which might explain the reduced ability of neuroleptics to induce catalepsy. On the other hand, they have a smaller population of higher affinity agonist binding sites, which might explain why such animals are supersensitive to dopamine agonists inducing stereotyped behaviour.

There was no change in the number of [^{3}H] piflutixol binding sites, alone or in the presence of a high concentration of sulpiride under any conditions; nor was there any change in the dissociation constant of this ligand in animals receiving either *cis*- or *trans*-flupenthixol (Table 2). This suggests that chronic neuroleptic intake does not lead to any change in the number of D-1 receptors. Also of interest is the fact that, despite *cis*-flupenthixol acting at D-1 receptors, there appeared to be no change in the dissociation constant for this ligand due to presence of drug in the tissue.

The apparent lack of change in [^{3}H] pifluxitol binding must be looked at in the light of the increase in dopamine sensitive adenylate cyclase actively in the same animals. Our evidence suggests that it may not be essential to have an actual change in the number of recognition sites for the adenylate cyclase enzyme, but that changes in receptor occupancy may be of greater importance in governing the level of adenylate cyclase activation by dopamine. Alternatively, the increased activation of adenylate cyclase may not represent any change in the recognition site, but rather a change in some component of the adenylate

Table 2 — The effect of administration of *cis*-flupenthixol (0.8–1.2 mg/kg/day) or *trans*-flupenthixol (0.9–1.2 mg/kg/day) to rats for 18 months on specific striatal binding of [^{3}H] spiperone, [^{3}H] N n-propylnoramorphine or [^{3}H] piflutixol and dopamine-stimulated adenylate cyclase activity.

Drug treatment	[^{3}H] spiperone B_{max} (pmoles/g tissue)	[^{3}H] spiperone K_D (nM)	[^{3}H] N n-propylnorapomorphine B_{max} (pmoles/g tissue)	[^{3}H] N n-propylnorapomorphine K_D (nM)	[^{3}H] piflutixol B_{max} (pmoles/g tissue)	[^{3}H] piflutixol K_D (nM)	[^{3}H] piflutixol (plus sulpiride 3×10^{-6}M) B_{max} (pmoles/g tissue)	[^{3}H] piflutixol (plus sulpiride 3×10^{-6}M) K_D (nM)	Adenylate cyclase (dopamine 50 μM) (pmoles/2.5min/ 2 mg tissue)
Control	10.2 ± 0.8	0.14 ± 0.02	—	—	84.6 ± 3.3	0.29 ± 0.02	59.7 ± 4.5	0.53 ± 0.04	12.8 ± 4.2
cis-Flupenthixol	14.0 ± 1.0[a]	0.85 ± 0.10[a]	2.4 ± 0.4[a]	0.52 ± 0.15[a]	89.4 ± 4.0	0.40 ± 0.02	53.6 ± 6.5	0.47 ± 0.08	31.8 ± 2.9[a]
trans-Flupenthixol	9.2 ± 0.8	0.25 ± 0.06	7.5 ± 1.7	1.29 ± 0.46	71.6 ± 3.2	0.26 ± 0.02	72.9 ± 8.8	0.57 ± 0.20	—

[a] $p < 0.05$ compared to control animals or animals receiving *trans*-flupenthixol.
Specific binding was defined using 100 μM dopamine ([^{3}H] spiperone), 1 μM (+)-butaclamol ([^{3}H] N n-propylnorapomorphine), 1 μM *cis*-flupenthixol ([^{3}H] piflutixol).

cyclase system beyond the recognition area. Neuroleptic drugs are known to induce changes in calmodulin and protein phosphorylation that could be of importance in governing the level of cyclic AMP formation (Lau and Gnegy, 1982).

5. STRIATAL DOPAMINE FUNCTION FOLLOWING WITHDRAWAL FROM CHRONIC NEUROLEPTIC TREATMENT

Since the primary objective of this work is to look at changes in dopamine function in relation to tardive dyskinesia, it was important to discover the effect of neuroleptic withdrawal following prolonged administration. Tardive dyskinesias often first appear, or are exacerbated, following drug withdrawal, and the movements may persist following drug removal. We examined the effects of drug withdrawal of animals either treated with trifluoperazine for a period of 12 months (Clow et al., 1980b), or with cis-flupenthixol for 18 months (unpublished data).

Following withdrawal from trifluoperazine treatment (Fig. 1) there was no further increase in apomorphine-induced stereotyped behaviour, [^{3}H] spiperone binding to striatal preparations, or dopamine sensitive adenylate cyclase activity in striatal tissue. The increase in apomorphine-induced stereotypy persisted for one month following drug withdrawal, but by 3 months and thereafter the animals showed a response which was not different from that observed in age-matched control animals. The increased number of [^{3}H] spiperone binding sites persisted for 3 months following drug withdrawal, but by 6 months had returned to normal. In contrast, the increase in the dissociation constant (K_D) had returned to control values 2 weeks following drug withdrawal and remained at this level thereafter. The change in dopamine stimulation of striatal adenylate cyclase, however, persisted throughout the drug withdrawal period of 6 months used in this experiment. The reason for this persistent change is not clear, but may be of relevance to the continuation of tardive dyskinesia following drug withdrawal.

The persistence of enhanced adenylate cyclase activity following drug withdrawal was confirmed in the subsequent cis-flupenthixol study. This change was still apparent 6 months following drug withdrawal, but by 12 months after drug removal dopamine sensitive adenylate cyclase activity in striatum had returned to control values. After 6 months' withdrawal from cis-flupenthixol administration we compared also changes in [^{3}H] spiperone and [^{3}H] piflutixol binding (both in the absence and presence of sulpiride) (Table 3). At this time, however, there was no change in the B_{max} for either [^{3}H] spiperone or [^{3}H]-pifluxitol, and there was no apparent difference in dissociation constant for the treated and control groups for either ligand. Absence of change must again be looked at in the light of the continuing increase of dopamine stimulation of striatal adenylate cyclase at this time. Our observations once more raise the possibility that either there are changes in receptor occupancy which are of

Table 3 — The effect of a 6-month withdrawal period following continuous administration of cis-flupenthixol (0.8–1.2 mg/kg/day) for 18 months compared to age-matched control animals on striatal [³H] spiperone and [³H] pifluxitol binding and dopamine stimulated adenylate cyclase activity (AC).

Drug treatment (mg/kg/day)	[³H] spiperone Bmax (pmoles/g)	[³H] spiperone K_D (nM)	[³H] pifluxitol Bmax (pmoles/g)	[³H] pifluxitol K_D (nM)	[³H] pifluxitol + sulpiride (30 μM) Bmax (pmoles/g)	[³H] pifluxitol + sulpiride (30 μM) KD (nM)	AC (pmoles/2.5 min/2 mg tissue)
Control (distilled water alone)	14.3 ± 1.2	0.28 ± 0.04	80.3 ± 6.2	0.37 ± 0.05	56.6 ± 3.7	0.41 ± 0.03	12.5 + 3.6
cis-flupenthixol (0.8–1.2 mg/kg/day)	14.5 ± 0.9	0.17 ± 0.02	84.5 ± 1.7	0.33 ± 0.03	64.5 ± 4.6	0.45 ± 0.06	26.9 ± 1.4[a]

[a] $p < 0.05$ compared to age-matched control animals.
Murugaiah et al., unpublished observations.

critical importance in determining the level of enzyme activity, or that some change in adenylate cyclase function has occurred beyond the dopamine recognition site. Whatever this change, it is of obvious importance to determine how this persistence is brought about for it may provide an answer to a number of prolonged effects occurring following neuroleptic withdrawal.

6. REGIONAL CHANGE IN CEREBRAL DOPAMINE FUNCTION IN RESPONSE TO CHRONIC NEUROLEPTIC ADMINISTRATION

The data we have presented so far has been restricted to changes in cerebral dopamine function occurring *within the striatum*. Dopamine fibres ascend to innervate not only the striatum, but also mesolimbic and mesocortical brain areas. We have examined changes occurring in [^{3}H] spiperone binding in the mesolimbic area (nucleus accumbens plus tuberculum olfactorium) as well as in striatum. In this area there was an increase in the number of [^{3}H] spiperone binding sites (Fig. 6), again suggesting the development of adaptive changes to the presence of neuroleptic drug in this brain area. However, we have no behavioural evidence to indicate that this was a functional supersensitivity, nor

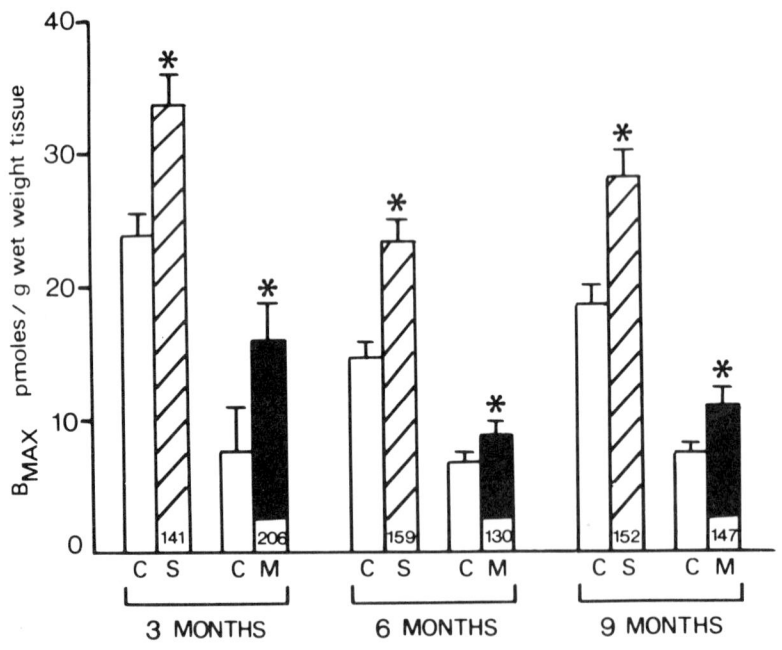

Fig. 6 – The effect of continuous administration of trifluoperazine dihydrochloride (4.5–5.6 mg/kg/day) in drinking water to rats for 9 months on the number of specific [^{3}H] spiperone binding sites (B_{max}) in striatal (S) and mesolimbic (M) tissue preparations compared to age-matched control (C) animals. The values inset in the columns represent the percentage of respective control values. *$p < 0.05$ compared to control animals.

have we any evidence that the increased number of receptors is of significance to the otherwise intact animal. Dopamine receptors may remain adequately blocked in the presence of neuroleptic drug in the mesolimbic area.

As far as the cortical dopamine systems are concerned, there is no evidence from any long-term neuroleptic experiments to indicate changes in dopamine function. It is difficult to measure receptor numbers within this region, and there is no easily measurable behavioural change associated with cortical dopamine systems. Evidence from short-term neuroleptic administration experiments suggest that tolerance does develop to the acute actions of neuroleptic drugs although effects may persist longer in this region than in subcortical areas (Scatton, 1977; Wheeler and Roth, 1980; Bacopoulos, 1981). However, until the appropriate long-term experiments are carried out, no judgement can be made as to the effects of chronic neuroleptic treatment on cortical dopamine systems.

There is some evidence to suggest that not all dopamine receptor systems respond in the same manner to neuroleptic drugs. After 12 months' trifluoperazine or thioridazine (Dyer *et al.*, 1981), intake there was no difference in serum prolactin levels from those observed in animals receiving the same dose of neuroleptic drug as an acute bolus (Fig. 7). This suggests that dopamine recep-

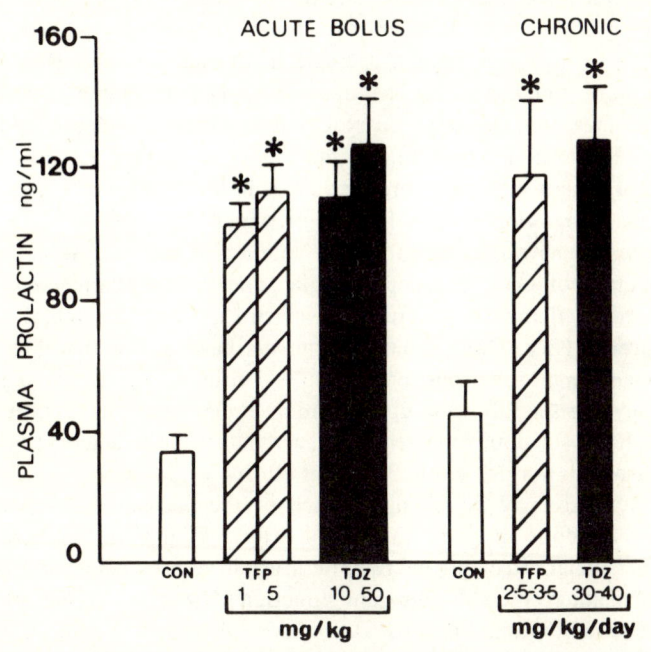

Fig. 7 — The effect on plasma prolactin levels of an acute oral bolus of trifluoperazine (TFP) (1 or 5 mg/kg 3 h previously) or thioridazine (TDZ) (10 or 50 mg/kg 3 h previously) and chronic oral administration of trifluoperazine (2.5–3.5 mg/kg/day) of thioridazine (30–40 mg/kg/day) for 12 months, compared to age-matched control animals. *$p < 0.05$ compared to age-matched control animals.

tors in the pituitary and/or hypothalamus do not respond in the same way as forebrain dopamine receptors. Perhaps this is not surprising because it is already known that their controlling mechanisms are markedly different from those occurring in either the striatum or mesolimbic dopamine containing areas of the brain (Annunziato, 1979).

7. CONCLUSIONS

It would appear from the data that we have presented, that striatal dopamine receptors can adapt to the continued administration of neuroleptic drugs so as to produce a normal or exaggerated response. This would appear to be of functional significance in the otherwise intact animal for it is associated which changes in striatal cholinergic function compatible with the development of dopamine receptor supersensitivity. Whether or not the nature of the increased numbers of dopamine receptors is identical to that of the endogenous receptor population remains unclear, but there is both behavioural and biochemical evidence to suggest that perhaps some fundamental change does occur so as to render them insensitive to the presence of neuroleptic drugs. Clearly, this does not affect all striatal dopamine receptor populations in the same way. While an increased number of D-2 receptors labelled by [^{3}H] spiperone was evident, this was accompanied by a corresponding decrease in the number of [^{3}H] N,n-propylnorapomorphine binding sites. The number of D-1 adenylate cyclase linked receptors, as measured by [^{3}H] piflutixol binding with and without supliride, did not alter, despite an increase in striatal adenylate cyclase activity. How the increase in adenylate activity is brought about in the absence of any measurable change in the recognition site for the enzyme is unclear.

These results have relevance to the production of tardive dyskinesias in man resulting from prolonged neuroleptic intake. The clinical pharmacology of this disorder suggests that such abnormal movements are due to cerebral dopamine receptor over-activity. Our animal experiments have shown the emergence of striatal dopaminergic supersensitivity, despite continued drug intake. Persistence of tardive dyskinesia following drug withdrawal also might be mirrored by the persistent change in dopamine sensitive adenylate cyclase activity; other parameters returned to normal within 3 months of drug withdrawal.

The data also may be of significance to the mechanism by which antipsychotic neuroleptic drugs produce their effects in man on long-term administration. Beneficial effects of neuroleptic drugs have been associated with their action in blocking cerebral dopamine function. However, in the striatum at least this does not occur on chronic drug administration to rodents. If neuroleptics are acting by blocking cerebral dopamine receptors then they must be doing this in another dopamine containing area of the brain. Prime candidates for this obviously are the mesolimbic and mesocortical dopamine containing systems. At the present time there is insufficient evidence to decide whether the changes that occur in the striatum also occur in these areas. There seems to be no reason why this should not be the case for, in sub-acute experiments, tolerance develops to the acute dopamine receptor blocking properties of neuroleptics in all areas. Nevertheless, it cannot be presumed that this will be the case.

In conclusion, the acute pharmacological profile of neuroleptic drugs does not reflect the adaptive changes which occur in response to continued chronic neuroleptic administration at least in animals. The profile of action of these drugs must be extended to include not only initial dopamine receptor blockade, but to explain the development of functional dopamine receptor supersensitivity at least within the rat striatum. The major finding from these experiments is that dopamine receptors in brain are capable of adapting to the continued presence of neuroleptic drugs so as to produce a normal or, indeed, an enhanced pharmacological response.

ACKNOWLEDGEMENTS

This study was supported by the Wellcome Trust, the Medical Research Council and the Research Funds of the Bethlem Royal and Maudsley Hospitals and King's College Hospital.

REFERENCES

Anden, N.-E., Butcher, S. G. and Corrodi, H. (1970) Receptor activity and turnover of dopamine and noradrenaline after neuroleptics. *Eur. J. Pharmacol.* **11**, 303–314.

Annunziato, L. (1979) Regulation of the tuberinfundibular and nigrostriatal systems. *Prog. Neuroendocrinol.* **29**, 66–76.

Asper, H., Bagglioni, M., Burki, H. R. and Lauener, G. (1973) Tolerance phenomena with neuroleptics. Catalepsy, apomorphine stereotypies and striatal dopamine metabolism in the rat after single and repeated administration of loxapine and haloperidol. *Eur. J. Pharmacol.* **22**, 287–294.

Bacopoulos, N. G. (1981) Biochemical mechanism of tolerance to neuroleptic drugs; regional differences in rat brain. *Eur. J. Pharmacol.* **70**, 585–586.

Burt, D. R., Creese, I. and Snyder, S. H. (1977) Antischizophrenic drugs: Chronic treatment elevates dopamine receptor binding. *Science* **196**, 326–328.

Carlsson, A. and Lindqvist, M. (1963) Effect of chlorpromazine or haloperidol on formation of 3-methoxytyramine and normetanephrine in mouse brain. *Acta Pharmacol. Toxicol.* **20**, 140–144.

Choi, R. L. and Roth, R. H. (1978) Development of supersensitivity of apomorphine-induced increase in acetylcholine levels and stereotypy after chronic fluphenazine treatment. *Neuropharmacology* **17**, 59–64.

Clow, A., Jenner, P. and Marsden, C. D. (1979a) Changes in dopamine-mediated behaviour during one year's neuroleptic administration. *Eur. J. Pharmacol.* **57**, 365–375.

Clow, A., Jenner, P. Theodorou, A. and Marsden, C. D. (1979b) Striatal dopamine receptors become supersensitive while rats are given trifluoperazine for six months. *Nature* **278**, 59–61.

Clow, A., Theodorou, A., Jenner, P. and Marsden, C. D. (1980a) Changes in rat striatal dopamine turnover and receptor activity during one year's neuroleptic administration. *Eur. J. Pharmacol.* **63**, 135–144.

Clow, A., Theodorou, A., Jenner, P. and Marsden, C. D. (1980b) Cerebral dopamine function in rats following withdrawal from one year of continuous neuroleptic administration. *Eur. J. Pharmacol.* **63**, 145–157.

Clow, A., Theodorou, A., Jenner, P. and Marsden, C. D. (1980c) A comparison of striatal and mesolimbic dopamine function in the rat during 6 months trifluoperazine administration. *Psychopharmacology* **69**, 227–233.

Consolo, S., Ladinsky, H., Samanin, R., Bianchi, S. and Ghezzi, D. (1978) Supersensitivity of the cholinergic response to apomorphine in the striatum following denervation or disuse supersensitivity of dopaminergic receptors in the rat. *Brain Res.* **155**, 45–54.

Creese, I. and Sibley, D. R. (1979) Radioligand binding studies: evidence for multiple dopamine receptors. *Comm. Psychopharmacol.* **3**, 385–395.

Dyer, R. G., Murugaiah, K., Theodorou, A., Clow, A., Jenner, P. and Marsden, C. D. (1981) During one year's neuroleptic treatment in rats, striatal dopamine receptor blockade decreases but serum prolactin levels remain elevated. *Life Sci.* **28**, 167–174.

Feuerstein, C. Demenge, P., Barrette, G., Silice, C., Guerin, B. and Mouchet, P. (1981a) Long-term effects of nigro-striatal ^{3}H haloperidol binding. *Eur. J. Pharmacol.* **76**, 457–459.

Feuerstein, C., Demenge, P., Caron, P., Barrette, G., Guerin, B., and Mouchet, P. (1981b) Supersensitivity time course of dopamine antagonist binding after nigrostriatal denervation: Evidence for early and drastic changes in the rat corpus striatum. *Brain Res.* **226**, 221–234.

Hyttel, J. (1981) Flupenthixol and dopamine receptor selectivity. *Psychopharmacology* **75**, 217.

Janssen, P. A. J., Niemegeers, C. J. E. and Schellekens, K. H. L. (1965) Is it possible to predict the clinical effects of neuroleptic drugs (major tranquillizers) from animal data? I. Neuroleptic activity spectra for rats. *Arz. Forsch.* **15**, 104–112.

Kebabian, J. W. and Calne, D. B. (1979) Multiple receptors for dopamine. *Nature* **277**, 93–96.

Klawans, H. L. (1973) The pharmacology of extrapyramidal movement disorders, in *Monographs in Neural Science* (ed. Cohen, M. M.). Karger, Basel.

Lau, Y.-S. and Gnegy, M. E. (1982) Chronic haloperidol treatment increased calcium-dependent phosphorylation in rat striatum. *Life Sci.* **30**, 21–28.

Marsden, C. D. and Jenner, P. (1980) The pathophysiology of extrapyramidal side-effects of neuroleptic drugs. *Psychol. Med.* **10**, 55–72.

Martres, M. P., Costentin, J., Baudry, M., Marcais, H., Protais, P. and Schwartz, J. C. (1977) Long-term changes in the sensitivity of pre-and postsynaptic dopamine receptors in mouse striatum evidenced by behavioural and biochamical studies. *Brain Res.* **136**, 319–337.

Moller-Nielsen, I., Christensen, A. V. and Fjalland, B. (1976) Receptor blockade and receptor supersensitivity following neuroleptic treatment, in *Antipsychotic Drugs: Pharmacodynamics and Pharmacokinetics* (eds. Sedvell, G., Uvnas, B. and Zotterman, B.) pp. 257–260. Pergamon Press, Oxford.

Muller, P. and Seeman, P. (1977) Brain neurotransmitter receptors after long-term haloperidol: dopamine, acetylcholine, serotonin, α-adrenergic and naloxone receptors. *Life Sci.,* **21**, 1751–1758.

Murugaiah, K. Theodorou, A., Mann, S., Clow, A., Jenner, P. and Marsden, C. D. (1982a) Chronic continuous administration of neuroleptic drugs alters cerebral dopamine receptors and increases spontaneous dopaminergic action in the striatum. *Nature* **296**, 570–572.

Murugaiah, K., Mann, S., Theodorou, A. E., Jenner, P. and Marsden, C. D. (1982b) Increased striatal acetylcholine after 14 months *cis*-flupenthixol treatment in rats suggests functional supersensitivity of dopamine receptors. *Life Sci.,* **21**, 181–188.

Niemegeers, C. J. E. and Janssen, P. A. J. (1979) A systematic study of the pharmacological activity of dopamine antagonists. *Life Sci.* **24**, 2201–2216.

Sayers, A. C., Burki, H. R., Ruch, W. and Asper, H. (1975) Neuroleptic-induced hypersensitivity of striatal dopamine receptors in the rat as a model of tardive dyskinesias. Effects of clozapine, haloperidol, loxapine and chlorpromazine. *Psychopharamcology* **41**, 97–104.

Scatton, B. (1977) Differential regional development of tolerance to increase in dopamine turnover upon repeated neuroleptic administration. *Eur. J. Pharmacol.* **46**, 363–369.

Sethy, V. H. (1976) Effects of chronic treatment with neuroleptics on striatal acetylcholine concentrations. *J. Neurochem.* **27**, 325–326.

Sethy, V. H. and van Woert, M. H. (1974) Modification of striatal acetylcholine concentration by dopamine receptor agonists and antagonists. *Res. Comm. Clin. Path. Pharmac.* **8**, 13–28.

Stadler, H., Lloyd, K. G., Gadae-Ciria, M. and Bartholini, G. (1973) Enhanced striatal acetylcholine release by chlorpromazine and its reversal by apomorphine. *Brain Res.* **55**, 476–480.

Tarsy, D. and Baldessarini, R. J. (1973) Pharmacologically induced behavioural supersensitivity to apomorphine. *Nature* **245**, 262–263.

Tarsy, D. and Baldessarini, R. J. (1974) Behavioural supersensitivity to apomorphine following chronic treatment with drugs which interfere with synaptic function of catecholamines. *Neuropharmacology* **13**, 927–940.

von Voigtlander, P. F., Losey, E. G. and Triezenberg, H. J. (1975) Increased sensitivity to dopaminergic agents after chronic neuroleptic treatment. *J. Pharmacol. Exp. Ther.* **193**, 88–94.

Wheeler, S. C. and Roth, R. H. (1980) Tolerance to fluphenazine and supersensitivity to apomorphine in central dopaminergic systems after chronic fluphenazine decanoate treatment. *Arch. Pharmacol.* **312**, 151–159.

9

Ligand-binding studies in brains of schizophrenics

F. Owen, A. J. Cross and T. J. Crow

Division of Psychiatry, Clinical Research Centre, Watford Road, Harrow, Middlesex HA1 3UJ, U.K.

Abbreviation used

The key to abbreviations used throughout the text is as follows: ADTN – 2-amino-6,7-dihydroxy-1,2,3,4-tetrahydronapthalene; B_{max} —maximum specific binding; DA – dopamine; GABA – γ-aminobutyric acid; H.Ch. – Huntington's Chorea; 5HT – 5-hydroxytryptamine, serotonin; K_D – dissociation constant; LSD – lysergic acid diethylamide; QNB – quinuclidinyl benzilate; Mr 2266 – [(−)-α-5,9-diethyl-2-(3-furylmethyl)-2′-hydroxy-6,7-benzomorphan] ; WB-4101 – (2,6-dimethoxyphenoxyethyl)aminomethyl-1,4-benzodioxane.

1. INTRODUCTION

The observation by Carlsson and Lindqvist (1963) that neuroleptic drugs interact with central dopaminergic mechanisms and that dopamine (DA) releasing drugs such as amphetamine (Randrup and Munkvad, 1966) could induce, in mentally normal individuals a psychosis closely resembling paranoid schizophrenia (Connell, 1958) led to the suggestion that some aberration in dopaminergic systems was involved in the aetiology of schizophrenia (Randrup and Munkvad, 1967, 1972). The resulting DA hypothesis of schizophrenia stated that schizophrenia was associated with excessive dopaminergic function in the central nervous system. The hypothesis was strengthened by the demonstration that neuroleptics were also effective in inhibiting the DA induced stimulation of adenylate cyclase (Miller *et al.*, 1974, Clement-Cormier *et al.*, 1974) and at displacing the high affinity binding of ligands to the DA receptor (Seeman *et al.*, 1975). However, subsequent studies of the concentrations of DA and its major end products 3,4-dihydroxyphenylacetic acid (DOPAC) and homovanillic acid (HVA) have produced no evidence of an increase in DA turnover in schizophrenic brains (Crow *et al.*, 1980). In addition the activities of enzymes responsible for the biosynthesis and inactivation of DA are similar in schizophrenic and control brains (Crow et al., 1979).

Earlier, Bowers (1974) and Crow et al. (1976) had suggested that the aberration in dopaminergic transmission in schizophrenic brains may be located at the post-synaptic receptor. With the advent of ligand binding techniques suitable for studying DA receptors (Seeman et al., 1975) it became possible to evaluate this possibility. The substance of this chapter is a description of the experiments carried out in the Division of Psychiatry at the Clinical Research Centre to investigate possible abnormalities in DA receptors in schizophrenic brains. During the course of the study it became clear that there were at least two forms of the DA receptor designated D1 and D2 (Kebabian and Calne, 1979). D1 receptors mediate the stimulation adenylate cyclase and do not bind butyrophenones with high affinity. D2 receptors bind butyrophenones with high affinity and do not stimulate adenylate cyclase. Seeman (1980) has suggested two further binding sites designated D3 and D4 (see also Chapter 4). Our initial study using [^{3}H]spiperone as ligand demonstrated a highly significant increase in D2 receptors in schizophrenic brains. Subsequent studies were carried out to determine the specificity of this change in receptors in terms of the other proposed forms of the DA receptor and also in terms of other putative neurotransmitter receptors. Further studies were also aimed at determining the effect of neuroleptic medication on the increase in D2 receptors observed in schizophrenic brains, since it has now been well established in animal experiments that D2 receptors increase in number in response to chronic neuroleptic administration although to a much smaller extent than that reported in schizophrenic brains (for review see Seeman, 1980, and Chapter 8). Lastly experiments were carried out on brain samples from patients whose schizophrenic symptoms had been assessed in life in an attempt to relate D2 receptor changes with the clinical features of the disease.

2. MATERIALS AND METHODS

2.1 Materials

(a) *Patients' brains*

Brains from patients with a hospital diagnosis of schizophrenia were received from a number of psychiatric hospitals. Brains were only included in the study if there was sufficient information in the case notes to satisfy the criteria for the 'nuclear syndrome' according to the 'syndrome check list' described by Wing et al. (1974) and the disease could also be described as 'definite' or 'probable' schizophrenia according to the criteria of Feighner et al. (1972).

Brains from patients dying with Huntington's Chorea (H.Ch.) or Alzheimer's disease were obtained from Drs L. L. Iversen and M. Rosser, MRC Neurochemical Pharmacology Unit, Cambridge. The clinical diagnoses of H.Ch. and Alzheimer's disease were confirmed neuropathologically.

(b) *Control brains*
Control brains were obtained from a district general hospital. Mean age, sex ratio, time between death and storage at 4 °C and time at 4 °C prior to autopsy and time deep frozen were similar in patients and controls as shown in Table 1. This sample was used for all the studies described here, although the results presented in Fig. 8 were obtained with a larger series of brains that were equally well-matched for the various collection and storage parameters.

Table 1 — Collection and storage details of schizophrenic and control brains.

	Schizophrenics ($n = 19$)	Controls ($n = 19$)
Age (years)	66 ± 14	73 ± 13
M/F	12/7	11/8
Time at room temperature (h)	2.5 ± 1.3	2.6 ± 0.8
Time at 4 °C (h)	44 ± 28	51 ± 30
Duration of storage (days)	397 ± 282	493 ± 177

Values as mean ± S.D.

(c) *Dissection and storage*
At autopsy brains were placed in expanded polystyrene moulds that retained approximate shape and were deep frozen at −40 °C. At dissection brains were sliced, still frozen, into coronal sections about 0.5 cm thick and the required areas dissected out according to the atlas of DeArmond *et al.* (1974). Samples were then diced and mixed and stored over liquid nitrogen.

(d) *Experimental animals*
Animal experiments were carried out, where appropriate, to supplement the work on human tissue and in all cases male Sprague Dawley rats were used (200 g weight approximately at start of experiment).

(e) *Radiochemicals*
Ligands used and source of purchase were as follows: [^{3}H] spiperone, [^{3}H]-ADTN, [^{3}H] WB 4101, [^{3}H] dihydroalprenolol, [^{3}H] GABA, [^{3}H] 5HT, [^{3}H]-LSD, [^{3}H] QNB, [^{3}H] diazepam and [^{3}H] etorphine were obtained from the Radiochemical Centre, Amersham. [^{3}H] clonidine was purchased from New England Nuclear, West Germany and [^{3}H] flupenthixol was a gift from Dr J. Hytell, Lundbeck A/S, Copenhagen.

2.2 Methods

(a) *Membrane preparations*

For binding studies both rat and human brain membranes were prepared by homogenising the tissue in 40 volumes of ice-cold 50 mM Tris-HCl buffer using a Polytron PT10, setting 6 for 15 seconds. The buffer was at pH 7.4 except for [^{3}H] dihydroalprenolol binding where the pH was 8.0. Homogenates were centrifuged at 50,000g for 20 minutes, the supernatant discarded and the pellets resuspended in 40 volumes of buffer. The suspensions were centrifuged at 50,000g for 20 minutes, the supernatants discarded, the pellets again resuspended in 40 volumes of buffer and stored frozen at −40 °C until assay.

(b) *Ligand binding techniques*

The receptors studied and the ligands and techniques used are presented in Table 2. In all experiments, on both human and animal tissue, samples were coded and assays carried out blind and in triplicate.

Table 2 — Techniques and ligands employed to study receptors in post-mortem brains.

Receptor	Ligand	References
Dopamine	[^{3}H] spiperone	Owen *et al.* (1978)
	[^{3}H] ADTN	Cross *et al.* (1979)
	[^{3}H] flupenthixol	Cross and Owen (1980)
Serotonin	[^{3}H] 5HT, [^{3}H] LSD	Bennett and Snyder (1976)
Muscarinic acetylcholine	[^{3}H] QNB	Yamamura and Snyder (1974)
β-noradrenergic	[^{3}H] dihydroalprenolol	Bylund and Snyder (1976)
α-noradrenergic	[^{3}H] WB4101, [^{3}H] clonidine	Greenberg *et al.* (1976)
GABA	[^{3}H] GABA	Enna and Snyder (1976)
Benzodiazepine	[^{3}H] diazepam	Möhler and Okada (1977)
Opiate	[^{3}H] etorphine	Owen *et al.* (in preparation)

3. RESULTS

3.1 Post-mortem stability of receptors

A preliminary experiment was carried out in the rat to determine the stability of neurotransmitter receptors under post-mortem conditions similar to those to which human tissue was subjected. Seventy rats were decapitated and the brains immediately removed from 10 animals and frozen. The remaining 60 heads were maintained at room temperature for 3 hours and then for varying lengths of time, up to 96 hours, at 4 °C. The results using [^{3}H] spiperone in striatal membranes and [^{3}H] GABA, [^{3}H] 5HT and [^{3}H] QNB in cerebral cortical membranes are shown in Fig. 1. Apart from a small (about 15%) decrease in [³H] GABA binding, the binding of the other ligands did not decrease significantly with increasing time at 4 °C. This strongly suggested that receptor binding studies in human brains would yield meaningful results.

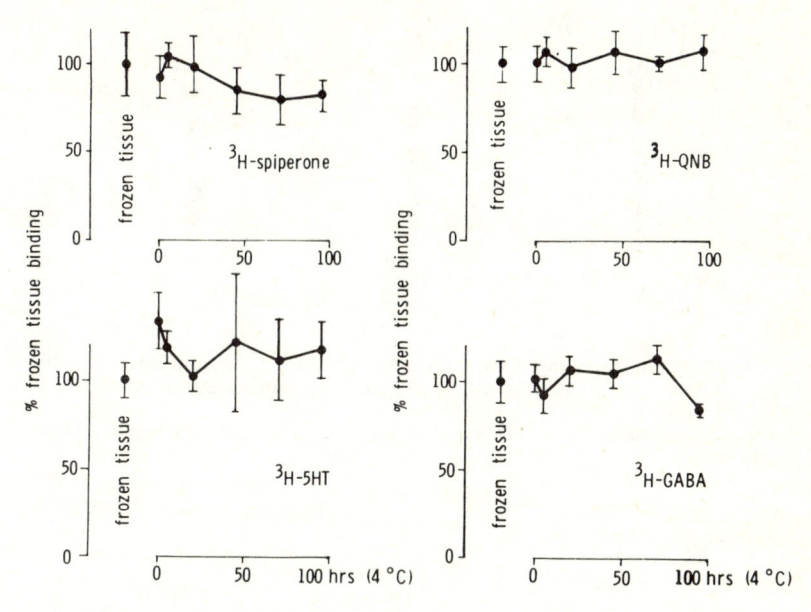

Each point represents mean ± SD for 10 rats.

Fig. 1 – Post-mortem stability of neurotransmitter receptors in the rat.

3.2 Binding of ligands to DA receptors in control and schizophrenic brains

(a) [³H] *Spiperone binding*

Using [³H] spiperone as ligand at 0.8 nM final concentration we assessed D2 receptors in membrane preparations of caudate, putamen and nucleus accumbens from schizophrenic and control brains. The results are presented in Fig. 2. There was a significant increase in [³H] spiperone binding in all three brain regions in samples from schizophrenic brains. This increase was about 55% in the caudate and putamen and 35% in the nucleus accumbens.

(b) *Scatchard analysis of* [³H] *spiperone binding*

In order to characterise further the increase in [³H] spiperone binding in schizophrenic brains we carried out a Scatchard analysis (Scatchard, 1949) of saturation data obtained by using concentrations (0.2, 0.3, 0.4, 0.8, 1.2 nM) of [³H]-spiperone in samples of caudate only. There was a 105% increase in maximum specific binding (B_{max}) or receptor number as shown in Fig. 3 ($p < 0.001$).

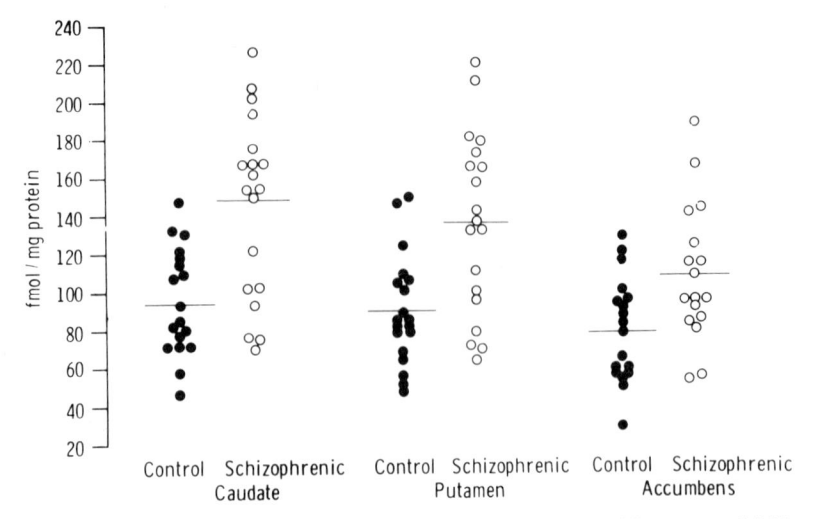

168 **Ligand-binding Studies in Brains of Schizophrenics** [Ch. 9

The differences between controls and schizophrenics are significant, $p < 0.001$ in caudate and putamen and $p < 0.01$ in nucleus accumbens (Students t test).

Fig. 2 – Specific [^{3}H] spiperone binding in control and schizophrenic brain samples.

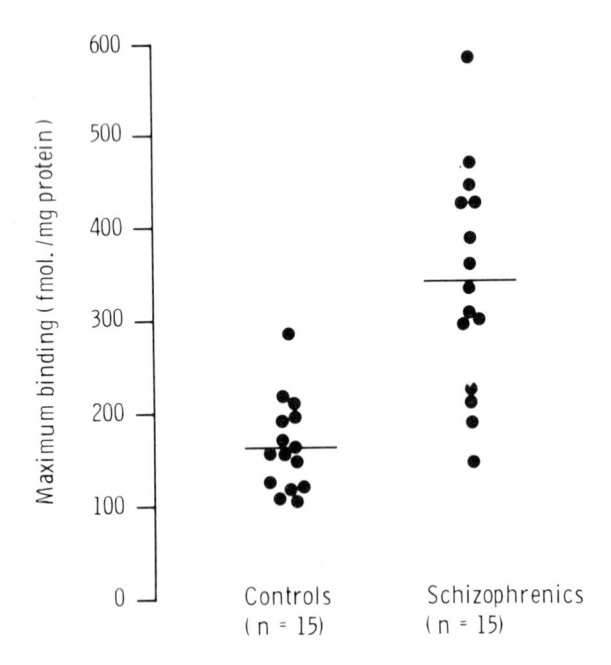

Fig. 3 – Maximum specific [^{3}H] spiperone binding in caudate samples from controls and schizophrenics.

Surprisingly there was also a large increase in the dissociation constant (K_D) in the samples from schizophrenics. However a closer analysis revealed that the increase in K_D was peculiar to those schizophrenics who had received neuroleptic medication up till death. It occurred to us that since neuroleptics are highly lipophilic with extremely high membrane: buffer partition coefficients that the drugs would not be removed from the membranes by the usual washing procedure and might compete with the tritiated ligand for the binding site and produce an apparent increase in K_D. We were able to confirm this suspicion in animal experiments the details of which have been published elsewhere (Owen et al., 1979).

(c) [^{3}H] *ADTN binding*
Under the conditions used to study DA agonist binding, i.e. with [^{3}H] ADTN at 7 nM and specific binding defined as that displaced by 1 μM apomorphine, binding would have been to the D3 binding site proposed by Seeman (1980). The results of applying this assay to samples of caudate and putamen from schizophrenic and control brains are illustrated in Fig. 4. There was no difference between controls and schizophrenics in [^{3}H] ADTN binding in either brain region.

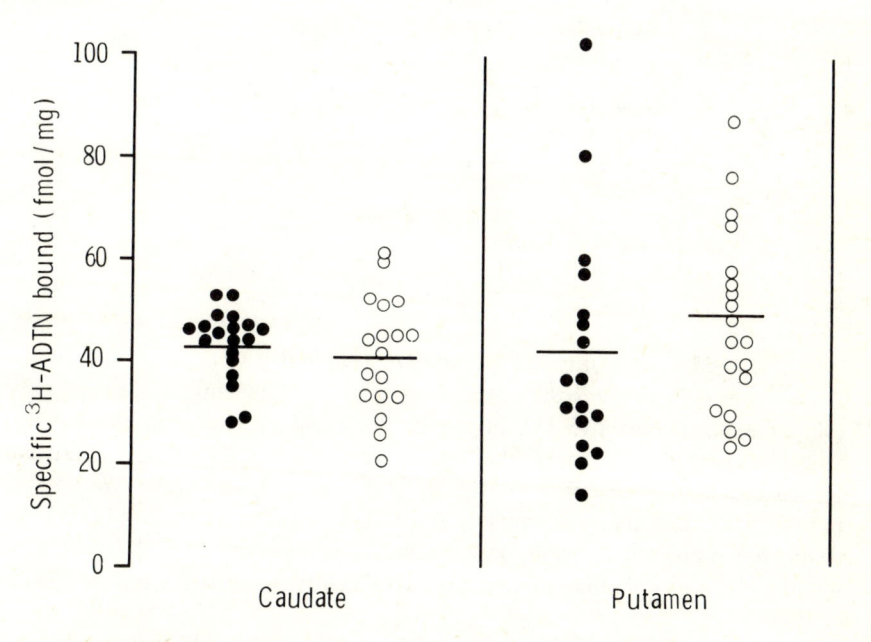

●, controls; ○, schizophrenics.

Fig. 4 — Specific [^{3}H] ADTN binding in control and schizophrenic brain samples.

(d) [³H] Flupenthixol binding

Hyttel (1978) proposed that [³H] flupenthixol was a specific ligand for those DA receptors linked to adenylate cyclase, i.e. the D1 receptors, although subsequently it became clear that flupenthixol binds with high affinity to both D1 and D2 binding sites (Cross and Owen, 1980). Initially we determined B_{max} and K_D values from Scatchard plots of [³H] flupenthixol binding in samples of caudate from controls and schizophrenics. The results are given in Table 3. There was a significant increase in B_{max} values in both drug-free and drug-treated schizophrenics. In addition, consistent with the findings with [³H] spiperone as ligand, there was a significant increase in K_D only in those patients receiving drugs.

Table 3 — Maximum [³H] flupenthixol binding in control and schizophrenic caudate samples

	K_D (nM)	B_{max} (fmol/mg protein)
Controls ($n = 8$)	3.7 ± 1.2	370 ± 124
Schizophrenics on drugs ($n = 7$)	7.6 ± 2.0[b]	637 ± 212[b]
Schizophrenics off drugs ($n = 8$)	4.6 ± 1.5	566 ± 130[a]

[a] $p < 0.02$.
[b] $p < 0.01$. Students t test (2-tailed).
(Values as mean ± S.D.)

(e) D1 and D2 components of [³H] flupenthixol binding

A further experiment was then carried out with [³H] flupenthixol as ligand. By using the highly selective D2 antagonist domperidone (Laduron and Leysen, 1979) we were able to resolve the binding of [³H] flupenthixol into its D1 and D2 components as described previously (Cross and Owen, 1980). We applied this modified technique to controls and drug-free schizophrenics only and the results are presented in Table 4. Only the D2 component of [³H] flupenthixol binding in schizophrenic brains was significantly increased compared with controls.

Our studies on the binding of ligands to the DA receptor therefore indicated that only the D2 receptors, i.e. those receptors that bind butyrophenones with high affinity, are increased in brains of schizophrenics.

Table 4 — D1 and D2 components of [³H] flupenthixol binding in control and schizophrenic caudate samples.

	D1	D2
Controls ($n = 9$)	104 ± 16	57 ± 9
Schizophrenic ($n = 6$) (drug-free)	133 ± 11	109 ± 16[a]

[a] $p < 0.02$ vs controls. Student's t test (2-tailed).
[³H] flupenthixol at 2 nM and D2 component defined by 100 nM domperidone.
Values as fmol/mg; mean ± SEM.

3.3 Binding of ligands to receptors other than DA

The binding of ligands to neurotransmitter receptors other than those for DA have also been studied to assess further the specificity of the increase in D2 receptors in brains of schizophrenics. The ligands and techniques employed have been outlined in Table 2. In samples of caudate and putamen (and also frontal cortex in the case of noradrenaline receptors) we were unable to demonstrate any significant difference between controls and schizophrenics in the binding of the ligands listed for receptors other than DA, i.e. the binding of ligands to 5HT, muscarinic acetylcholine, noradrenaline, GABA, benzodiazepine and opiate receptors were similar in controls and schizophrenics. The majority of this work has been published in detail elsewhere (Owen *et al.*, 1981).

Our study on opiate receptors is more recent and the results do not confirm the significant reduction in [³H] naloxone binding in samples of caudate from schizophrenic brains reported by Reisine and colleagues (1979). In the present study with [³H] etorphine (0.4 nM) as ligand and the selective 'μ' agonist, [D-Ala², MePhe⁴, Gly-ol⁵] enkephalin (Chapter 2 and Kosterlitz and Patterson, 1981) at 100 nM to define [³H] etorphine binding to 'μ' binding sites and by difference, using the benzomorphan Mr 2266 at 1.5 μM to define non-specific binding, the binding of [³H] etorphine was resolved into its 'μ' and '$\delta + \kappa$' components. The results in samples of caudate are presented in Fig. 5. There was no difference in total specific [³H] etorphine binding or in its 'μ' or '$\delta + \kappa$' components between controls and schizophrenics

Of the receptors studied in control and schizophrenic brains significant changes were observed only in DA receptors.

3.4 Effect of neuroleptic medication on the results
(a) [³H] *ADTN Binding in the Rats after Chronic Neuroleptic administration*
The fact that [³H] ADTN binding was not increased in samples from schizophrenic brains (Fig. 3) was of interest since Muller and Seeman (1977) reported

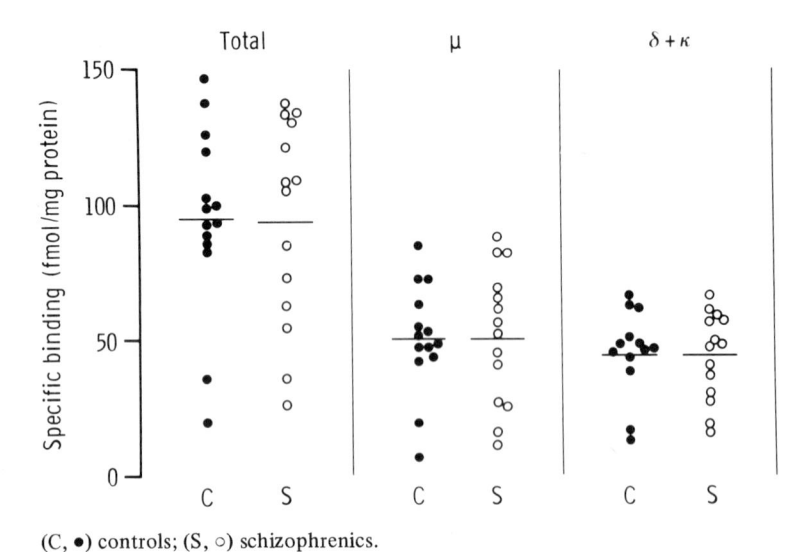

[Figure: scatter plot showing Specific binding (fmol/mg protein) on y-axis (0-150), with three paired groups Total, μ, and δ+κ, each with C (controls, ●) and S (schizophrenics, ○) columns.]

(C, ●) controls; (S, ○) schizophrenics.

Fig. 5 – Specific [³H] etorphine binding in control and schizophrenic brain samples.

a significant increase in the binding of [³H] apomorphine (which has binding characteristics similar to [³H] ADTN) in striatal and mesolimbic regions of rat brain after chronic haloperidol administration. This finding was confirmed using [³H] ADTN as ligand in rats after 6 weeks' administration of haloperidol (1.5 mg/kg in the drinking water). The results of the binding of [³H] ADTN and [³H] spiperone to striatal membranes are depicted in Fig. 6. Both [³H] ADTN and [³H] spiperone binding were significantly increased in the drug-treated animals, suggesting perhaps that in our particular sample of brains from schizophrenics the influence of neuroleptic induced DA receptor supersensitivity might be small.

(b) [³H] *Spiperone binding in drug-free and drug-treated schizophrenics*
Although generally there was a paucity of information in the patients' case notes it was possible to determine the drug status of the majority of the schizophrenics. The result of an analysis of neuroleptic medication against maximum specific [³H] spiperone binding values is presented in Table 5. Two patients who apparently had never received neuroleptics, had [³H] spiperone binding values that were significantly elevated above controls. Similarly 5 patients who had not received neuroleptics for at least one year prior to death had binding values significantly higher than controls. The highest binding values were observed in the group definitely receiving medication up to death.

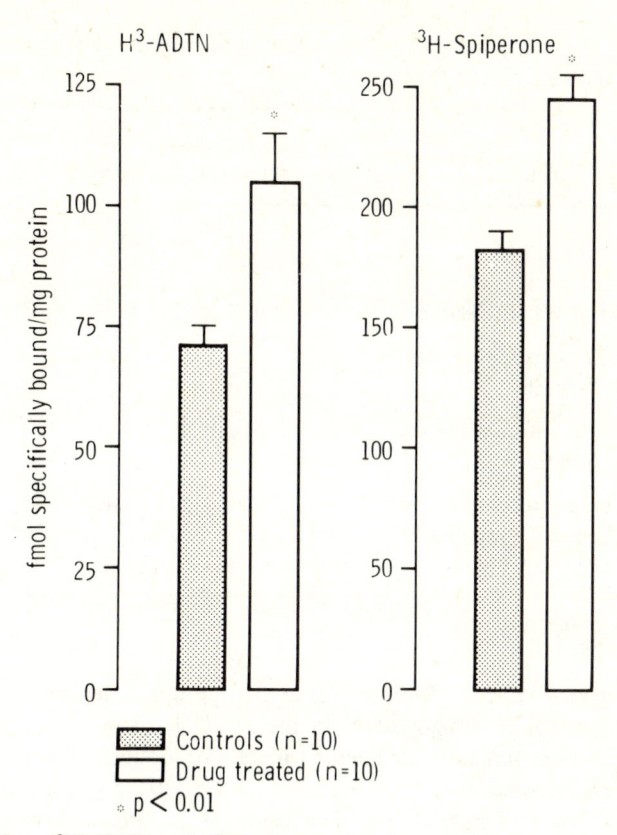

Fig. 6 – [³H]ADTN and [³H]spiperone binding in rat striatum after chronic haloperidol administration.

(c) [³H] *Spiperone binding in samples from medicated and unmedicated patients dying with Huntington's Chorea or Alzheimer's Disease*

In an attempt to determine the extent to which neuroleptic-induced DA receptor supersensitivity contributed towards the increased D2 receptors observed in schizophrenic brains we have also assessed D2 receptors in samples of brain from two groups of non-schizophrenic patients some of whom had received neuroleptics up to the time of death and others who had not. These were firstly samples from patients dying with Huntington's Chorea (H.Ch.) and secondly samples from patients dying with Alzheimer's disease. For both of these neuropsychiatric disorders a number of biochemical changes are known to occur in the brain (for reviews see Bird (1976), Perry (1979)). In relation to the H.Ch. samples, therefore, we carried out a preliminary experiment using the rat striatal kainic acid lesion model of H.Ch. (McGeer *et al.*, 1978) with the aim of determining whether or not the lesions impair the ability of DA receptors to proliferate in response to neuroleptic medication. After kainic acid lesions of the

Table 5 — Relationship between maximum specific [³H] spiperone binding and neuroleptic medication.

Group	No.	B_{max} (fmol/mg protein)	p
Controls	15	167.0 ± 50.4	—
Schizophrenics:			
Never had neuroleptics	2	296.9 , 214.8	<0.05
No medication < 1 year before death	5	249.7 ± 68.1	<0.01
Uncertain if medicated up to death	6	380.9 ± 90.6	<0.001
Definitely receiving medication up to death	4	390.3 ± 163.4	<0.001
All	15	339.7 ± 119.6	<0.001

Values as mean ± S.D.

right striatum rats were allocated to one of two groups either receiving or not receiving haloperidol (1.5 mg/kg/day) in the drinking water. A further set of unlesioned rats were similarly allocated to these two groups. After six weeks of drug treatment followed by a 7-day washout period [³H] spiperone binding was carried out as described in Section 2. The results of this preliminary experiment are presented in Table 6. The results demonstrate that after striatal kainic acid lesions in the rat, [³H] spiperone binding increases in response to chronic neuroleptic administration to the same extent as unlesioned animals. This would suggest that patients with H.Ch., treated with neuroleptics, would develop DA

Table 6 — The effect of chronic haloperidol treatment on [³H] spiperone binding in striatum of lesioned, unlesioned and control rats.

	Drug-free	Drug-treated
Unlesioned rats	166 ± 20.6 (10)	202 ± 22.9 (9) ↑21% $p<0.05$
Lesioned rats		
Left striatum	137.2 ± 20.4 (9)	188.6 ± 21.0 (8) ↑37% $p<0.001$
Right striatum	89.8 ± 23.6 (9)	119 ± 23.7 (8) ↑33% $p<0.001$

Numbers in brackets refer to numbers of rats. Results (mean ± S.D.) expressed as fmol/mg protein.

receptor supersensitivity to the same extent as normal individuals. The results for [³H] spiperone binding (0.8 nM final concentration) in samples of caudate from neuroleptic-treated patients dying with H.Ch. and also from those with no history of ever having received such medication are illustrated in Fig. 7. There was no significant difference between the two groups in the binding of [³H] spiperone.

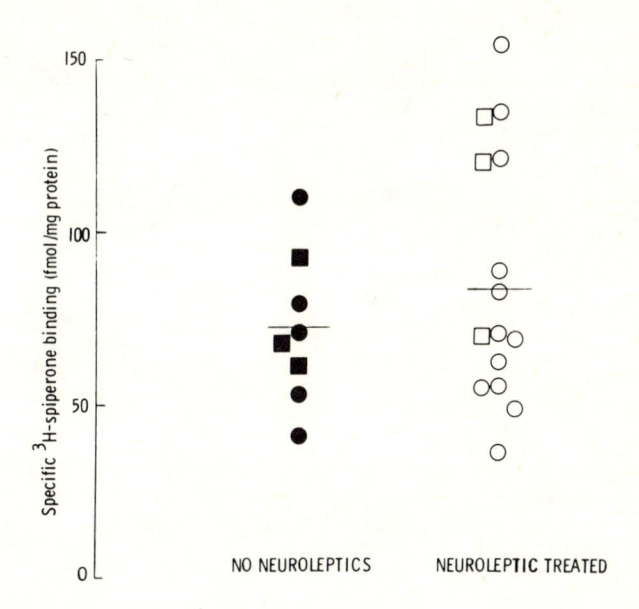

■ ▫, tetrabenazine treated; [³H] spiperone at 0.8 nM.

Fig. 7 − [³H] Spiperone binding in caudate samples from patients with Huntington's Chorea, treated and untreated with neuroleptics.

(d) [³H] *Spiperone binding in samples from medicated and unmedicated patients dying with Alzheimer's disease*

In a similar study of samples from patients dying with Alzheimer's disease, [³H] spiperone binding was carried out in samples of putamen. In this study patients were defined as drug-free provided they had not received neuroleptics for one month prior to death. The results for [³H] spiperone binding in these samples are presented in Fig. 8. Again there was no significant difference in [³H] spiperone binding in samples from individuals who were drug-free or drug-treated up to the death.

3.5 Relationship between D2 receptors and schizophrenic symptoms

There is rarely sufficient information in a patient's case notes to assess satisfactorily the severity of the predominant features of schizophrenia. For any

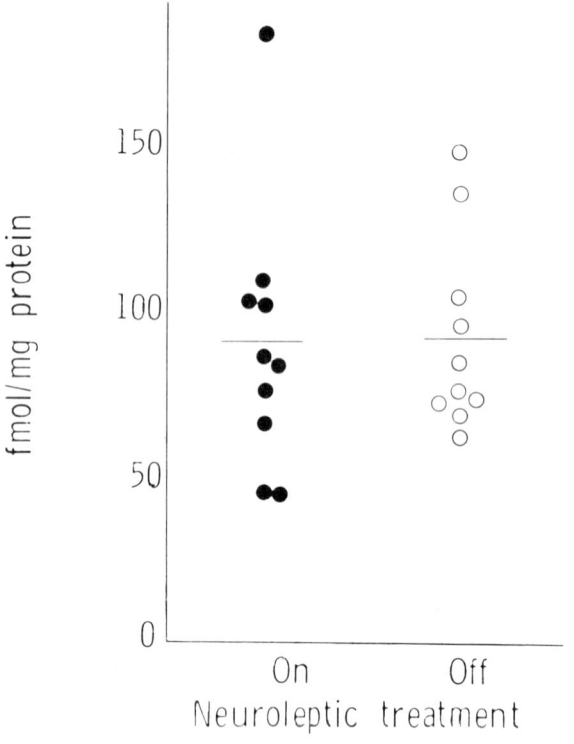

[³H] Spiperone at 0.8 nM.

Fig. 8. — [³H] Spiperone binding in putamen samples from patients with Alzheimer's disease, treated and untreated with neuroleptics.

realistic attempt to relate biochemical measures in post-mortem brain tissue with the symptomatology of schizophrenia it was necessary to accumulate a series of brains from patients whose psychoses had been carefully assessed in life. A large number of schizophrenics have now been so assessed using the rating scale devised by Krawiecka and colleagues (1977) and we now have maximum specific [³H] spiperone binding values on 14 of these patients. In a much larger series of brains than used previously we have assessed D2 receptors (B_{max} [³H] spiperone binding values) in samples of putamen and nucleus accumbens and the results are presented in Fig. 9. The results were qualitatively similar to our earlier finding — there was a highly significant increase in D2 receptors in schizophrenic brains. Within this larger series of schizophrenic brains were brains from patients who had been clinically rated during life. The severity of the positive and negative symptoms of schizophrenia were then correlated with maximum specific [³H]-spiperone binding values and the results are given in Table 7. The positive features of schizophrenia were positively and significantly correlated with D2

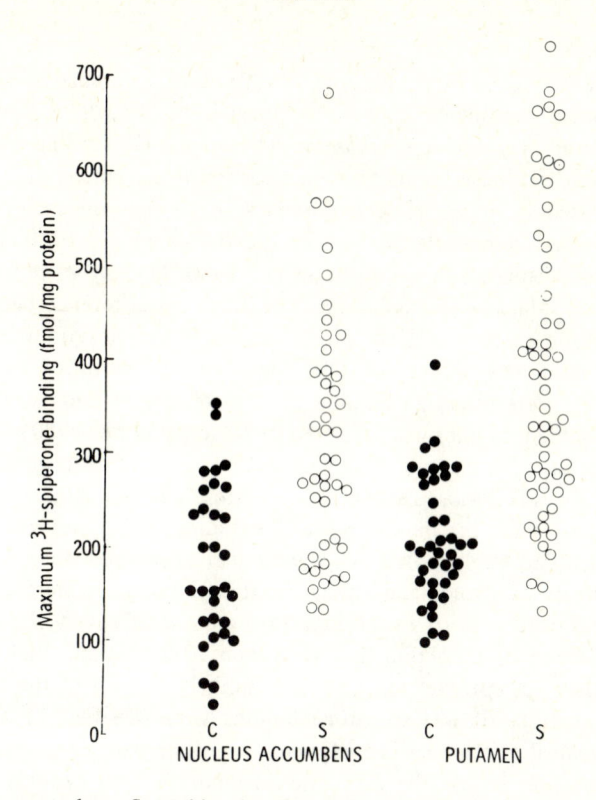

(●, c – controls; ○, S – schizophrenics). Values (mean ± SD, fmol/mg of protein) were: putamen, controls ($n = 39$) 210 ± 66.3, schizophrenics ($n = 53$) 388.1 ± 159.7; accumbens, controls ($n = 32$) 180.1 ± 85.9, schizophrenics ($n = 44$) 316.5 ± 129.7. $p < 0.001$ in each case.

Fig. 9 – Maximum specific [^{3}H] spiperone binding in nucleus accumbens and putamen of controls and schizophrenics.

receptor numbers whereas the negative symptoms were inversely and non-significantly correlated.

Table 7 – Relationship between D2 receptors and symptomatology in schizophrenics assessed before death.

	r	p
[^{3}H] spiperone binding (B_{max}) vs positive symptoms	0.698	<0.01
[^{3}H] spiperone binding (B_{max}) vs negative symptoms	−0.369	N.S.

($n = 14$).

4. DISCUSSION

Apart from the report by Reynolds *et al.* (1980) of no difference in B_{max} values for [^{3}H]spiperone binding in putamen samples from controls and neuroleptic treated schizophrenics and a decrease in [^{3}H]spiperone binding in drug-free schizophrenics, there now seems to be a measure of agreement that D2 receptors are increased in brains of schizophrenics, given that the majority of the patients had received neuroleptics during life (Owen *et al.*, 1978, Lee *et al.*, 1978, Lee and Seeman, 1980, Reisine *et al.*, 1980, Mackay *et al.*, 1980). An earlier report by Mackay and colleagues (1978) of no difference between controls and schizophrenics in [^{3}H]spiperone binding in samples of nucleus accumbens using low concentrations of [^{3}H]spiperone was modified to increased B_{max} values for [^{3}H]spiperone binding in caudate and nucleus accumbens of samples from schizophrenics in a later report (Mackay *et al.*, 1980). The failure to demonstrate an increase in [^{3}H]spiperone binding in schizophrenic brains in the earlier study was almost certainly due to residual neuroleptics competing with [^{3}H]spiperone for the binding sites. The findings of Reynolds *et al.* (1980) are more difficult to explain although the number of samples studied was small. The suggestion by Reynolds and colleagues that the major difference between their study and other reports in the literature was the relatively short death to storage deep-frozen or dissection time (less than 10 hours) and that this might account for their exceptional results seems unlikely in view of the remarkable post-mortem stability of [^{3}H]spiperone binding sites (see Fig. 1). Whether or not methodological differences can explain these discrepancies is unclear but of obvious importance is that the B_{max} values reported by Reynolds *et al.*, are 2–4 times higher than those reported by Mackay *et al.* (1980).

Up to the present time it would seem that the change in receptors in schizophrenic brains may be specific to D2 receptors — [^{3}H]ADTN binding to the D3 site, and the D1 site assessed by the residual specific binding of [^{3}H]flupenthixol after displacement by domperidone were both no different from controls. The binding of ligands to several other neurotransmitter receptors in Table 2 and Section 3.3 also revealed no significant differences between controls and schizophrenics. Of interest with respect to these non-dopaminergic receptors are the negative findings with [^{3}H]LSD and [^{3}H]etorphine as ligands. Previously Bennett *et al.* (1979) have reported a decrease in [^{3}H]LSD binding in frontal cortex of schizophrenics. In the present study we were unable to demonstrate any significant difference between controls and schizophrenics in [^{3}H]LSD binding in samples of caudate or putamen. In a separate study, but with samples used in the present study Whitaker *et al.* (1981) by contrast to Bennett and colleagues reported an increase in [^{3}H]LSD binding in a small number of samples of frontal cortex from drug-free but not drug-treated schizophrenics. This result requires further investigation. Neuroleptics are effective displacers of high affinity [^{3}H]LSD binding and Whitaker *et al.* (1981) pointed out that residual neuroleptics in membrane preparations from drug-treated schizophrenics might compete for [³H] LSD binding sites and form the basis of an explanation of the decrease in [³H] LSD binding in schizophrenic brains reported by Bennett and colleagues.

The only other non-dopaminergic receptor change reported in schizophrenic brain was the study of opiate receptors by Reisine *et al.* (1980) who used [³H] naloxone as ligand. We were unable to replicate this finding using a more elaborate technique, as described in Section 3.3 to differentiate the binding of [³H] etorphine into its components binding to 'μ' and also '$\delta + \kappa$' opiate receptors.

The outstanding question arising from studies of D2 receptors in schizophrenic brains is whether or not the reported increases are solely due to the effects of chronic neuroleptic treatment. On the one hand Mackay *et al.* (1980) reported that the increase in D2 receptors in their samples of schizophrenic brains was a result of neuroleptic medication: patients who had been drug-free for at least 1 month prior to death had [³H] spiperone binding values similar to controls. On the other hand Owen *et al.* (1978) reported increased D2 receptors in patients who had not received neuroleptics for at least 1 year prior to death and two patients who had never received neuroleptics. In addition Lee and Seeman (1980) who used both [³H] haloperidol and [³H] spiperone as ligands reported increased D2 receptors in caudate samples from 11 schizophrenics who had never received neuroleptics. It is not easy to reconcile these two conflicting views but the results of experiments included in this chapter lend strong support to the view that the increase in D2 receptors reported in schizophrenic brains is not solely due to the result of neuroleptic medication. Although [³H]-ADTN binding is increased in rats after chronic neuroleptic treatment (Fig. 5) the binding of this ligand to schizophrenic brain is no different from controls (Fig. 3). This is consistent with the report by Lee and Seeman (1978) of no change in [³H] apomorphine binding (which has similar binding characteristics to [³H] ADTN) in schizophrenic brains. Moreover, Marsden and Jenner (1980) (see also Chapter 8) have reported that after chronic neuroleptic treatment to rats and withdrawal from the drug that D2 receptors have returned to control values at a time when the DA stimulation of adenylate cyclase remains significantly elevated. Considering the close relationship between DA stimulated adenylate cyclase and [³H] flupenthixol binding (Hyttel, 1978; Cross and Owen, 1980) the result of Marsden and Jenner seems qualitatively opposite to that observed in schizophrenic brains. Consistent with this argument is the report by Carenzi *et al.* (1975) of normal DA stimulation of adenylate cyclase in caudate samples from schizophrenics.

The finding in the present study that [³H] spiperone binding is not increased in samples from neuroleptic-treated patients dying with H.Ch. (Fig. 7) or Alzheimer's disease (Fig. 8) suggests that the degree to which D2 receptors increase in number, in man, in response to neuroleptic medication may be small. There are two major objections to these studies and these are that the dose of neuroleptics administered, in general, is lower and for a shorter period of time in the case of patients with H.Ch. and Alzheimer's disease compared with schizophrenics. However, when [^{3}H] spiperone binding values in the initial study (Fig. 2) were grouped into those from patients who had received low doses of neuroleptics (similar to those received by the H.Ch. and Alzheimer patients) and high doses, the binding values were similar in the two groups. Also it seems unlikely that duration of treatment is an important determinant of the degree of DA receptor supersensitivity achieved. It has been demonstrated in rats (Owen *et al.*, in preparation) receiving chronic neuroleptic medication that the full extent of D2 receptor increase is achieved in 7 days with no further increase up to 9 months. The large increases in D2 receptors observed in some schizophrenics may not, therefore, be simply attributable to neuroleptic treatment. This seems to be borne out in our most recent study where the increase in D2 receptors was shown to be significantly correlated with the positive symptoms of schizophrenia.

Obviously, the relationship between D2 receptors and schizophrenia and the extent of the effect of neuroleptics on the results could be clarified by a biochemical analysis of a large series of brains from schizophrenics who were definitely drug-free. This, however, may not prove a feasible strategy and future experiments to evaluate the significance of D2 receptor increases in schizophrenia may have to await the development of *in vivo* techniques, perhaps based on positron emission, similar to those used by Comar *et al.* (1979a, 1979b) in baboons and humans.

ACKNOWLEDGEMENTS

We would like to acknowledge Dr G. Slavin and Dr A. B. Price of Northwick Park Hospital, Harrow, Miss M. Williams of Shenley Hospital, Herts, and Dr J. A. N. Corsellis, Mr R. Alston and Mrs R. Brown of Runwell Hospital, Essex, for their assistance in the collection of the post-mortem material. We are also grateful to Professor H. W. Kosterlitz and Dr S. J. Paterson of Marischal College, Aberdeen, for their expert guidance in the study of opiate receptors.

REFERENCES

Bennett, J. P. and Snyder, S. H. (1976) Serotonin and lysergic acid diethylamide binding in rat brain membranes: relationship to post synaptic serotonin receptors. *Mol. Pharmacol* 12, 373–389.

Bennett, J. P., Enna, S. J., Bylund, D. B., Gillin, J. C., Wyatt, R. J. and Snyder, S. H. (1979) Neurotransmitter receptors in frontal cortex of schizophrenics. *Arch. Gen. Psychiatry* 36, 927–934.

Bird, E. D. (1976) Biochemical studies on γ-aminobutyric acid metabolism in Huntington's Chorea, in *Biochemistry and Neurology* (Bradford, H. F. and Marsden, C. D., eds.) pp. 83–90. Academic Press, London.

References

Bowers, M. B. (1974) Central dopamine turnover in schizophrenic syndromes. *Arch. Gen. Psychiatry* **31**, 50–54.

Bylund, D. B. and Snyder, S. H. (1976) Beta-adrenergic receptor binding in membrane preparations from mammalian brain. *Mol. Pharmacol.* **12**, 568–580.

Carenzi, A., Gillin, C., Guidotti, A., Schwartz, M., Trabucchi, M. and Wyatt, J. (1975) Dopamine-sensitive adenyl cyclase in human caudate nucleus: a study in control subjects and schizophrenic patients. *Arch Gen Psychiatry* **32**, 1056–1059.

Carlsson, A. and Lindqvist, M. (1963) Effect of chlorpromazine and haloperidol on the formation of 3-methoxy-tyramine and normetanephrine in mouse brain. *Acta Pharmac. Tox.* **20**, 140–144.

Clement-Cormier, Y. C., Kebabian, J. W., Petzold, G. L. and Greengard, P. (1974) Dopamine-sensitive adenylate cyclase in mammalian brain: A possible site of action of antipsychotic drugs. *Proc. Natl. Acad. Sci. USA* **71**, 1113–1117.

Comar, D., Maziere, M., Godot, J. M., Berger, G., Soussaline, F., Menini, Ch. Arfel, G. and Naquet, R. (1979a) Visualisation of ^{11}C-flunitrazepam displacement in the brain of the live baboon. *Nature* **280**, 329–331.

Comar, D., Zarifian, E., Verhas, M., Soussaline, F., Maziere, M., Berger, Cr., Loo, H., Cuche, H., Kellershon, C. and Deniker, P. (1979b) Brain distribution and kinetics of ^{11}C-Chlorpromazine in schizophrenics: positron emission tomography studies. *Res. Psychiatry* **1**, 23–29.

Connell, P. H. (1958) *Amphetamine Psychosis*, Maudsley Monograph No. 5. Chapman and Hall, London.

Cross, A. J. and Owen, F. (1980) Characteristics of ^{3}H-*cis*-flupenthixol binding to calf brain membranes. *Eur. J. Pharmacol.* **65**, 341–347.

Cross, A. J., Crow, T. J. and Owen, F. (1979) The use of ADTN (2-amino,6,7-dihydroxy, 1,2,3,4-tetrahydronapthalene) as ligand for brain dopamine receptors. *Br. J. Pharmacol.* **66**, 87–88P.

Crow, T. J., Deakin, J. F. W., Johnstone, E. C. and Longden, A. (1976) Dopamine and schizophrenia. *Lancet* **ii**, 563.

Crow, T. J., Baker, H. F., Cross, A. J., Joseph, M. H., Lofthouse, R., Longden, A., Owen, F., Riley, G. J., Glover, V. and Killpack, W. S. (1979) Monamine mechanisms in chronic schizophrenia: post-mortem neurochemical findings. *Br. J. Psychiat.* **134**, 249–256.

Crow, T. J., Owen, F., Cross, A. J., Johnstone, E. C., Joseph, M. H. and Longden, A. (1980) The dopamine receptor as the site of the primary disturbance in schizophrenia, in *Enzymes and Neurotransmitters in Mental Disease*. (Usdin, E. Sourkes, T. L. and Youdim, M. B. H., eds.) pp. 559–574. John Wiley, Chichester.

DeArmond, S. J., Fusco, H. M. and Dewey, M. M. (1974) *Structure of the Human Brain*. Oxford University Press, New York.

Enna, S. J. and Snyder, S. H. (1976) Influence of ions, enzymes and detergents on γ-amino-butyric acid receptor binding in synaptic membranes of rat brain. *Mol. Pharmacol.* **13**, 442–453.

Feighner, J. P., Robins, E., Guze, S. B., Woodruff, R. A., Winokur, G. and Munoz, R. (1972) Diagnostic criteria for use in psychiatric research. *Arch. Gen. Psychiatry* **26**, 57–63.

Greenberg, D. A., U'Prichard, D. C. and Snyder, S. H. (1976) Alpha-noradrenergic receptor binding in mammalian brain: differential labelling of agonist and antagonist states. *Life Sci.* **19**, 69–76.

Hyttel, J. (1978) Effects of neuroleptics on ^{3}H-haloperidol and ^{3}H-cis(z)-flupenthixol binding and on adenylate cyclase activity in vitro. *Life Sci.* **23**, 551–556.

Kebabian, J. W. and Calne, D. B. (1979) Multiple receptors for dopamine. *Nature* **277**, 93–96.

Kosterlitz, H. W. and Paterson, S. J. (1981) Tyr-D-Ala-Gly-MePhe-NH(CH$_2$)$_2$OH is a selective ligand for the μ-opiate binding site. *Br. J. Pharmac.* **73**, 299P.

Krawiecka, M., Goldberg, D. and Vaughan, M. (1977) A standardised psychiatric assessment for rating chronic psychotic patients. *Acta Psychiat. Scand.* **55**, 299–308.

Laduron, P. M. and Leysen, J. W. (1979) Domperidone, a specific in vitro dopamine antagonist, devoid of in vivo central dopaminergic activity. *Biochem. Pharmac.* **28**, 2161–2165.

Lee, T. and Seeman, P. (1980) Elevation of brain neuroleptic/dopamine receptors in schizophrenia. *Am. J. Psychiatry* **137**, 191–197.

McGeer, E. G., Olney, J. W. and McGeer, P. L. (1978) *Kainic Acid as a Tool in Neurobiology.* Raven Press, New York.

Mackay, A. V. P., Doble, A., Bird, E. D., Spokes, E. G. Quik, M. and Iversen, L. L. (1978) ^{3}H-Spiperone binding in normal and schizophrenic post-mortem human brain. *Life Sci.* **23**, 527–532.

Mackay, A. V. P., Bird, E. D., Spokes, E. G., Rossor, M. and Iversen, L. L. (1980) Dopamine receptors and schizophrenia: drug effect or illness. *Lancet* **ii**, 915–916.

Marsden, C. D. and Jenner, P. (1980) The pathophysiology of extrapyramidal side-effects of neuroleptic drugs. *Psychol. Med.* **10**, 55–72.

Miller, R. J., Horn, A. S. and Iversen, L. L. (1974) The action of neuroleptic drugs on dopamine-stimulated adenosine 3',5'-monophosphate production in neostriatum and limbic forebrain. *Mol. Pharmacol.* **10**, 759–766.

Möhler, H. and Okada, T. (1977) Properties of ^{3}H-diazepam binding to benzodiazepine receptors in rat cerebral cortex. *Life Sci.* **20**, 2101–2110.

Muller, P. and Seeman, P. (1977) Brain neurotransmitter receptors after long-term haloperidol: dopamine, acetylcholine, serotonin, α-noradrenergic and naloxone receptors. *Life Sci.* **21**, 1751–1758.

Owen, F., Cross, A. J., Crow, T. J., Longden, A., Poulter, M. and Riley, G. J. (1978) Increased dopamine receptor sensitivity in schizophrenia. *Lancet* **ii**, 223–226.

Owen, F., Cross, A. J., Poulter, M. and Waddington, J. L. (1979) Change in the characteristics of ^{3}H-spiperone binding to rat striatal membranes after acute chlorpromazine administration: effects of buffer washing of membranes. *Life Sci.* **25**, 385–390.

Owen, F., Cross, A. J., Crow, T. J., Lofthouse, R. and Poulter, M. (1981) Neurotransmitter receptors in brain in schizophrenia. *Acta Psychiat. Scand. Suppl.* **63**, 20–26.

Perry, E. K. (1979) Correlations between psychiatric neuropathological and biochemical findings in Alzheimer's disease, in *Alzheimer's Disease: Early recognition of potentially reversible deficits* (Glen, A. I. M. and Whalley, L. J. eds.) pp. 27–32. Churchill Livingstone, Edinburgh.

Randrup, A. and Munkvad, I. (1966) On the role of catecholamines in the amphetamine excitatory response. *Nature* **211**, 540.

Randrup, A. and Munkvad, I. (1967) Stereotyped activities produced by amphetamine in several animal species and man. *Psychopharmacologia* **11**, 300–310.

Randrup, A. and Munkvad, I. (1972) Evidence indicating an association between schizophrenia and dopaminergic hyperactivity in the brain. *Orthomol. Psychiatr.* **1**, 2–7.

Reisine, T. D., Rossor, M., Spokes, E., Iversen, L. L. and Yamamura, H. I. (1980) Opiate and neuroleptic receptor alterations in human schizophrenic brain tissue. *Adv. Biochem. Psychopharmacol* **21**, 443–450.

Reynolds, G. P., Reynolds, L. M., Riederer, P., Jellinger, K. and Gabriel, E. (1980) Dopamine receptors and schizophrenia, drug effect or illness. *Lancet* **ii**, 1251.

Scatchard, G. (1949) The attraction of proteins for small molecules and ions. *Ann. N.Y. Acad. Sci.* **51**, 660–672.

Seeman, P. (1980) Brain dopamine receptors. *Pharmacol. Rev.* **32**, 230–287.

Seeman, P., Chau-Wong, M., Tedesco, J. and Wong, K. (1975) Brain receptors for antipsychotic drugs and dopamine: direct binding assays. *Proc. Nat. Acad. Sci. USA* **72**, 4376–4380.

Whitaker, P. M., Crow, T. J. and Ferrier, I. N. (1981) Tritiated LSD binding in frontal cortex in schizophrenia. *Arch. Gen. Psychiatry* **38**, 278–280.

Wing, J. K., Cooper, J. E. and Sartorius, N. (1974) *The Measurement and Classification of Psychiatric Symptoms: An Instruction Manual for the PSE and Cetego Programs.* Cambridge University Press, London.

Yamamura, H. I. and Snyder, S. H. (1974) Muscarinic cholinergic binding in rat brain. *Proc. Nat. Acad. Sci. USA* **71**, 1725–1729.

10

Pituitary Gonadotrophin-Releasing-Hormone Receptors: Physiological Regulation

R. N. Clayton

Department of Medicine, University of Birmingham, Edgbaston, Birmingham B15 2TH, U.K.

Abbreviations used

GnRH	–	gonadotrophin releasing hormone
GnRH-A	–	gonadotrophin releasing hormone agonist analogue
GnRH-R	–	gonadotrophin releasing hormone receptor
LH	–	luteinising hormone
M.E.L.	–	median eminence lesion

1. INTRODUCTION

It is now well established that the tonic and pulsatile release of pituitary gonadotrophic hormones is regulated by the hypothalamic gonadotrophin releasing hormone (GnRH) (Fink, 1976). GnRH is synthesised in the perikarya of neurones located in various hypothalamic nuclei (differing somewhat between species) and is then transported along axons which terminate in the median eminence region of the hypothalamus. Quanta of GnRH are released from the storage granules located in the terminals of these axons into the hypophyseal portal vasculature for transport to the pars distalis of the pituitary (Ben-Jonathan *et al.*, 1973). Release of GnRH into the portal vessels is affected by a number of neurotransmitters of the central catecholaminergic system, and can be induced experimentally by depolarisation of the axon terminals *in vivo* and *in vitro* (Fink, 1979; Barraclough *et al.*, 1982). Under physiological circumstances, the secretion of GnRH is pulsatile rather than continuous and there is increasing evidence that modulation of pulse frequency, rather than the pulse amplitude, plays a critical, if not a major, role in the regulation of gonadotrophin secretion, and hence the reproductive state of the organism, (Yen *et al.*, 1972; Knobil, 1980).

Introduction

Because of the inaccessibility of the hypothalamo-hypophyseal portal vasculature relatively few data are available about changes in GnRH secretion under different circumstances. Nevertheless, it seems that the portal plasma concentration of GnRH is increased following removal of negative steroid feedback (Sarkar and Fink, 1979; Sherwood and Fink, 1980). Additionally, reduction in GnRH secretion and pulse frequency is implied by the finding of a dramatic decline in the number of serum gonadotrophin peaks in lactating females (Bohnet et al., 1976).

From the foregoing it is evident that hormonal regulation of hypothalamic GnRH secretion represents the final integrated response arising from interaction of many feedback signals within the central nervous system. However, it is well recognised that gonadal steroid hormones can influence gonadotrophin secretion by acting directly upon the anterior pituitary to modulate not only tonic secretion of the glycoproteins but also gonadotroph responsiveness to the releasing-hormone stimulus (Drouin et al., 1976; Hsueh et al., 1979). Therefore, it becomes important to determine the relative contribution of direct pituitary and indirect (via the hypothalamus and endogenous GnRH secretion) sites of action of sex-steroid hormones.

There is now compelling evidence that the initial interaction of GnRH with gonadotrophs occurs with specific receptors located within the plasma membrane of the cells (Clayton and Catt, 1981a). Subsequently, a series of intracellular events is activated which ultimately results in secretion of stored gonadotrophins, and probably leads to stimulation of their synthesis (Conn et al., 1981). Any or all of these steps have the potential for regulation by steroid hormones, as well as other intra- and extra-cellular signals.

This review is concerned with the changes that result in the initial step of GnRH interaction with its receptor sites following *in vivo* manipulation of the sex-steroid hormone environment. In addition, evidence is presented that GnRH itself regulates the number of its own receptor sites, a phenomenon which probably contributes to changes in the sensitivity of the gonadotroph response to the hypothalamic hormone.

1.1 GnRH receptor measurement

Initial attempts to characterise and quantitate pituitary binding sites for GnRH employed radio-iodinated GnRH ($[^{125}I]$ GnRH) as the ligand, and either crude rat pituitary membranes or enriched plasma membrane fractions from bovine anterior pituitaries as the source of receptors. These assays were highly unsatisfactory for a number of reasons, as outlined in Table 1 and reviewed in detail elsewhere (Clayton and Catt, 1981a). The major problem with $[^{125}I]$ GnRH was the presence of a high proportion of low-affinity ($K_d \approx 1$ μM) GnRH binding sites which are considered of dubious physiological significance, given that the concentration of GnRH in the portal plasma is $<$ 1 nM. This, and the other problems, were completely resolved with the availability for radiolabelling of

Table 1 — Comparison of various radioligands for the GnRH ligand-binding assay

	[^{125}I] GnRH	[^{125}I] GnRH-A[a]
Low affinity binding ($K_a \approx 10^6$ M^{-1})	High capacity 75% of specific B/T	ABSENT
High affinity binding ($K_a \approx 10^9$ M^{-1})	Low capacity < 25% of specific B/T	100% B/T
Receptor affinity (K_a)	3×10^8 M^{-1}	$2-6 \times 10^9$ M^{-1}
Ligand degradation at 4 °C	30–40%	ABSENT
Receptor source	Enriched bovine pituitary membranes best	Crude 10,800 × g rat pituitary membranes *or* individual pituitary homogenate
Suitability for physiological studies	USELESS	EXCELLENT

[a] Any GnRH agonist analogues with D-amino acid in position 6 (except DLys^6) and Pro^9 NEthylamide.

synthetic super-active, slowly degradable, agonist analogues (Clayton et al., 1979) of the natural decapeptide. When used as the ligand, [^{125}I] (DSer(tBu)^6)-des-Gly^{10}GnRH NEthylamide ([^{125}I] GnRH-A), or other analogues with D-amino acid substitutions in position 6 and glycinamide^{10} replaced by ethylamide, interacts with a single class of high-affinity binding sites (apparent $K_a \approx 5 \times 10^9$ M^{-1}) in pituitary membrane preparations and cultured anterior pituitary cells (Naor et al., 1981). The advantages of these ligands are summarised in Table 1. Thus, ligand-binding assays based on the use of these analogues have become established as the method of choice for accurate measurement of GnRH receptors. In addition, it should be emphasised that this methodology enables the quantitation of GnRH receptor content of single rat pituitary glands thus permitting their correlation with individual serum and pituitary hormone levels. The detailed characterisation of pituitary GnRH receptors (GnRH-R) has been reviewed elsewhere (Clayton and Catt, 1981a), and will not be further elaborated here. However, two points are worth re-emphasising: (i) the close correlation between the receptor affinity and biological activity of a range of agonist and antagonist analogues strongly supports the notion that sites identified with these binding assays are responsible for mediation of GnRH action *in vivo*; (ii) GnRH receptors are only identified in those tissues in which the decapeptide has a biological effect. Indeed even within the pituitary GnRH-R are restricted to the gonadotrophs (Naor et al., 1982). Having established and validated the method for GnRH-R measurement studies were performed to evaluate GnRH-R in a number of physiological circumstances. Studies to date have been performed largely in rats or hamsters.

2. PHYSIOLOGICAL REGULATION OF GnRH RECEPTORS

2.1 Estrous cycle

Gonadotrophin release in response to exogenously adminstered GnRH is maximal on the afternoon of pro-estrous in rats, just prior to the spontaneous pre-ovulatory LH surge (Aiyer et al., 1974; Gordon and Reichlin, 1974). This is attributable, in part, to the 'sensitising' effects of increasing circulating concentrations of 17β-estradiol, which generally start rising sometime late on metestrous in animals with 4-day cycle lengths. One of the mechanisms by which estrogens can 'sensitise' the pituitary is by increasing the readily releasable pool of gonadotrophin (Pickering and Fink, 1979), possibly by stimulation of its synthesis. However, other mechanisms may contribute, such as stimulation of cellular hormone release mechanisms. In this regard the possible contribution of an increase in the pituitary content of GnRH receptors during pro-estrous has been investigated.

A number of studies have shown that GnRH-R values on pro-estrous are double the nadir values measured on estrous and early metestrous Fig. 1. (Savoy-Moore et al., 1980; Clayton et al., 1980; Adams and Spies, 1981; Ferland et al., 1981). Furthermore, GnRH-R increase between the evening of metestrous and diestrous II morning, prior to increased sensitivity of the gland to GnRH, and at a time when serum estradiol concentrations begin to rise (Savoy-Moore et al., 1980). Thus, increasing GnRH-R per se is not the sole mechanism responsible for the enhanced pre-ovulatory sensitivity to GnRH. Nevertheless, the result of this increase in GnRH-R could serve to increase pituitary gonadotroph responsiveness to increased endogenous GnRH stimulation occurring on the afternoon of pro-estrous (Sarkar et al., 1976). Exogenous estradiol given to ovariectomised monkeys increases pituitary GnRH-R, beginning 12 h after exposure and reaching a maximum at 36 h, just at the beginning of the E_2 induced LH surge. Thus, in a primate, estrogen sensitisation of pituitary gonadotrophin secretion is mediated, in part, by an increase in GnRH-R (Adams et al., 1981). It remains to be established whether estradiol itself, or some other factor, is responsible for the GnRH-R increase, and whether this operates directly on pituitary gonadotrophs or via the hypothalamus to stimulate endogenous GnRH secretion which then induces its own pituitary receptors (vide infra). It should be emphasised that the changes in GnRH-R could not be accounted for by altered receptor affinity.

Another intriguing and consistent observation was that GnRH-R declined to nadir values coincident with the time of the serum LH rise (Savoy-Moore et al.,

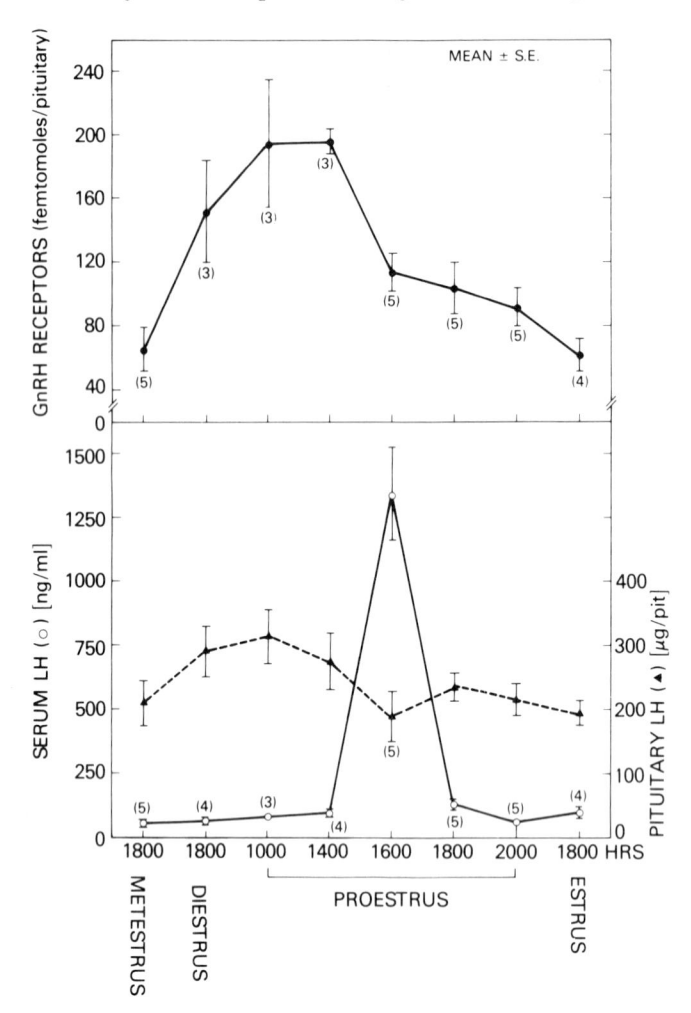

Fig. 1 – Pituitary GnRH receptor content (upper panel) of glands obtained from adult female rats at different stages of a 4-day estrous cycle. Data are mean ± SE of the number of animals shown in parentheses. Bottom panel shows serum LH (o) ng/ml RP-1 and pituitary LH content (▲).

1980; Clayton et al., 1980; Adams and Spies, 1981). Initially this was interpreted as arising from loss of GnRH receptors due to increased occupancy during the LH surge since (i) endogenous GnRH secretion is maximal at this time in the cycle, (ii) blockade of the spontaneous LH surge, and the antecedent endogenous GnRH rise, with pentobarbitone in the early afternoon of pro-estrous prevents this GnRH receptor fall (Barkan et al., 1981) and (iii) 30 min following 225 ng of GnRH intravenously a significant reduction in GnRH-R was observed, though this was 15 min after the serum LH peak (Savoy-Moore *et al.*, 1981). However, GnRH-R reduction was not achieved by the successive intravenous administration of the decapeptide to pentobarbitone blocked rats, in amounts capable of eliciting LH release (Barkan *et al.*, 1981). This casts some doubt on the original interpretation that the acute GnRH receptor fall is a consequence of increased endogenous GnRH secretion. Currently, an acceptable explanation for the abrupt fall in GnRH-R, coincident with the LH surge, awaits further experimentation. The unlikely explanation of an ultra-short feedback of gonadotrophins on GnRH-R is discounted since GnRH-R are elevated following ovariectomy when gonadotrophin values are as high as those at the time of the LH surge.

2.2 Post-partum

In contrast to pro-estrous in rats, or the late follicular phase of the primate menstrual cycle, when the pituitary is 'super-sensitive' to GnRH, the immediate post-partum period is characterised by anestrous and anovulation respectively (Smith and Neill, 1977; Duchen and McNeilly, 1980). Not only is this physiological state associated with reduced gonadotrophin responsiveness to GnRH (Lu *et al.*, 1976), but pulsatility of LH secretion is also lost, implying inhibition of hypothalamic GnRH secretion. Currently, it is uncertain whether absent GnRH secretion during the immediate post-partum period is a direct consequence of the accompanying hyperprolactinaemia or arises from neural input generated by the suckling reflex. It was not surprising, therefore, to find that pituitary GnRH receptor values in glands from lactating rats were only 50% of those in glands from metestrous females, a nadir time during the estrous cycle (Clayton *et al.*, 1980). Thus, a situation of diminished pituitary responsivity to GnRH is accompanied by reduced GnRH-R thereby supporting the concept of a positive correlation between GnRH-R, basal serum gonadotrophin levels, and responsiveness of the gonadotroph to the releasing hormone. It has recently been shown that experimental hyperprolactinaemia can attenuate the GnRH-R and serum LH increase following gonadectomy (Clayton and Bailey, 1982). The implications of this observation are discussed later.

2.3 Ovariectomy

Further support for a physiological role of GnRH receptor regulation comes from studies in ovariectomised rats. GnRH-R are significantly elevated by the second or third day following ovariectomy, though two-fold increases are not apparent until the major increase in serum gonadotrophins, occurring at 4 or 5 days (Clayton and Catt, 1981b, Frager *et al.*, 1981). This receptor increase is prevented by the daily administration of estradiol, progesterone, or estradiol plus progesterone commencing at the time of ovariectomy (Clayton and Catt, 1981b). This provides additional evidence for gonadal steroid regulation of GnRH-R though the site of this interaction, *vis-à-vis* the pituitary, hypothalamus or a combination of the two, has yet to be determined. The GnRH-R increase following ovariectomy is much slower than that following orchidectomy (see Section 2.4), a fact which may be attributable to a significant negative feedback contribution from adrenal steroid production (Rabii and Ganong, 1976). It would be interesting to determine the time course of the GnRH-R increase in combined ovariectomised/adrenalectomised animals.

2.4 Orchidectomy

The removal of androgen negative feedback in adult male rats is associated with

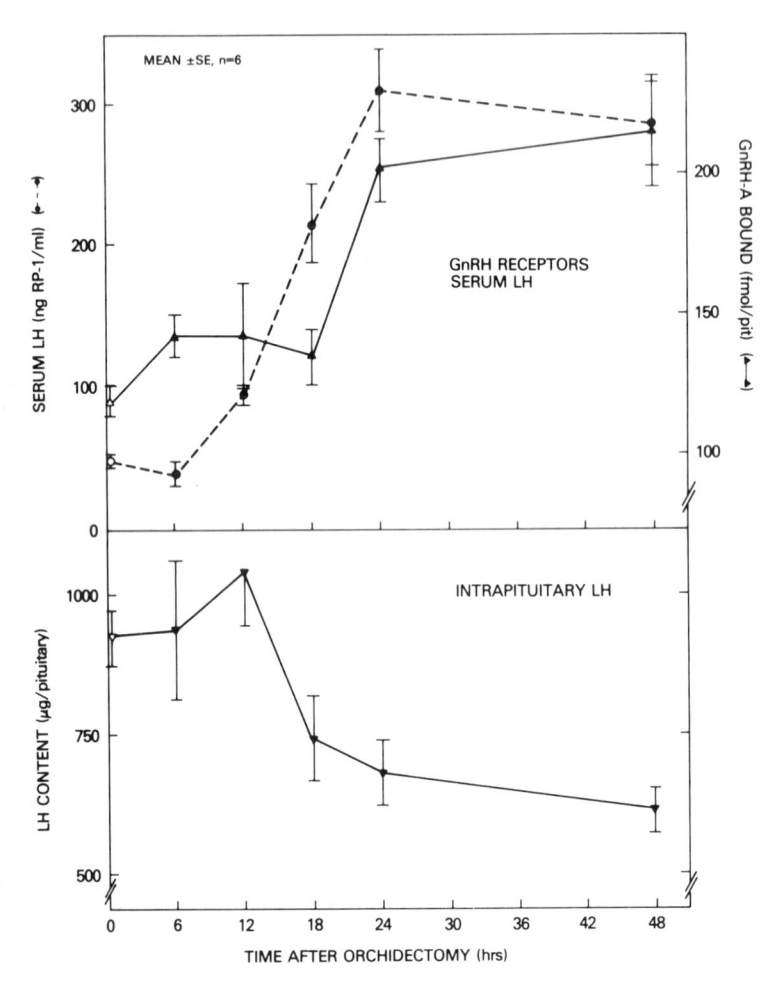

Fig. 2 — GnRH receptor content (▲) and serum LH (ng/ml RP-1 ●) following orchidectomy of adult male rats (upper panel). Pituitary LH is shown in lower panel. Data are mean ± SE of 6 animals at each time.

a significant increase in serum LH between 6 and 12 h post-orchidectomy. However, GnRH-R do not rise significantly until between 18 and 24 h after the operation, indicating a time lag of about 12 h between the gonadotrophin and receptor responses (Fig. 2). A doubling of GnRH-R is consistently observed by 24 h with values increasing only slightly thereafter for the first 10 days post-orchidectomy, (Clayton and Catt, 1981b; Frager et al., 1981; Conne et al., 1982). However, by 14 days post-castration GnRH-R increase by about another 40%, at a time when serum LH values also increase further. Between 14 and 30 days post-orchidectomy there is no further GnRH-R or serum LH increase (Clayton et al., 1982). Replacement of testosterone, dihydrotestosterone, estradiol-17β, but not progesterone, commencing at the time of orchidectomy, prevent both GnRH-R and serum LH increases measured 5 days later. In addition, the elevated GnRH-R values of chronically orchidectomised animals are reduced to those of intact controls following insertion of silastic implants containing testosterone propionate. However, this result takes at least a week to attain with implants that produce near-physiological serum testosterone levels. With supra-physiological serum testosterone values the effect occurs earlier (Clayton and Catt, 1981b). These data indicate that, in the male as in the female, GnRH-R are regulated by physiologically relevant gonadal steroids. Again the site of negative androgen action on GnRH-R is not clear, although a direct action at the pituitary level to suppress gonadotrophin secretion (Hsueh et al., 1979), and GnRH-R levels has been reported (Giguere et al., 1981). This does not preclude an additional level of negative steroid feedback at the hypothalamic level (*vide infra*), and a combination of the two sites of action appears likely.

Another good example of the negative androgen feedback action of GnRH-R is demonstrated during male sexual development (Fig. 3). From high concentrations during the first 25 days of post-natal life GnRH-R fell progressively over the 15-day period during which serum testosterone values steadily increased, to reach mature adult values by 50 days of age. There was a corresponding fall in basal serum LH values. The period of 30–50 days is the stage of puberty and maturation of spermatogenesis in this strain of rats. It remains to be determined whether the decrease in GnRH-R concentration reflects a reduction in the high proportion of gonadotroph cells found in immature rat pituitaries (Denef et al., 1978) or a decrease in the number of receptors per cell in a static population of gonadotrophs. Likewise, it is as yet unclear whether the high pre-pubertal GnRH-R concentration is a reflection of high levels of endogenous GnRH secretion during this time. Whatever the mechanism for the high pituitary GnRH receptor content this finding is consistent with the greater sensitivity of immature pituitaries to GnRH *in vitro* and *in vivo* (Ojeda et al., 1977). A summary of GnRH receptor regulation by gonadal hormones is presented in Table 2.

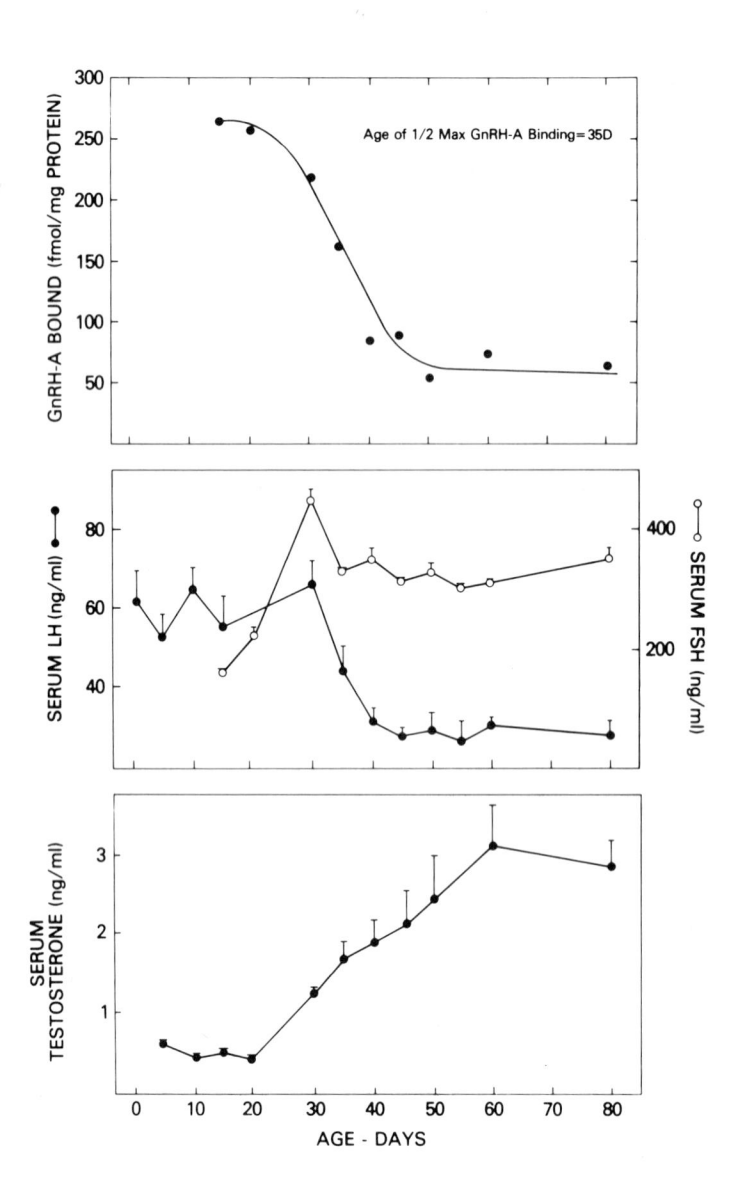

Fig. 3 – Pituitary GnRH receptor concentration (fmol/mg protein) during development of male rats. Receptor data derived from Scatchard analysis of GnRH-A binding to an homogenate of pooled pituitaries at each age, hence no estimate of the error. Serum gonadotrophin and testosterone values are the mean ± SE of not less than 5 separate determinations at each age.

Table 2 — Regulation of pituitary GnRH receptors by gonadal hormones in animals.

---

1. GnRH-R following orchidectomy
   - increase 2–3-fold within 24 h
   - show positive correlation with basal serum gonadotrophin levels
   - increase prevented by testosterone in dose-dependent manner and by 5α-dihydrotestosterone, estradiol and diethylstilbestrol, but not progesterone
2. GnRH-R following ovariectomy
   - 2-fold increase occurs slowly 2–3 days post-castration
   - increase appears just prior to major serum LH rise
   - increase prevented by estradiol, progesterone, and estradiol + progesterone
3. GnRH-R during 4-day estrous cycle
   - increase commences on metestrous evening coincident with rise in serum estradiol
   - levels high throughout diestrous and pro-estrous
   - dramatic fall in GnRH-R coincident with LH surge to nadir values
4. GnRH-R during development
   - males: levels high pre-pubertally and fall with increase in serum testosterone at time of puberty
   - females: high pre-pubertal levels fall *after* fall in gonadotrophins and serum estradiol
5. GnRH-R in lactating rats
   - 50% below values at nadir on estrous/metestrous
6. Induced hyperprolactinaemia
   - attenuated post-gonadectomy receptor and serum LH increase

---

3. MECHANISM OF PHYSIOLOGICAL REGULATION OF PITUITARY GnRH RECEPTORS

From the foregoing physiological data it is apparent that there is a good positive correlation between the pituitary GnRH receptor content and the inferred hypothalamic GnRH secretory activity (summarised in Table 3). This has led to the hypothesis that GnRH-R are regulated by the decapeptide itself (Clayton *et al.*, 1982a, 1982b; Pieper *et al.*, 1982). The concept of autoregulation of peptide hormone receptors has been confirmed for prolactin (Posner *et al.*, 1975) and angiotensin II receptors (Hauger *et al.*, 1978). In the latter instance this regulation may be of physiological importance to the maintenance of sodium balance. However, the following studies are the first to demonstrate that a CNS-derived peptide hormone can regulate the responsiveness of a target tissue

Table 3 — Positive correlation between pituitary GnRH-R and endogenous GnRH secretion

Physiological state	GnRH secretion	GnRH receptors
Gonadectomy — direct measurement of GnRH in hypophyseal portal plasma	↑	↑
Lactation — absent pulsatile LH secretion — low basal serum LH — decreased responsiveness to GnRH	↓	↓

via a process of receptor regulation. Of course, additional steps in the pathway leading to hormone secretion, beyond the membrane receptor site, might also be under direct positive control by GnRH.

The experiments performed to test the hypothesis of GnRH receptor autoregulation employed adult male rats, where the effects of various manipulations on the GnRH-R and serum LH responses 5 or 6 days after orchidectomy were analysed. This experimental model was chosen (1) to avoid cyclical variations in GnRH-R that would complicate data interpretation had female animals been used, and (2) because the responses are easily measurable and highly reproducible in males. The experimental approaches adopted and their rationale are summarised in Table 4.

Considering the first approach, the stereotactic placement of a destructive lesion in the floor of the 3rd ventricle (median eminence lesion (M.E.L.)), prior to orchidectomy, completely prevents the GnRH-R and LH increases 6 days after castration. That this was due solely to elimination of GnRH and no other hypothalamic factor(s) was confirmed by the restoration of castration responses if M.E.L.-bearing orchidectomised rats were replaced with GnRH three times daily (Clayton *et al.*, 1982a). Further, the effect of removing endogenous GnRH rapidly (within 8 days) reduced GnRH-R when their elevation had been well established 10 days post-castration (Fig. 4). Indeed GnRH-R values in animals bearing M.E.L. for longer than one week were 30% lower than those of intact controls, an observation also seen following treatment with a GnRH antiserum (Clayton *et al.*, 1982b). These data indicate that (i) GnRH is essential for the post-castration receptor and serum LH increases (ii) some exposure to GnRH is required in intact male rats to maintain GnRH-R at 'normal' values (*vide infra*).

If the potent GnRH antagonist analogue (N-acetyl Ala^1, DPhe^2, DTrp^{3,6})

Table 4 — Evidence that endogenous GnRH is essential for maintenance of pituitary GnRH receptors.

Rationale for approach	Experimental protocol	Results
1. Removal of GnRH at source	Mechanical destruction of medial basal hypothalamus (M.E.L.)	Placed prior to castration–prevents GnRH-R and serum LH increase. GnRH-R values 30% lower than intact controls. Effect of M.E.L reversed by exogenous GnRH. Reduced GnRH-R values when placed in chronically castrated males.
2. Counteract GnRH action at pituitary receptor sites	Continuous infusion of GnRH antagonist analogue	Prevents post-castration GnRH-R and serum LH rise. Reduced values below intacts due to receptor occupancy. Remaining receptors are 'functional'.
3. Abolition of endogenous GnRH action prior to arrival at pituitary	Immunoneutralisation by passive transfer of GnRH antibody	GnRH-R and serum LH increases post-orchidectomy prevented. GnRH-R values 30% < intact controls. GnRH-R remaining are fully functional. GnRH-R recovery following treatment with antiserum coincides with decline in GnRH binding capacity of the rat serum. GnRH-R increased by concurrent administration of agonist analogue which does not cross-react with anti-GnRH serum.

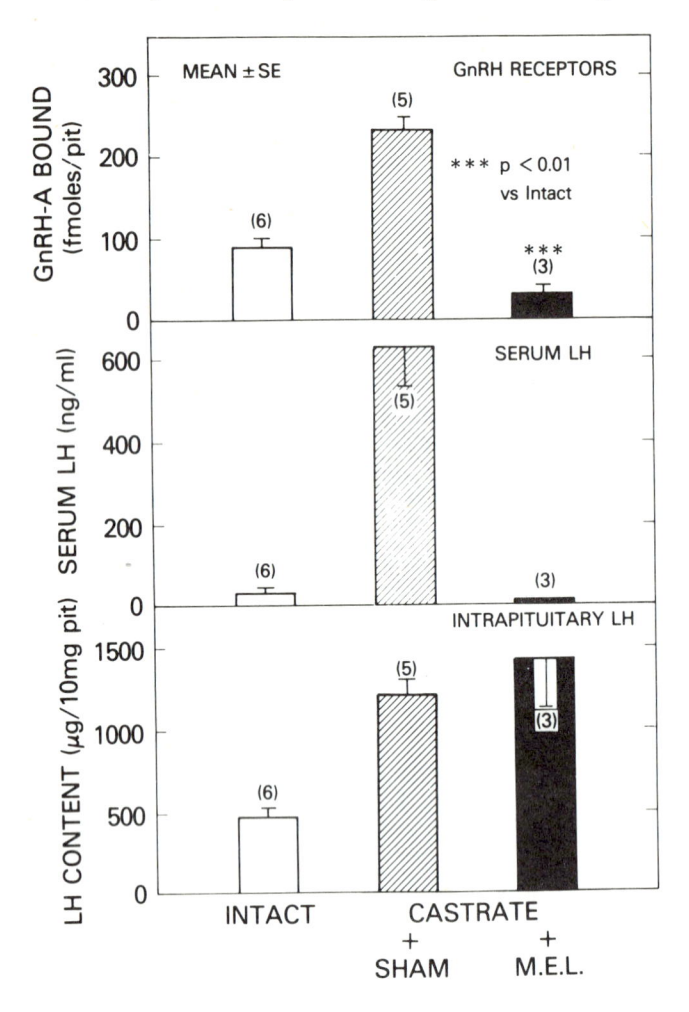

Fig. 4 – Effect of destructive lesion in the median eminence on pituitary GnRH receptors, serum LH and pituitary LH. Lesions were placed 8 days prior to sacrifice and 3 weeks after orchidectomy of adult rats. Values are mean ± SE of the number of animals shown in parentheses.

GnRH is infused continuously from intraperitoneal osmotic minipumps, commencing from the time of orchidectomy, this also prevents the post-castration GnRH-R and serum LH increase. Indeed, GnRH-R values measured are only 25% of intact controls, which almost certainly reflects receptor occupancy by the antagonist, rather than net receptor loss. This is indicated by experiments performed to dissociate antagonist-receptor complexes prior to GnRH receptor measurement. When such manoeuvres are performed, GnRH-R values from

Sec. 3]   Physiological Regulation of Pituitary GnRH Receptors   197

antagonist-infused orchidectomised animals are indistinguishable from those of intact controls, though the post-castration rise in GnRH-R has been prevented (Clayton *et al.*, 1982a). Since antagonist-infused animals can respond with a normal increment in LH release following exogenous GnRH (Clayton *et al.*, 1982a) the remaining receptors are clearly functional, in contrast to the situation when GnRH-R are reduced by a similar magnitude following high dose GnRH agonist infusion but the receptors are unresponsive to GnRH (Clayton,

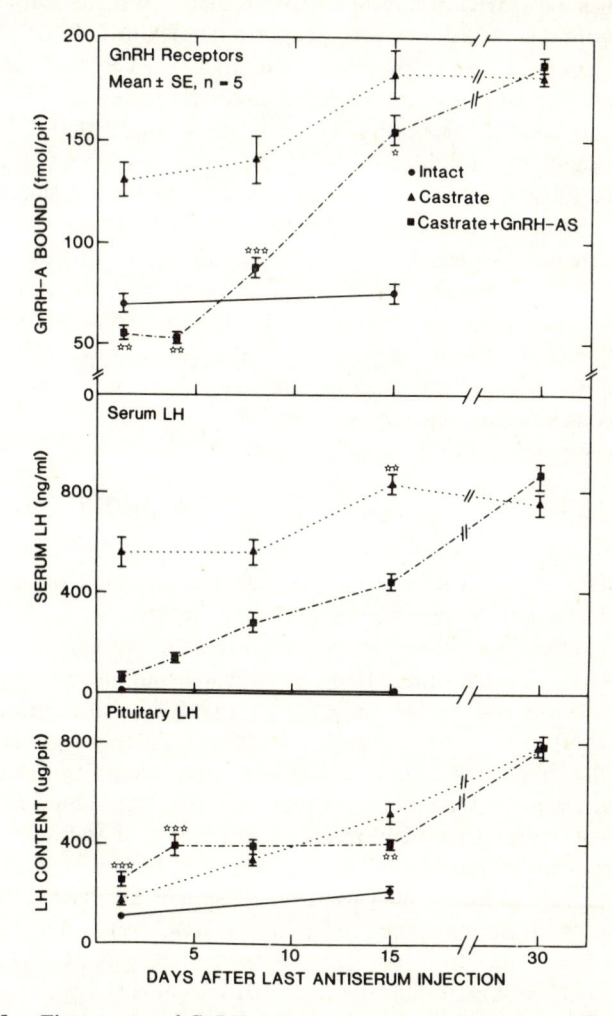

Fig. 5 – Time course of GnRH receptor and serum LH recovery following treatment of orchidectomised rats with 0.25 ml of GnRH antiserum for 6 days. Antiserum treatment commenced at the time of orchidectomy. Groups of 5 animals each were sacrificed at 1, 4, 8, 15 and 30 days after the last antiserum injection. *$p < 0.05$, **$p < 0.02$, ***$p < 0.01$ vs castrate or intact controls.

1982). These experiments show that GnRH interaction with its own receptors is a prerequisite for the increase occurring post-orchidectomy.

The third approach, immunoneutralisation of endogenous GnRH by daily subcutaneous injection of a high affinity antiserum, also prevents the post-orchidectomy responses and further reduces GnRH-R values, by 30% compared with intact controls, in chronically orchidectomised animals. A GnRH agonist analogue, which does not cross-react with this antiserum, can effectively induce GnRH-R, indicating that GnRH receptor autoregulation is shared with functionally similar peptides. When the rate of recovery of GnRH-R is determined following cessation of antiserum administration to orchidectomised rats was determined, it was found that more than 15 days were required before GnRH-R regained castrate control values (Fig. 5). This coincides with the time at which the binding capacity of the animal sera for [^{125}I] GnRH was much reduced.

From the foregoing experiments the results of which are summarised in Table 4, the conclusion is that endogenous GnRH is an absolute requirement for the maintenance and regulation of the concentration of its own receptors in the pituitary. In this manner the hypothalamus can exercise control over the sensitivity and maximum responsiveness of pituitary gonadotroph function. These conclusions are further supported by the data from studies in which exogenous GnRH or an agonist analogue are administered either intermittently or by continuous infusion (*vide infra*).

4. AUTOREGULATION OF GnRH-R BY GnRH AND AGONIST ANALOGUE

The possibility that the 'self-priming' effect of endogenous GnRH might be due to autoinduction of GnRH-R has been considered as a possible explanation for the increased pituitary responsiveness occurring on pro-estrous (*vide supra*). That this is feasible has been demonstrated by induction of GnRH receptors in adult male rats following either frequent subcutaneous injections of GnRH (Frager *et al.*, 1981) or once-daily injection of a GnRH agonist analogue (Clayton *et al.*, 1980). The degree of GnRH receptor induction by this means is slightly less than that seen following orchidectomy, suggesting the possibility that increased androgen secretion, resulting from the exogenous GnRH, might limit the receptor rise. This might imply an interaction of androgen with exogenous GnRH at the pituitary level, a view supported by the finding of a direct inhibitory effect of androgen on GnRH receptors and LH release in pituitary cells incubated with the steroid *in vitro* (Giguere *et al.*, 1981). Further support for this view is available from the observation that exogenous GnRH fails to enhance the acute post-castration rise in GnRH receptors (Frager *et al.*, 1981), perhaps an indication of maximal stimulation by endogenous GnRH. However, the exact site of negative steroid feedback *in vivo* vis-à-vis pituitary or hypothalamus can only be satisfactorily resolved in a model whereby the contribution from endogenous decapeptide is eliminated, and exogenous stimulation provided at a controlled frequency and magnitude. While such a primate model has been extensively studied (Knobil 1980), no rodent equivalent is available. The data from preliminary experiments with exogenous GnRH administration to animals with an intact hypothalamo-pituitary-gonadal axis are consistent with that described in the preceding section, and indicate that receptor induction is also feasible in the female. Whether this phenomenon is a significant physiological component of the rat estrous cycle has yet to be determined.

A number of studies in rodents and primates have clearly shown that the method by which GnRH is administered is crucial for determination of pituitary gonadotroph responsiveness (Belchez et al., 1978). Continuous infusion of the decapeptide causes an initial abrupt rise in serum LH, which then declines progressively as the infusion continues. During the infusion period, sudden increments of GnRH to well above the infusion rate, fail to elicit an LH peak, though administration of other secretagogues is effective in this regard (Smith and Vale, 1981). These data indicate that the pituitary refractoriness or 'desensitisation' induced by GnRH cannot be accounted for by 'exhaustion' of pituitary gonadotrophin stores, and imply a defect in the GnRH-stimulated LH release mechanism. Possible abnormalities include (1) GnRH-induced receptor loss (or 'down-regulation'), possibly resulting from increased processing of occupied receptors in a manner analogous to that described for many other hormones and protein molecules (for review see Catt et al., 1979) or (2) uncoupling of occupied hormone receptors from the intracellular events requiring activation prior to release of gonadotrophins.

To obtain some insight into the mechanism of desensitisation, pituitary refractoriness was induced in adult male rats continuously infused with either GnRH or an agonist analogue (GnRH-A). Continuous infusion of low doses of GnRH to intact rats, surprisingly, increased GnRH-R by 30–40% and achieved serum GnRH values, measured with a highly specific GnRH antiserum, of about 40 pg/ml (< 0.3 nM). These values for GnRH are very similar to those measured in pituitary stalk plasma of rats (Fink, 1976; Eskay et al., 1979; Sarkar and Fink, 1979; Sherwood and Fink, 1980). Basal serum LH values were unchanged by these levels of serum GnRH. Similarly, low doses of GnRH-agonist, producing serum levels of GnRH-A of < 50 pg/ml, also induced larger ($\sim 70\%$) GnRH-R increases in intact (but not castrated) animals (Clayton, 1982) and basal serum LH levels were doubled in the intact animals with GnRH-R increases.

The results of different doses of GnRH infusion to intact or castrated rats are summarised in Table 5. Qualitatively identical results were obtained with GnRH agonist analogue infusions (Clayton, 1982).

Increasing the infusion dose of GnRH or GnRH-A caused a dose-dependent reduction in GnRH-R and basal serum LH levels when administered to chronically (2-week) orchidectomised rats for 6 days. Highest doses of the peptides reduced GnRH-R to values well below those of intact control animals. This reduction could not be completely accounted for by *in vivo* receptor occupancy,

Table 5 — Effect of continuous infusion (for 6 days) of GnRH (or a GnRH agonist analogue) upon GnRH receptors and gonadotrophin levels in intact and castrated male rats.

Infusion regimen	Intact animals	Acute castration	Chronic castration
Low dose GnRH (serum levels <40 pg/ml)			
GnRH-R	↑ 30–40%	↔	↔
Basal LH	↔	↔	↔
Pituitary LH	↔	↔	↔
Intermediate dose GnRH (serum levels ≈ 260 pg/ml)			
GnRH-R	↔	→ to 50% of castrate control	→ to intact control values
Basal LH	Doubled	→ to 40% of castrate control	→ to 40% of castrate control
Pituitary LH	→ 50%	→ to 50% of castrate control	→ to 50% of castrate control
High dose GnRH (serum levels 1600 pg/ml)			
GnRH-R	30% of intact control	→ to 70% of intact control	→ to 50% of intact control
Basal LH	Doubled	→ to 25% of castrate control	→ to 60% of castrate control
Pituitary LH	10% of control	→ to 10% of castrate control	→ to 30% of intact control

↔ = no change from appropriate control group.

since intact control values were not completely restored following dissociation of occupied receptors prior to receptor analysis (Clayton, 1982). Nevertheless, receptor occupancy accounted for about 65% of the measured reduction in GnRH-R, with net receptor loss contributing the remainder. Despite this high degree of receptor occupancy, both basal serum LH and pituitary LH content fell in the castrated animals. This suggests that the occupied receptors were functionally dissociated from the intracellular mechanisms required for normal hormone release and synthesis. The view is supported by the observation that a high dose of GnRH (10 μg s.c.), given towards the end of the GnRH or GnRH-A infusion, completely failed to raise serum LH levels. However, such a GnRH injection, to animals at the end of high dose GnRH antagonist infusion produced a brisk serum LH increase (Clayton, 1982). The dose of antagonist infused produced a similar extent of receptor occupancy (about 80%). Thus, the gonadotrophs's intracellular release mechanisms can clearly distinguish between receptors occupied by a GnRH agonist or antagonist analogue.

Agonist-induced pituitary desensitisation, with 'uncoupling' of receptors from the release mechanism might arise from major conformational changes in the gonadotroph surface membrane. Hormone-receptor complexes are probably internalised in a manner analogous to occupied insulin, $\alpha 2$ macroglobulin and epidermal growth factor receptors. In support of this hypothesis is the observation that GnRH receptors in cultured gonadotrophs first aggregate into clusters and then migrate to the pole of the cell within a very short time of exposure to the ligand (Naor *et al.*, 1981; Hazum *et al.*, 1980). Such processing of H-R complexes might account for receptor loss when the cells are exposed to 'supraphysiological' ligand concentrations, but whether a similar mechanism, applies at physiological concentrations of GnRH remains to be determined.

To date the evidence suggests that at extremes of circulating GnRH concentrations its pituitary receptors can be either induced (low GnRH) or markedly reduced (high GnRH). Qualitatively similar biphasic receptor autoregulation has been obtained with GnRH and analogue treatment of pituitary cells *in vitro* (Loumaye and Catt, 1982). It seems likely that the GnRH receptor number is maintained *in vivo* within a fairly narrow range, and depends on a balance between occupancy-induced processing and the rate of reappearance of vacant receptors on the cell surface. Processing of hormone-receptor complexes and desensitisation may represent a generalised mechanism by which target cells protect themselves from excessive stimulation.

While receptor 'down-regulation' is one factor in pituitary refractoriness to GnRH, it is apparent that the pituitary LH content also falls dramatically in these circumstances. In intact animals, this might possibly be caused by the 2–3-fold higher serum LH levels. This seems unlikely since serum LH values are 10-fold greater 6 days after orchidectomy and the pituitary LH content is not reduced, if anything it may be slightly elevated. Thus, the very low pituitary LH content of glands continuously exposed to exogenous GnRH indicates that reduced gonadotrophin stores may also be a manifestation of receptor loss and/or uncoupling. This implies that GnRH is required for a normal rate of gonadotrophin synthesis as well as release, and that this action of the decapeptide is also mediated through the same receptors.

While such *in vivo* experiments shed some light into the mechanisms of GnRH receptor function, detailed *in vitro* analysis of receptors in cultured pituitary cells is required for a direct and precise correlation between GnRH-R numbers and cell responsiveness. Analysis of GnRH-R in cultured pituitary cells may prove difficult to interpret, since these appear to lose a majority of their binding sites as secretory processes recover over the first few days in culture following enzymatic dispersion (Naor *et al.*, 1980). However, one report indicates that culture of pituitary cells in the presence of androgens is associated with diminished responsiveness to GnRH and a reduction in GnRH-R (Giguere *et al.*, 1981).

5. CONCLUSIONS

The availability of superagonist analogues of GnRH for radio-iodination has enabled the development of accurate and reproducible receptor binding assays for GnRH. GnRH receptors can now be determined in single pituitary glands. GnRH receptor numbers change in many physiological circumstances, without alteration in their affinity. A positive correlation exists between pituitary GnRH receptors, basal gonadotrophin secretion, and levels of GnRH in hypophyseal portal plasma. The concentration of pituitary GnRH receptors is regulated by GnRH itself, and thus their measurement reflects the degree of prior exposure of the pituitary to endogenous GnRH. Positive regulation of GnRH-R by low concentrations of the peptide may contribute to the 'self-priming' effect of the hormone observed during the estrous cycle. Conversely, gonadotroph desensitisation consequent upon repeated exposure to high concentrations of GnRH is the result of GnRH receptor loss, 'uncoupling' of occupied receptors from the release machinery, and depletion of pituitary LH available for release. Thus, under physiological conditions, the gonadotroph response to GnRH represents a finely regulated balance between GnRH receptor induction and processing, both events being regulated primarily from the hypothalamus. In this manner, hypothalamic GnRH secretion forms the final pathway from the central nervous system in the control of gonadotrophin secretion, and hence reproductive function.

ACKNOWLEDGEMENTS

This work was supported, in part, by grants from the Medical Research Council. I am grateful to Miss L. C. Bailey and Mr K. Schumacher for many of the hormone assays.

REFERENCES

Adams, T. E. and Spies, H. G. (1981) Binding characteristics of gonadotrophin-releasing hormone receptors throughout the estrous cycle of the hamster. *Endocrinology* **108**, 2245–2253.

Adams, T. E., Norman, R. L. and Spies, H. G. (1981) Gonadotrophin-releasing hormone receptor binding and pituitary responsiveness in estradiol primed monkeys. *Science* **213**, 1388–1390.

Aiyer, M. S., Fink, G. and Grieg, F. (1974) Changes in the sensitivity of the pituitary gland to luteinising hormone releasing factor during the estrous cycle of the rat. *J. Endocrinol.* **60**, 47–64.

Barkan, A., Regiani, S., Dguncan, J. A. and Marshall, J. C. (1981) Pituitary GnRH receptors during the LH surge in female rats: effects of pentobarbital blockage and GnRH injections. *Abs. No. 820 Program of the 63rd Mtg. of the Endocrine Society, Cincinnati, Ohio. June 17–19 1981.*

Barraclough, C. A. and Wise, P. M. (1982) The role of catecholamines in the regulation of pituitary luteinising hormone and follicle-stimulating hormone secretion. *Endocrine Reviews* **3**, 91–119.

Belchetz, P. E., Plant, T. M., Nakai, Y., Keogh, E. J. and Knobil, E. (1978) Hypophyseal responses to continuous and intermittent delivery of hypothalamic gonadotrophin-releasing hormone. *Science* **202**, 631–633.

Ben-Jonathan, N., Mical, R. S. and Porter, J. C. (1973) Superfusion of hemipituitaries with portal blood I. LRF secretion in castrated and diestrous rats. *Endocrinology* **93**, 497–503.

Bohnet, H. G., Dahlen, H. G., Wuttke, W. and Schneider, H. P. G. (1976). Hyperprolactinaemic anovulatory syndrome. *J. Clin. Endo. Metab.* **42**, 132–143.

Catt, K. J., Harwood, J. P., Aguilera, G. and Dufau, M. L. (1979) Hormonal regulation of peptide receptors and target cell responses. *Nature* **280**, 109–116.

Clayton, R. N. (1982) Gonadotrophin releasing hormone modulation of its own pituitary receptors: evidence for biphasic regulation. *Endocrinology* **111**. 152–161.

Clayton, R. N. and Bailey, L. C. (1982) Hyperprolacteinaemia attentuates the gonadotrophin releasing hormone receptor response to gonadectomy. *J. Endocrinol.* **95**, 267–274.

Clayton, R. N. and Catt, K. J. (1981a) Gonadotrophin releasing hormone receptors: characterisation, physiological regulation and relationship to reproductive function. *Endocrine Reviews* **2**, 186–209.

Clayton, R. N. and Catt, K. J. (1981b) Regulation of pituitary gonadotrophin releasing hormone receptors by gonadal hormones. *Endocrinology* **108**, 887–895.

Clayton, R. N., Shakespear, R. A., Duncan, J. A. and Marshall, J. C. with Appendix by P. J. Munson and D. Rodbard (1979) Radioiodinated nondegradable gonadotrophin releasing hormone analogs: new probes for the investigation of pituitary gonadotrophin releasing hormone receptors. *Endocrinology* **105**, 1369–1381.

Clayton, R. N., Solano, A. R., Garcia-Vela, A., Dufau, M. L. and Catt, K. J. (1980) Regulation of pituitary receptors for gonadotrophin releasing hormone during the rat estrous cycle. *Endocrinology* **107**, 699–706.

Clayton, R. N., Channabasavaiah, K., Stewart, J. M. and Catt, K. J. (1982a) Hypothalamic regulation of pituitary gonadotrophin-releasing hormone receptors I. effects of hypothalamic lesions and a gonadotrophin releasing hormone antagonist. *Endocrinology* **110**, 1108–1115.

Clayton, R. N., Popkin, R. M. and Fraser, H. M. (1982b) Hypothalamic regulation of pituitary GnRH receptors II: effects of gonadotrophin-releasing hormone immunoneutralisation. *Endocrinology* **110**, 1116–1123.

Conn, P. M., Marian, J., McMillan, M., Stern, J., Rogers, D., Hamby, M., Penna, A. and Grant, E. (1981) Gonadotrophin releasing hormone action in the pituitary a three step mechanism. *Endocrine Reviews* **2**, 174–185.

Conne, B. S., Scaglioni, S., Lang, U., Sizanenko, P. C. and Aubert, M. L. (1982) Pituitary receptor sites for gonadotrophin-releasing hormone: effect of castration and substitutive therapy with sex-steroids in the male rat. *Endocrinology* **110**, 70–79.

Denef, C., Hautekeete, E., de Wolf, A. and Banderschuren, B. (1978) Pituitary basophils from immature male and female rats: distribution of gonadotrophs and thyrotrophs as studied by unit gravity sedimentation. *Endocrinology* **103**, 724–735.

Drouin, J. and Labrie, F. (1976) Selective effect of androgens on LH and FSH release in anterior pituitary cells in culture. *Endocrinology* **98**, 1528–1534.

Drouin, J., Legace, L. and Labrie, F. (1976) Estradiol induced increase of the LH responsiveness to LH releasing hormone (LHRH) in rat anterior pituitary cell cultures. *Endocrinology* **99**, 1477–1481.

Duchen, M. R. and McNeilly, A. S. (1980) Hyperprolactinaemia and long-term lactational amenorrhoea. *Clin. Endocrinol.* **12**, 621–627.

Eskay, D. L., Mical, R. S. and Porter, J. C. (1979) Relationship between luteinising hormone releasing hormone concentration in hypophysial portal blood and luteinising hormone release in intact castrated and electrochemically-stimulated rats. *Endocrinology* **100**, 263–270.

Ferland, L., Marchetti, B., Sequin, C., Lefebvre, F. A., Recres, J. T. and Labrie, F. (1981) Dissociated changes of pituitary luteinising hormone releasing hormone (LHRH) receptors and responsiveness to the neurohormone induced by 17β-estradiol and LHRH in vivo in the rat. *Endocrinology* **109**, 87–93.

Fink, G. (1976) The development of the releasing factor concept. *Clin. Endocrinol.* **5** (Suppl.) 245s–260s.

Fink, G. (1979) Neuroendocrine control of gonadotrophin secretion. *Brit. Med. Bull.* **35**, 155–159.

Frager, M. S., Pieper, D. R., Tonetta, S., Duncan, J. A. and Marshall, J. C. (1981) Pituitary gonadotrophin-releasing hormone receptors: effects of castration, steroid replacement and the role of GnRH in modulating receptors in the rat. *J. Clin. Invest.* **67**, 615–623.

Giguere, V., Lefebvre, F. A. and Labrie, F. (1981) Androgens decrease LHRH binding sites in rat anterior pituitary cells in culture. *Endocrinology* **108**, 350–352.

Gordon. J. H. and Reichlin, S. (1974) Changes in pituitary responsiveness to luteinising hormone releasing factor during the rat estrous cycle. *Endocrinology* **94**, 974–978.

Hauger, R. L., Aguilera, G. and Catt, K. J. (1978) Angiotensin II regulates its receptor sites in the adrenal glomerulosa zone. *Nature* **271**, 176–178.

Hazum, E., Cuatrecasas, P., Marian, J. and Conn, P. M. (1980) Receptor mediated internalisation of fluorescent gonadotrophin releasing hormone by pituitary gonadotrophs. *Proc. Natl. Acad. Sci. USA* **77**, 6692–6695.

Hsueh, A. J. W., Erickson, G. F. and Yen, S. S. C. (1979) The sensitising effect of estrogens and catechol estrogen on cultured pituitary cells to luteinising hormone-releasing hormone: its antagonism by progestins. *Endocrinology* **104**, 807–813.

Knobil, E. (1980) The neuroendocrine control of the menstrual cycle. *Rec. Prog. Horm. Res.* **36**, 53–88.

Lagace, L., Massicotte, J. and Labrie, F. (1980) Acute stimulatory effects of progesterone on luteinising hormone and follicle stimulating hormone release in rat anterior pituitary cells in culture. *Endocrinology* **106**, 684–689.

Loumaye, E. and Catt, K. J. (1982) Homologous regulation of gonadotrophin releasing hormone receptors in cultured pituitary cells. *Science* **215**, 983–985.

Lu, K. H., Chen, H. T., Grandison, L., Huang, H. H. and Meites, J. (1976) Reduced luteinising hormone release by synthetic luteinising hormone releasing hormone (LHRH) in post-partum lactating rats. *Endocrinology* **98**, 1235–1240.

McNeilly, A. S. (1980) Prolactin and the control of gonadotrophin secretion in the female. *J. Reprod. Fertil.* **58**, 537–549.

Naor, Z., Clayton, R. N. and Catt, K. J. (1980) Characterisation of gonadotrophin-releasing hormone receptors in cultured rat pituitary cells. *Endocrinology* **107**, 1144–1152.

Naor, Z., Atlas, D., Clayton, R. N., Forman, D. S., Amsterdam, A. and Catt, K. J. (1981) Fluorescent derivative of gonadotrophin releasing hormone: visualisation of hormone-receptor interaction in cultured pituitary cells. *J. Biol. Chem.* **256**, 3049–3052.

Naor, Z., Childes, G. V., Leifer, A. M., Clayton, R. N., Amsterdam, A. and Catt, K. J. (1982) Gonadotrophin releasing hormone binding and activation of enriched population of pituitary gonadotrophs. *Mol. Cell. Endocrinol.* **25**, 85–97.

Ojeda, S. R., Jameson, H. S. and McCann, S. M. (1977) Developmental changes in pituitary responsiveness to luteinising hormone releasing hormone (LHRH) in the female rat: ovarian-adrenal influence during the infantile period. *Endocrinology* **100**, 440–451.

Peiper, D. R., Gala, R. R., Regiani, S. R. and Marshall, J. C. (1982) Dependence of pituitary gonadotrophin releasing hormone receptors on GnRH secretion from the hypothalmus. *Endocrinology* **110**, 749–753.

Pickering, A. J. M. C. and Fink, G. (1979) Variation in size of the readily releasable pool of luteinising hormone during the estrous cycle of the rat. *J. Endocrinol.* **83**, 53–59.

Posner, B. I., Kelly, P. A. and Friesen, H. G. (1975) Prolactin receptors in rat liver: possible induction by prolactin. *Science* **188**, 57–59.

Rabii, J. and Ganong, W. F. (1976) Response of plasma estradiol and plasma LH to ovariectomy, ovariectomy plus adrenalectomy and estrogen injection at various ages. *Neuroendocrinol.* **20**, 270–281.

Sarkar, D. K. and Fink, G. (1979) Effects of gonadal steroids on output of luteinising hormone in pituitary stalk blood from female rats. *J. Endocrinol.* **80**, 303–313.

Sarkar, D. K., Chiappa, S. A. and Fink, G. (1976) Gonadotrophin releasing hormone surge in proestrous rats. *Nature* **264**, 461–463.

Savoy-Moore, R., Schwartz, N. B., Duncan, J. A. and Marshall, J. C. (1980) Pituitary gonadotrophin releasing hormone receptors during the rat estrous cycle. *Science* **209**, 942–944.

Savoy-Moore, R. T., Schwartz, N. B., Duncan, J. A. and Marshall, J. C. (1981) Pituitary GnRH receptors on proestrus: effect of pentobarbital blockade of ovulation in the rat. *Endocrinology* **109**, 1360–1364.

Sherwood, N. M. and Fink, G. (1980) Effect of ovariectomy and adrenalectomy on luteinising hormone-releasing hormone in pituitary stalk blood from female rats *Endocrinology* **106**, 363–367.

Smith, M. S. and Neill, J. D. (1977) Inhibition of gonadotrophin secretion during lactation in the rat: relative contribution of suckling and ovarian steroids. *Biol. Reprod.* **17**, 255–261.

Smith, M. A. and Vale, W. W. (1981) Desensitisation to gonadotrophin releasing hormone observed in superfused pituitary cells on cytodex beads. *Endocrinology* **108**, 752–759.

Yen, S. S. C., Tsai, C. C., Vandenberg, G. and Rebar, N. (1972) Gonadotrophin dynamics in patients with gonadal dysgensis: A model for the study of gonadotrophin regulation. *J. Clin. Endo. Metab.* **35**, 897–904.

11

The role of membrane phospholipids in receptor transducing mechanisms

Michael J. Berridge

ARC Unit of Invertebrate Chemistry and Physiology, Department of Zoology, University of Cambridge, Downing Street, Cambridge, U.K.

Abbreviations used

Con A	—	concanavalin A
DG	—	diacylglycerol
DPI	—	diphosphoinositide
GTP	—	guanosine triphosphate
IDP	—	inositol diphosphate
IMP	—	inositol monophosphate
ITP	—	inositol triphosphate
PC	—	phosphatidylcholine
PDGF	—	platelet derived growth factor
PE	—	phosphatidylethanolamine
PI	—	phosphatidylinositol
TPI	—	triphosphoinositide

1. INTRODUCTION

The phospholipids which make up the membrane bilayer contribute in many ways to the operation of cell surface receptors. One obvious role is to provide a structural matrix in which these receptors are embedded. Another function of this matrix is to provide a physical barrier which means that chemical signals detected on the outside must somehow pass through this hydrophobic layer in order to influence processes inside the cell. A specific GTP-binding protein, which lies within this hydrophobic domain, functions to couple receptors on the outside of the membrane to adenylate cyclase located on the cytoplasmic surface and thus provides a good example of how information moves across this barrier. There are some indications that this adenylate cyclase system may have fairly specific requirements for particular phospholipids. For example, removal of phosphatidylinositol with a specific phospholipase C results in a reduction in the basal level of adenylate cyclase (Panagia et al., 1981). Other membrane enzymes, especially those concerned with ion transport, also have specific requirements for certain phospholipid species (Roelofsen, 1981). Such observations suggest that membrane phospholipids may not function just as a physical barrier but may play a rather specific role in various membrane functions including signal transduction.

Another reason for suspecting that membrane phospholipids play a dynamic role in receptor mechanisms is supported by the observation that certain phospholipids undergo pronounced metabolic changes upon activation of certain receptors. The two major changes which have been described are a methylation reaction whereby phosphatidylethanolamine (PE) is converted to phosphatidylcholine (PC) and a phosphatidylinositol response during which phosphatidylinositol (PI) is hydrolysed to diacylglycerol (DG). These metabolic alterations of membrane phospholipids have assumed great importance because they appear to be an integral part of the receptor mechanisms responsible for generating second messengers such as calcium, cyclic GMP and the prostaglandins. This link between phospholipid metabolism and second messenger formation may have very considerable clinical importance because such receptor mechanisms have been implicated in numerous phenomena including inflammation, hypertension, mental retardation resulting from galactosemia, control of cell growth and manic-depressive illness. These clinical aspects will be described later following a more detailed description of some basic aspects of the methylation and phosphatidylinositol responses.

2. METHYLATION RESPONSE

The basic observation underlying the methylation response is the sequential methylation of phosphatidylethanolamine (PE) to phosphatidyl-N-monomethylethanolamine and then to phosphatidylcholine (PC) by means of two lipid transmethylating enzymes which are embedded in the membrane. The orientation of these enzymes is such that during the conversion of PE to PC the lipid is flipped from the cytoplasmic to the outer membrane leaflet. This metabolic transformation of PE to PC is triggered by a variety of agonists such as the action of isoproterenol on reticulocytes (Hirata et al., 1979b), IgE and Con A on mast cells and leukemic basophils (Crews et al., 1980; Hirata et al., 1979a; Ishizaka et al., 1980, 1981), glucagon, vasopressin and angiotensin II on rat hepatocytes (Castaño et al., 1980, Alemany et al., 1981) and vasopressin on the rat pituitary (Prasad and Edwards, 1981). While there appears to be growing evidence for the existence of a receptor-mediated methylation response in many different cell types it is difficult to establish the functional importance of this response.

The first point to appreciate about this methylation response is that it is usually of very small magnitude and Vance and Kruijff (1980) have argued that the metabolic transformations are far too small to induce any functional change in the membrane. However, small changes in phospholipids within local domains may be sufficient to induce significant perturbations in the membrane. The fact that the methylation response is induced by a variety of hormones and related signal molecules has led to the suggestion that it may play some role in the generation of second messengers (Hirata and Axelrod, 1980; Ishizaka et al., 1980). The problem is to explain how the conversion of PE to PC results in the activation of adenylate cyclase or the opening of calcium channels. One idea is that as the phospholipids are flip-flopped within the membrane they may alter membrane fluidity which may then effect 'transmembrane signalling' (Hirata and Axelrod, 1980; Axelrod et al., 1981). In considering this possible involvement in 'transmembrane signalling' it has not been established yet whether the methylation response plays a direct role in generating second messengers or whether it functions indirectly to modulate other mechanisms. For example, the proposed change in fluidity may provide more favourable conditions for receptors to activate adenylate cyclase. Another point to consider is whether the methylation response might be activated by second messengers instead of being responsible for their generation. Alemany et al. (1981) have demonstrated that methylation is secondary to an increase in the intracellular level of calcium. In the case of blood platelets, the methylation response has no role to play in inducing aggregation (Cordasco et al., 1981; Randon et al., 1981) which suggests that there is no obligatory link between methylation of PE and the generation of second messengers.

An interesting aspect of the possible involvement of this methylation response in cellular control is that some processes may be regulated by a decrease rather than an increase in methylation. For example, the chemotactic factor f-Met-leu-phe induces a fall in phospholipid methylation (Hirata et al., 1979a; Pike et al., 1979). This decrease in methylation may be associated with the activation of phospholipase A_2 which uses PC as a substrate for the release of arachidonic acid.

Such observations have led to the idea that one function of the methylation response may be to provide PC to the outer leaflet of the membrane where it can be degraded by phospholipase A_2 (Hirata and Axelrod, 1980). It is clear that the methylation response has been implicated in a large number of membrane processes but much more needs to be done in order to establish the functional significance of this response.

3. THE PHOSPHATIDYLINOSITOL RESPONSE

The hydrolysis of phosphatidylinositol (PI) is an important metabolic event associated with the activation of a certain class of receptors which use calcium as a second messenger (Michell, 1975, 1979; Berridge, 1980, 1981; Michell and Kirk, 1981; Putney, 1981). Receptors may be classified on the basis of the transducing mechanism which they employ to carry out their effect on the cell. Some receptors, such as the nicotinic cholinergic receptor, are linked directly to ion channels to induce rapid phasic fluctuations in membrane potential. Other receptors function to generate second messengers such as the cyclic nucleotides and calcium, by activating specific biochemical pathways. There are those receptors (β-adrenergic, H_2 histaminergic, V_2 vasopressin) which act by stimulating adenylate cyclase whereas another class of receptors (α_1-adrenergic, H_1 histaminergic, V_1 vasopressin and muscarinic cholinergic) bring about a profound alteration in the metabolism of PI. Since the latter receptor class also act through calcium, it has been proposed that this PI response is responsible for generating this calcium signal and this aspect will be dealt with later following a description of agonist-dependent PI metabolism.

3.1 Phosphoinositide metabolism

The metabolism of PI is complicated by the fact that this phospholipid can be further phosphorylated to give the polyphosphoinositides (Hawthorne and White, 1975). There are kinases which add an extra phosphate group to the 4-position of the inositol headgroup of PI to form diphosphoinositide (DPI) and a further phosphate group to the 5-position to form triphosphoinositide (TPI) (Fig. 1). Although these polyphosphoinositides represent only about 10% of the PI, they turn over rapidly and may play a crucial role in the agonist-dependent PI response. The basis of this response is that agonists stimulate hydrolysis of the inositol phosphate headgroup. There are specific phosphodiesterases which cleave the diester bond to release the corresponding inositol phosphates (Fig. 1) with diacylglycerol remaining in the membrane. It is this hydrolytic cleavage which is stimulated by agonists. The problem is to determine which of the phosphoinositides is the immediate substrate for the action of the agonist. Previously it was thought that agonists acted to stimulate the hydrolysis of PI to liberate inositol-1,2-cyclic phosphate and diacylglycerol but more recent observations on cells which have been prelabelled with ^{32}P have shown that agonists stimulate a more rapid reduction in the levels of labelled DPI and TPI (Akhtar and Abdel-Latif, 1980; Michell et al., 1981) suggesting that the primary receptor-mediated event is the breakdown of these polyphosphoinositides. However, since all these phosphoinositides are in equilibrium with each other any reduction in TPI, for example, will result in a decline in DPI and PI as these are converted to TPI by the kinase enzymes. In this way, PI will be destroyed indirectly as it is used up to replace the TPI.

One way of trying to decide which of the phosphoinositides is used is to study the nature of the inositol phosphates which are released during receptor activation. In brain homogenates, acetylcholine stimulates the formation of inositol monophosphate (IMP), inositol diphosphate (IDP) and inositol tri-

The Phosphatidylinositol Response

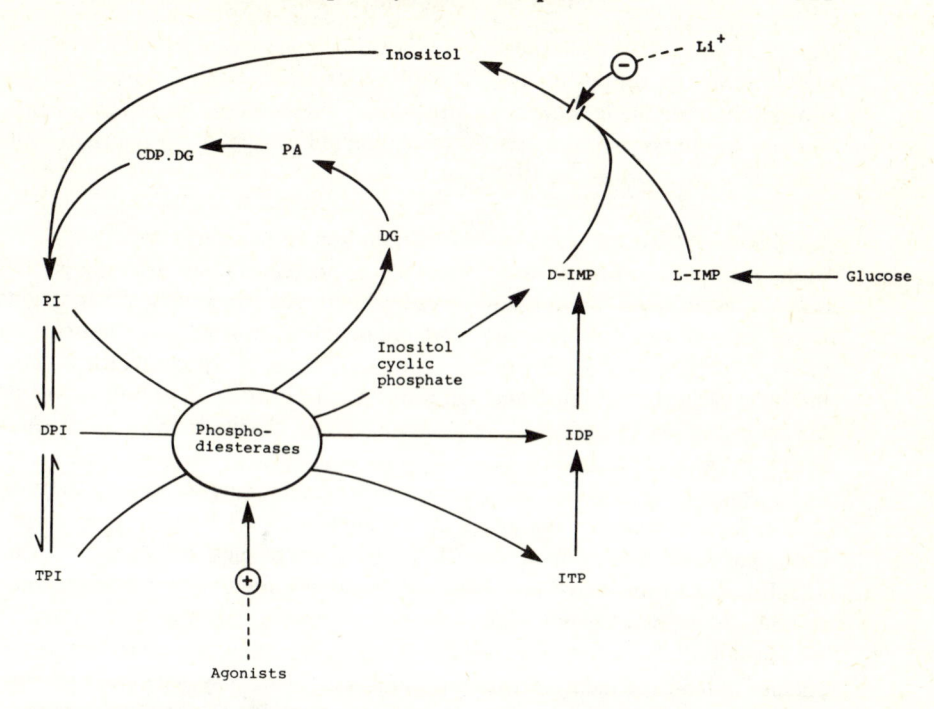

Fig. 1 – A summary of agonist-dependent phosphatidylinositol (PI) metabolism. PI exists in equilibrium with the polyphosphoinositides DPI (diphosphoinositide) and TPI (triphosphoinositide). Agonists stimulate phosphodiesterases which hydrolyse these phospholipids to diacylglycerol (DG) with the release of inositol cyclic phosphate from PI, inositol diphosphate (IDP) from DPI and inositol triphosphate (ITP) from TPI. These inositol phosphates are all converted to D-*myo*-inositol phosphate (D-IMP) which is then dephosphorylated to inositol. Glucose is converted to the L-enantiomer (L-IMP) which is dephosphorylated to inositol by the same phosphatase. Lithium inhibits this formation of inositol and thus interferes with the subsequent resynthesis of PI from DG. The latter is converted to phosphatidic acid (PA) and CDP-diacylglycerol (CDP.DG) which then recombines with inositol to reform PI.

phosphate (ITP) which is consistent with the possible involvement of the polyphosphoinositides (Durell *et al.*, 1968). When applied to iris smooth muscle, acetylcholine stimulates an increase in the release of ITP (Akhtar and Abdel-Latif, 1980) which further suggests that TPI might be the primary substrate for the agonist-dependent event. The ITP is converted to IDP which is further dephosphorylated to IMP and then to inositol (Fig. 1). The inositol-1-phosphatase, which is the final enzyme in this dephosphorylation sequence, is strongly inhibited by lithium (Hallcher and Sherman, 1980). By reducing the activity of this enzyme, lithium will severely curtail the turnover of inositol leading to a reduction in the level of inositol with a corresponding accumulation of IMP

(Sherman et al., 1981; Berridge et al., 1982). It has been argued that such a reduction in the supply of inositol could reduce the level of phosphatidylinositol in the membrane which, in turn, might reduce the effectiveness of those receptor mechanisms which use this phospholipid as part of a signal transducing mechanism (Berridge et al., 1982).

3.2 Relationship of phosphoinositide metabolism to calcium signalling

While the ability of agonists to stimulate an increase in the hydrolysis of PI is firmly established, there is still considerable doubt concerning the functional significance of this change in phospholipid metabolism. A unique feature of this PI response is that it is always associated with those receptors which function through calcium. The important question which remains unresolved is whether the hydrolysis of PI is directly responsible for the change in membrane permeability to calcium, as first proposed by Michell (1975), or whether it is a parallel event responsible for regulating some other membrane process (Cockcroft, 1981). So far, most of the evidence for implicating PI hydrolysis in calcium gating has been rather circumstantial. The pharmacological evidence has clearly established that these two processes are always triggered upon the activation of certain receptor types. However, there are a number of ways of establishing such a close relationship. For example, the two processes may be parallel but independent events, the hydrolysis of PI may gate calcium or the gating of calcium may then lead to the hydrolysis of PI. The sensitivity of the PI response to calcium thus becomes an important criterion for distinguishing between some of these alternatives. If the hydrolysis of PI is dependent upon calcium then it should be possible to abolish this response by stimulating cells in a calcium-free solution. Unfortunately, the sensitivity of the PI response to calcium seems to vary considerably from tissue to tissue. In some tissues the hydrolysis of PI is independent of calcium (Jones and Michell, 1975, 1978; Cockcroft and Gomperts, 1979; Jafferji and Michell, 1976; Kirk et al., 1978; Fain and Berridge, 1979). On the other hand, there are other tissues, such as neutrophils (Cockcroft et al., 1980a, 1980b), where calcium seems to be responsible for initiating the PI response. Such contradictory observations are confusing and have led Cockcroft et al. (1980a, 1980b) to question the notion that the hydrolysis of PI has anything to do with calcium signalling. However, it may be somewhat premature to reject the calcium hypothesis on the grounds of its variable sensitivity to calcium because the effect of calcium on PI metabolism is extremely complicated and open to a number of alternative interpretations. In addition to affecting the enzymes responsible for phosphoinositide hydrolysis, calcium can also influence the synthesis of PI (Berridge and Fain, 1979; Egawa et al., 1981; Gibson and Brammer, 1981; Smith and Hauser, 1981). Because calcium can influence so many aspects of PI metabolism it is difficult to interpret experiments where the availability of calcium is altered. Another complicating factor arises from the fact that the hydrolysis of PI is not only mediated

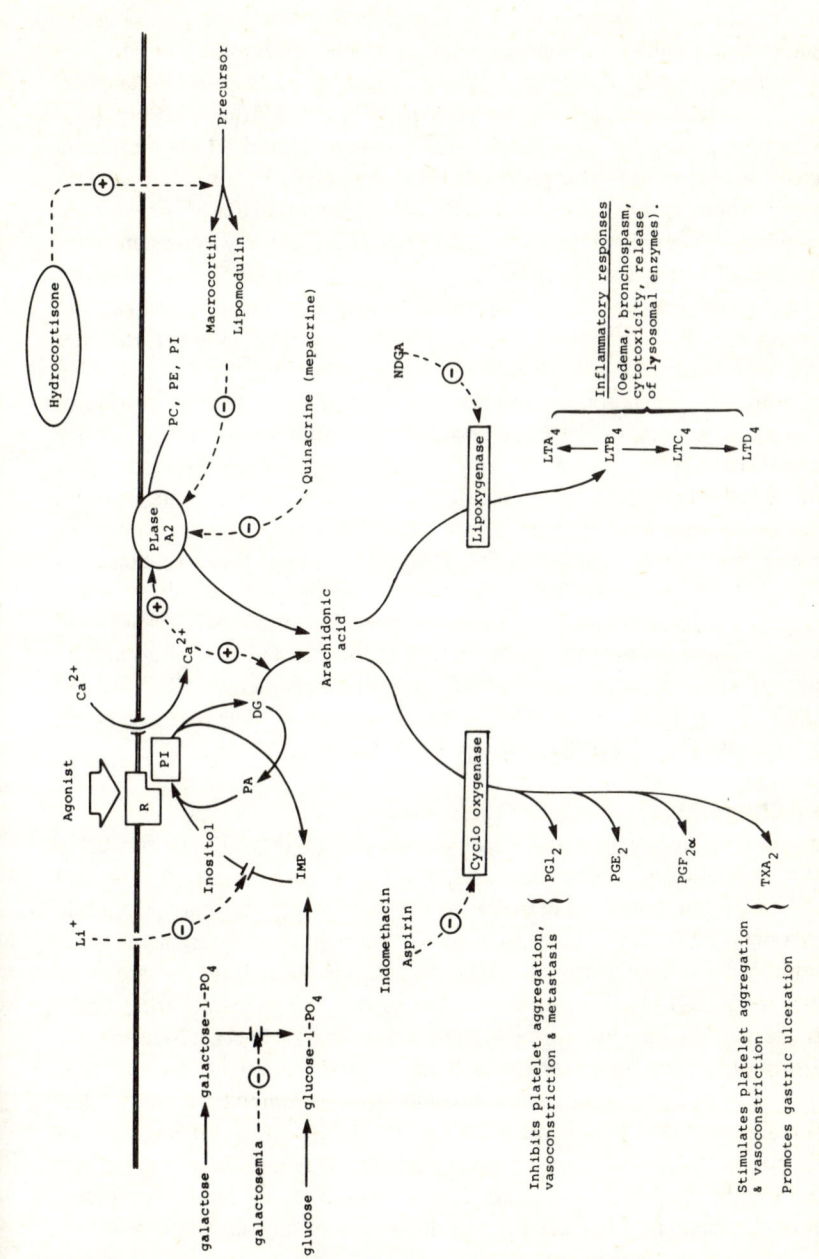

Fig. 2 – The relationship between the PI response and arachidonic acid metabolism. Agonists acting on specific receptors (R) stimulate the hydrolysis of phosphatidylinositol (PI) to diacylglycerol (DG) and inositol-1-phosphate (IMP). A diacyglycerol lipase can remove arachidonic acid from DG. Arachidonic acid can also be released from various phospholipids by phospholipase A_2 (PLase A_2) which is inhibited by quinacrine. Both enzymes are dependent on calcium which may enter the cell as part of the PI response. Metabolism of arachidonic acid via the cyclooxygenase system is inhibited by aspirin and by indomethacin whereas the lipoxygenase pathway is blocked by nordihydroguaiaretic acid (NDGA). Some of the products of arachidonic metabolism are prostacyclin (PGI_2), prostaglandin E_2 (PGE_2), prostaglandin F_2 ($PGF_{2\alpha}$), thromboxane A_2 (TXA_2) and a series of leukotrienes (LTA_4, LTB_4, LTC_4 and LTD_4).

by a phosphodiesterase (Fig. 1) to liberate diacyglycerol but can also be the substrate for phospholipase A_2 which is a calcium-dependent enzyme (Fig. 2). It is conceivable, therefore, that there is an initial calcium-independent PI response responsible for generating a calcium signal which then activates phospholipase A_2 to induce a further breakdown of PI and other phospholipids (Fig. 2). Such a sequential calcium-independent event followed by a calcium-dependent breakdown of phospholipids may be important in cells such as neutrophils, blood platelets and mast cells which are particularly active in releasing arachidonic acid. Studies on blood platelets are entirely consistent with this hypothesis because the phosphodiesterase (phospholipase C) was relatively independent of calcium in contrast to phospholipase A_2 which required calcium (Billah et al., 1980). Calmodulin antagonists also reveal that these two enzymes may be regulated independently of each other with phospholipase A_2 functioning primarily to release arachidonic acid (Walenga et al., 1981). In order to resolve the calcium dependency of the PI response, future studies must aim to clearly establish which enzyme system is being activated by the agonist and which enzyme is activated secondarily as a result of a change in calcium concentration.

In the final analysis, the most direct way of showing whether or not PI hydrolysis is responsible for calcium signalling is to find out the molecular basis of how the two process are linked. While considerable progress has been made in characterising the hydrolysis of PI, there has been less progress in establishing the nature of the calcium gate. One idea is that the ionophore is phosphatidic acid which is generated during the hydrolysis of PI (Michell et al., 1977; Putney et al., 1980; Salmon and Honeyman, 1980; Barritt et al., 1981. Harris et al., 1981; Ohsako and Deguchi, 1981; Putney, 1981; Putney et al., 1980).

3.3 The multifunctional role of the PI response

The receptors which function to hydrolyse phosphatidylinositol are also those responsible for releasing arachidonic acid, for activating guanylate cyclase and for activating protein kinase C (Berridge, 1981). The PI response may play a role in the two main pathways for releasing arachidonic acid (Fig. 2). It may provide the calcium signal necessary to activate phospholipase A_2 or it may play a more direct role by generating diacylglycerol which is then the substrate for a lipase which strips off the arachidonic acid. The latter is then metabolised via the cylclooxygenase or the lipoxygenase systems to generate a wide range of very potent substances (e.g prostaglandins, prostacyclins, thromboxanes and leukotrienes) (Wolfe, 1982). These eicosanoids have a wide spectrum of action including inflammation, induction of labour, dysmenorrhea, sodium excretion and control of the circulatory system. The clinical significance of the prostaglandins is immediately apparent from the therapeutic action of aspirin which is widely used for its analgesic, antirheumatic and antipyretic properties. Aspirin exerts its therapeutic action by inhibiting the cyclooxygenase enzyme which generates the prostaglandins, prostacyclins and thromboxanes (Fig. 2).

The therapeutic action of glucocorticoids in suppressing both acute and chronic inflammation may also depend upon an action on arachidonic acid metabolism. The steroid acts on target cells, such as the neutrophils, to stimulate the release of proteins such as *lipomodulin* (Hirata, 1981) or *macrocortin* (Flower, 1981) which act to inhibit phospholipase A_2 and thus curtail the release of arachidonic acid (Fig. 2). Whether or not these inhibitory proteins will also prevent the release of arachidonic acid from diacylglycerol has not been established.

The release of arachidonic acid and its conversion to various metabolites may also be responsible for the stimulation of guanylate cyclase (Takai *et al.*, 1981). The receptors which act by stimulating the hydrolysis of PI are also capable of increasing the formation of cyclic GMP. However, the way in which receptors are linked to guanylate cyclase has not been resolved. Takai *et al.* (1981) suggest that the hydrolysis of PI may induce a cascade of events one of which is the release of arachidonic acid which may be further metabolised to yield not only the prostaglandins etc., as described above, but also some metabolite capable of activating guanylate cyclase.

Another aspect of this cascade which begins with the hydrolysis of PI is that diacylglycerol may activate a new species of calcium-dependent protein kinase (Takai *et al.*, 1979; Kishimoto *et al.*, 1980). Diacylglycerol may thus function as a second messenger operating within the plane of the membrane to activate protein kinase C which then alters membrane proteins. Protein kinase C has a widespread occurrence in tissues from many different phyla (Kuo *et al.*, 1980). The proteins which are phosphorylated by this kinase are different from those which are the substrate for the calmodulin-sensitive protein kinase (Wrenn *et al.*, 1980; Kotoh *et al.*, 1981). In blood platelets, the DG-sensitive kinase phosphorylates a 40K protein which participates in the release of serotonin (Takai *et al.*, 1981).

The hydrolysis of PI may thus have a multifunctional role responsible for generating a whole battery of intracellular signals. Not only has it been implicated in calcium signalling but it also plays a role in the release of arachidonic acid, in the activation of guanylate cyclase and in the stimulation of protein kinase C.

4. CLINICAL SIGNIFICANCE OF THE PI RESPONSE

At present there is no indication of any clinical disorder which has been attributed to derangement of the biochemical pathways responsible for PI metabolism. However, there are numerous hints in the literature suggesting that phosphatidylinositol may play a role in pathogenesis. One way that the phosphoinositides may contribute to pathogenesis is to function as receptors for various toxins. There is some evidence to suggest that TPI may contribute to the action of diptheria toxin (Friedman *et al.*, 1982). If cells are pre-incubated with a monoclonal antibody directed against TPI the cytotoxic effect of the toxin is greatly reduced. Perhaps of greater significance, however, is the role of the phosphoinositides in the action of receptors that control many vital physiological processes. The best way of approaching this problem is to consider some examples of how changes in the PI response might contribute to various clinical disorders by inducing subtle alterations in these normal cellular control processes.

4.1 Immediate hypersensitivity

The PI response plays an important role in the immediate hypersensitive reaction which is responsible for various allergic reactions such as hay fever, asthma and anaphylactic shock. Some of the control processes responsible for this allergic reaction are summarised in Fig. 3. The binding of an antigen with receptors on the surface of mast cells triggers a classical PI response (Cockroft and Gomperts, 1979) which seems to be responsible for the release of histamine. In addition,

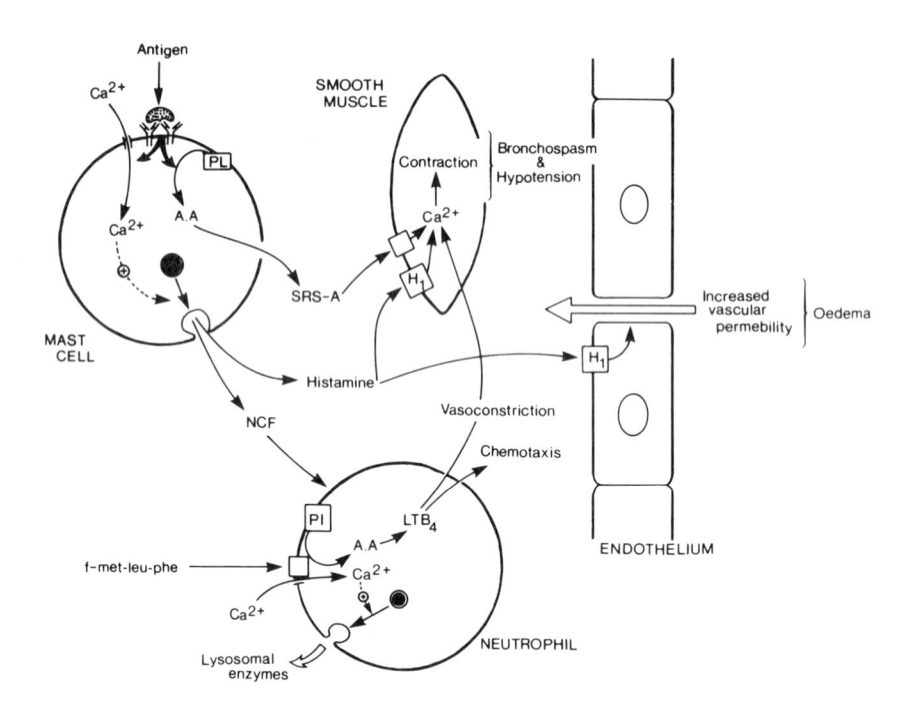

Fig. 3 – Cellular interactions responsible for immediate hypersensitive reactions. Antigen interacting with mast cell receptors induces the release of histamine, neutrophil chemotactic factor (NCF) and arachidonic acid (AA). The latter is converted to the slow-reacting substances of anaphylaxis (SRS-A) which act together with histamine to contract smooth muscle. Histamine also acts on the endothelium to increase vascular permeability. PL-phospholipids ,PI-phosphatidylinositol; LTB_4-leukotriene B_4.

there is also a release of arachidonic acid which is then metabolised via the lipoxygenase system to generate slow-reacting substances of anaphylaxis (SRS-A) which are a mixture of leukotrienes. Histamine and SRS-A stimulate smooth muscle to contract thus leading to bronchospasm and hypotension. Smooth muscle has H_1 histamine receptors which respond to histamine by eliciting a classical PI response. Neighbouring endothelial cells also have H_1 receptors which hydrolyse PI when they respond to histamine with an increase in vascular permeability leading to oedema (Fig. 3). Neutrophils also play an important role in local inflammatory responses. They respond to specific signals, such as f-met-leu-phe, with a large breakdown of PI much of which is metabolised to arachidonic acid and then to the leukotrienes which are particularly important in inflammation (Lewis and Austen, 1981).

4.2 Aggregation of blood platelets

The PI response also plays a role in the aggregation of blood platelets (Fig. 4). If a blood vessel is damaged various signals are produced such as collagen and ADP which interact with blood platelets to induce a PI response leading to aggregation and clot formation. During this PI response there is a precipitous fall in the level of PI and a corresponding rise in the level of diacylglycerol (Rittenhouse-Simmons, 1979; 1980; Prescott and Majerus, 1981). The latter then activates protein kinase C, as described earlier, and also functions as a substrate for a lipase which liberates arachidonic acid (Fig. 4). Arachidonic acid is converted to PGG_2 some of which is metabolised to thromboxane A_2 which helps to induce aggregation and release. Thromboxane A_2 may also leave the platelets to stimulate contraction of smooth muscles and has been implicated in the pathogenesis of gastric ulceration (Whittle *et al.*, 1981). Some of the PGG_2 escapes from the platelets and is taken up by the surrounding endothelium where it is converted into prostacyclin (PGI_2) which is an extremely potent inhibitor of platelet aggregation. Prostacyclin diffuses back to the platelet where it activates adenylate cyclase to produce the cyclic AMP responsible for limiting platelet aggregation. There is thus an antagonism between thromboxane A_2 which promotes aggregation and prostacyclin which attempts to maintain the *status quo*, and a yin-yang balance between these arachidonic acid metabolites may be important in the development of atherosclerosis and arterial thrombi (Moncada and Vane, 1979). Prostacyclin may also be an extremely important antimetastatic agent because it prevents tumour cells from establishing new colonies (Honn *et al.*, 1981).

The prostacyclin synthesised by endothelial cells also plays a pivotal role in regulating the contractile activity of vascular smooth muscle and hence contributes to the vascular tone which determines blood pressure. Noradrenaline released from adrenergic nerves interacts with α_1-receptors to induce smooth muscle contraction (Fig. 4). These receptors give a classical PI response which

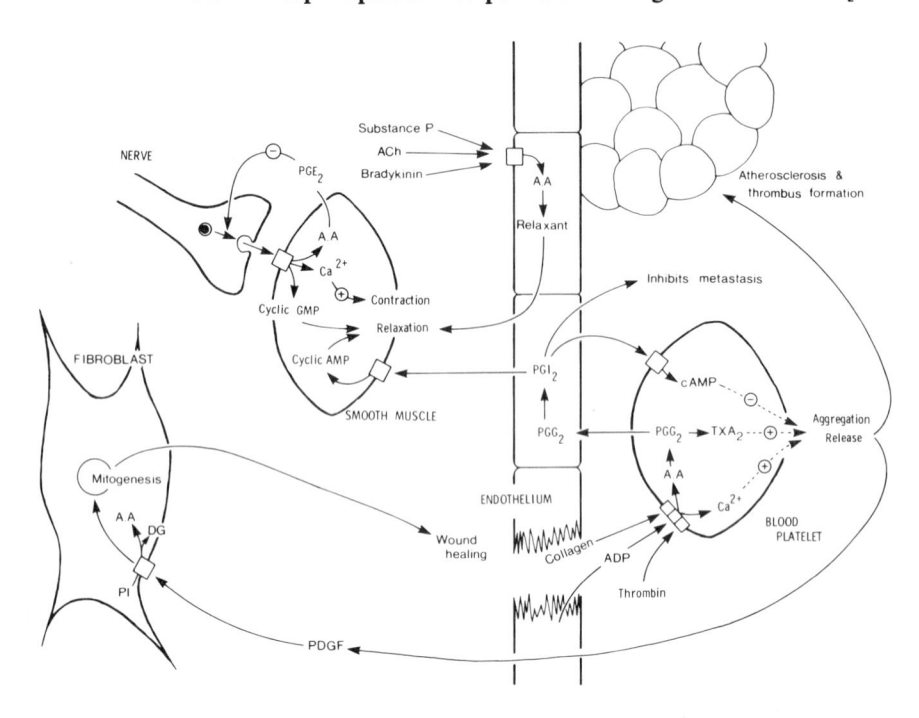

218   Membrane Phospholipids in Receptor Transducing Mechanisms   [Ch.

Fig. 4 — The role of the PI response in platelet aggregation and some related events (see text for further details). AA — arachidonic acid; ACh — acetylcholine; PGE_2 — prostaglandin E_2; PGG_2 — prostaglandin G_2; PGI_2 — prostacyclin: TXA_2 — thromboxane A_2.

may contribute not only to the calcium signal responsible for contraction but also stimulates the formation of cyclic GMP and the release of arachidonic acid. The latter is metabolised to prostaglandin E_2 which may feedback to regulate release of noradrenaline from the nerve endings (Westfall, 1977; Vanhoutte et al., 1981). These smooth muscle receptors thus display a classical multifunctional PI response which is mainly responsible for contraction but also regulates other aspects of the neuro-effector system. Prostaglandins coming from the endothelial cells can contribute to relaxation. Endothelial cells may also respond to agonists such as substance P, acetylcholine and bradykinin to produce another arachidonic acid metabolite which causes smooth muscle relaxation (Furchgott and Zawadzki, 1980; Furchgott, 1981). It is evident that the PI response plays a central role in regulating smooth muscle tone and may thus function in the control of blood pressure. It has been suggested that an alteration in polyphosphoinositide metabolism may play a role in the development of hypertension (Kiselev et al., 1981).

4.3 Cellular proliferation

The PI response may also play an important role in the regulation of cellular proliferation (Hoffmann *et al.*, 1974; Ristow *et al.*, 1975; Ristow *et al.*, 1980). During the release reaction, blood platelets secrete a platelet derived growth factor (PDGF) which stimulates neighbouring fibroblasts to divide and thus contribute to wound healing (Fig. 4). This mitogenic protein was found to induce a profound hydrolysis of PI leading to a phasic elevation of diacylglycerol and a large increase in the release of arachidonic acid (Habernicht *et al.*, 1981). The release of arachidonic acid is particularly interesting because there are suggestions that it may be involved in the initiation of DNA synthesis by rat liver cells (Boynton and Whitfield, 1980). An enhanced turnover of arachidonyl-phosphatidylinositol has also been reported in L-cells following stimulation with antibody (Shearer and Richards, 1981), and in a human epidermoid carcinoma cell line in response to epidermal growth factor (Sawyer and Cohen, 1981). As described earlier, two mechanisms are available for releasing arachidonic acid (Fig. 2). Habernicht *et al.* (1981) consider that it is released as part of the PI response whereas Shier (1980) suggests that the arachidonic acid released in response to PDGF results from activation of phospholipase A_2. The synthesis of PGE_2 by transformed BALB/3T3 cells in response to bradykinin and thrombin is also thought to result from the activation of phospholipase A_2 (Hong and Deykin, 1981). Although there are clear indications that the PI response is induced when a variety of cells are stimulated to grow and divide, the precise function of the hydrolysis of PI has not been established. There is an intriguing report that Duchenne-dystrophic human skin fibroblasts respond to Con A with a larger change in phosphatidylinositol metabolism than that seen in normal cells (Rounds *et al.*, 1980).

4.4 Neural disorders and mental illness

Another area where alterations in PI metabolism may be of clinical importance is in the central nervous system because PI responses are an integral component of many neural receptor mechanisms (Hawthorne and Pickard, 1979; Michell, 1981; Subramanian *et al.*, 1981). An interesting aspect of inositol metabolism in the brain is that all the inositol used for the synthesis of PI is made within the brain because the blood—brain barrier is impermeable to this polyhydric alcohol (Margolis *et al.*, 1971; Barkai 1981). Glucose is converted to *myo*-inositol-1-phosphate which is then dephosphorylated to *myo*-inositol by a phosphatase enzyme. This synthetic pathway for supplying inositol can be interfered with in various ways which may lead to changes in neural activity which are of clinical importance. It has been suggested that the neurotoxic effects of galactosemia may result from the accumulation of galactose-1-phosphate and a decrease in the level of free inositol (Warefield and Segal, 1978). Galactosemia also leads to changes in PI metabolism including a decrease in the PI response to acetylcholine (Warefield and Segal, 1978; Berry *et al.*, 1981). A change in the availability of inositol may thus impair the receptor mechanisms which operate through a PI response and thus lead to important changes in neural processing which could account for the mental retardation associated with galactosemia.

Studies on the mode of action of lithium provide further support for the idea that neural mechanisms might be altered through subtle changes in inositol metabolism. As described earlier, lithium has a potent inhibitory effect on the *myo*-inositol-1-phosphatase which converts IMP to inositol (Fig. 1). Since, the enzyme is responsible for dephosphorylating both the D-enantiomer (produced from the breakdown of PI) as well as the L-enantiomer (produced by synthesis from glucose), lithium will block the formation of free inositol which declines markedly in the brain while the level of IMP builds up (Allison and Stewart, 1971; Allison *et al.*, 1976; Sherman *et al.*, 1981; Berridge *et al.*, 1982). Berridge *et al.* (1982) have suggested that such a decline in the level of free inositol will reduce the synthesis and hence the level of PI in the membrane. By lowering membrane PI, lithium may reduce the effectiveness of those receptor mechanisms which use this phospholipid. A natural extension of this hypothesis is that manic-depressive illness may depend upon the hyperactivity of those receptor mechanisms which use PI hydrolysis as a transducing mechanism.

These few examples illustrate the potential importance of phospholipid metabolism not only in the control of a wide range of different physiological events but also suggest that many clinical disorders may stem from alterations in those receptor mechanisms which use PI as a transducing mechanism. There clearly is an urgent need to find out more about these phospholipid events and how they are influenced by hormones and neurotransmitters.

REFERENCES

Akhtar, R. A. and Abdel-Latif, A. A. (1980) Requirement for calcium ions in acetylcholine-stimulated phosphodiesteratic cleavage of phosphatidyl-*myo*-inositol 4,5-bisphosphate in rabbit iris smooth muscle. *Biochem. J.* **192**, 783–791.

Alemany, S., Varela, I. and Mato, J. M. (1981) Stimulation by vasopressin and angiotensin of phospholipid methyltransferase in isolated rat hepatocytes. *FEBS Letters* **135**, 111–114.

Allison, J. H. and Stewart, M. A. (1971) Reduced brain inositol in lithium-treated rats. *Nature New Biology* **233**, 267–268.

Allison, J. H., Blisner, M. E., Holland, W. H., Hipps, P. P. and Sherman, W. R. (1976) Increased brain *myo*-inositol 1-phosphate in lithium-treated rats. *Biochem. Biophys. Res. Commun.* **71**, 664–670.

Axelrod, J., Hirata, F., Crews, F. T., Ishizaka, T., Ishizaka, K., McGivney, A. and Siraganian, R. P. (1981) Lipids and the receptor-mediated release of histamine from cells, in *Chemical Neurotransmitters 75 years* (Stjärne, L., Hedqvist, P., Lagercrantz, H. and Wennalm, A. eds.) pp. 319–328. Academic Press, London.

Barkai, A. I. (1981) *Myo*-inositol turnover in the intact rat brain: increased production after D-amphetamine. *J. Neurochem.* **36**, 1485–1491.

Barritt, G. J., Dalton, K. A. and Whiting, J. A. (1981) Evidence that phosphatidic acid stimulates the uptake of calcium by liver cells but not calcium release from mitochondria. *FEBS Letts.* **125**, 137–140.

Berridge, M. J. (1980) Receptors and calcium signalling. *Trends Pharmacol. Sci.* **1**, 419–424.

Berridge, M. J. (1981) Phosphatidylinositol hydrolysis: a multifunctional transducing mechanism. *Mol. Cellular Endocrin.* **24**, 115–140.

Berridge, M. J., Downes, C. P. and Hanley, M. R. (1982) Lithium amplifies agonist-dependent phosphatidylinositol responses in brain and salivary glands. *Biochem. J.* **206**, 587–595.

Berry, G., Yandrasitz, J. R. and Segal, S. (1981) Experimental galactose toxicity: effects on synaptosomal phosphatidylinositol metabolism. *J. Neurochem.* **37**, 888–891.

Billah, M. M., Lapetina, E. G. and Cuatrecasas, P. (1980) Phospholipase A_2 and phospholipase C activities of platelets. *J. Biol. Chem.* **255**, 10227–10231.

Boynton, A. L. and Whitfield, J. F. (1980) Possible involvement of arachidonic acid in the initiation of DNA synthesis by rat liver cells. *Exptl. Cell Res.* **129**, 474–478.

Castaño, J. G., Alemany, S., Nieto, A. and Mato, J. M. (1980) Activation of phospholipid methytransferase by glucagon in rat hepatocytes. *J. Biol. Chem.* **255**, 9041–9043.

Cockroft, S. (1981) Does phosphatidylinositol breakdown control the Ca^{2+}-gating mechanism? *Trends Pharmacol. Sci.* **2**, 340–342.

Cockroft, S. and Gomperts, B. D. (1979) Evidence for a role of phosphatidylinositol turnover in stimulus-secretion coupling. Studies with rat peritoneal mast cells. *Biochem. J.* **178**, 681–687.

Cockroft, S., Bennett, J. P. and Gomperts, B. D. (1980a) Stimulus-secretion coupling in rabbit neutrophils is not mediated by phosphatidylinositol breakdown. *Nature* **288**, 275–277.

Cockroft, S., Bennett, J. P. and Gomperts, B. D. (1980b) f*MetLeuPhe*-induced phosphatidylinositol turnover in rabbit neutrophils is dependent on extracellular calcium. *FEBS Letts.* **110**, 115–118.

Cordasco, D. M., Segarnick, D. J. and Rotrosen, J. (1981) Human platelet phospholipid methylation. *Life Sci.* **29**, 2299–2309.

Crews, F. T., Morita, Y., Hirata, F., Axelrod, J. and Siraganian, R. P. (1980) Phospholipid methylation affects immunoglobulin E-mediated histamine and arachidonic acid release in rat leukemic basophils. *Biochem. Biophys. Res. Commun.* **93**, 42–49.

Durell, J., Sodd, M. A. and Friedal, R. O. (1968) Acetylcholine stimulation of the phosphodiesteratic cleavage of guinea pig brain phosphoinositides. *Life Sci.* **7**, 363–368.

Egawa, K., Sacktor, B. and Takenawa, T. (1981) Ca^{2+}-dependent and Ca^{2+}-independent degradation of phosphatidylinositol in rabbit vas deferens. *Biochem. J.* **194**, 129–136.

Fain, J. N. and Berridge, M. J. (1979) Relationship between hormonal activation of phosphatidylinositol hydrolysis, fluid secretion and calcium flux in the blowfly salivary gland. *Biochem. J.* **178**, 45–58.

Flower, R. (1981) Glucocorticoids, phospholipase A_2 and inflammation. *Trends Pharmacol. Sci.* **2**, 186–189.

Friedman, R. L., Iglewski, B. H., Roerdink, F. and Alving, C. R. (1982) Suppression of cytoxicity of diptheria toxin by monoclonal antibodies against phosphatidylinositol phosphate. *Biophys. J.* **37**, 23–24.

Furchgott, R. F. (1981) The requirement for endothelial cells in the relaxation of arteries by acetylcholine and some other vasodilators. *Trends Pharmacol. Sci.* **2**, 173–176.

Furchgott, R. F. and Zawadzki, J. V. (1980) The obligatory role of endothelial cells in the relaxation of arterial smooth muscle by acetylcholine. *Nature* **288**, 373–378.

Gibson, A. and Brammer, M. J. (1981) The influence of divalent cations and substrate concentration on the incorporation of *myo*-inositol into phospholipids of isolated bovine oligodendrocytes. *J. Neurochem.* **36**, 868–874.

Habernicht, A. J. R., Glomset, J. A., King, W. C., Nist, C., Mitchell, C. D. and Ross, R. (1981) Early changes in phosphatidylinositol and arachidonic acid metabolism in quiescent Swiss 3T3 cells stimulated to divide by platelet-derived growth factor. *J. Biol. Chem.* **256**, 12329–12335.

Hallcher, L. M. and Sherman, W. R. (1980) The effects of lithium ion and other agents on the activity of *myo*-inositol-1-phosphatase from bovine brain. *J. Biol. Chem.* **255**, 10896–10901.

Harris, R. A., Schmidt, J., Hitzemann, B. A. and Hitzemann, R. J. (1981) Phosphatidate as a molecular link between depolarization and neurotransmitter release in the brain. *Science* **212**, 1290–1291.

Hawthorne, J. N. and Pickard, M. R. (1979) Phospholipids in synaptic function. *J. Neurochem.* **32**, 5–14.

Hawthorne, J. N. and White, D. A. (1975) Myo-inositol lipids. *Vitamins and Hormones* **33**, 529–573.

Hirata, F. (1981) The regulation of lipomodulin, a phospholipase inhibitory protein, in rabbit neutrophils by phosphorylation. *J. Biol. Chem.* **256**, 7730–7733.

Hirata, F. and Axelrod, J. (1980) Phospholipid methylation and biological signal transmission. *Science,* **209**, 1082–1090.

Hirata, F., Axelrod, J. and Crews, F. T. (1979a) Concanavalin A stimulates phospholipid methylation and phosphatidylserine decarboxylation in rat mast cells. *Proc. Natl. Acad. Sci. USA* **76**, 4813–4816.

References

Hirata, F., Strittmatter, W. J. and Axelrod, J. (1979b) β-adrenergic receptor agonists increase phospholipid methylation, membrane fluidity, and β-adrenergic receptor adenylate cyclase coupling. *Proc. Natl. Acad. Sci. USA* **76**, 368–372.

Hoffmann, R., Ristow, H.-J., Pachowsky, H. and Frank, W. (1974) Phospholipid metabolism in embryonic rat fibroblasts following stimulation by a combination of the serum proteins S1 and S2. *Eur. J. Biochem.* **49**, 317–324.

Hong, S. L. and Deykin, D. (1981) The activation of phosphatidylinositol-hydrolyzing phospholipase A_2 during prostaglandin synthesis in transformed mouse BALB 3T3 cells. *J. biol. Chem.* **256**, 5215–5219.

Honn, K. V., Cicone, B. and Skoff, A. (1981) Prostacyclin: a potent antimetastatic agent. *Science* **212**, 1270–1272.

Ishizaka, T., Hirata, F., Ishizaka, K. and Axelrod, J. (1980) Stimulation of phospholipid methylation, Ca^{2+} influx, and histamine release by bridging of IgE receptors on rat mast cells. *Proc. Natl. Acad. Sci. USA* **77**, 1903–1906.

Ishizaka, T., Hirata, F., Sterk, A. R., Ishizaka, K. and Axelrod, J. (1981) Bridging of IgE receptors activates phospholipid methylation and adenylate cyclase in mast cell plasma membranes. *Proc. Natl. Acad. Sci. USA* **78**, 6812–6816.

Jafferji, S. S. and Michell, R. H. (1976) Stimulation of phosphatidylinositol turnover by histamine, 5-hydroxytryptamine and adrenaline in the longitudinal smooth muscle of guinea pig ileum. *Biochem. Pharmacol.* **25**, 1429–1430.

Jones, L. M. and Michell, R. H. (1975) The relationship of calcium to receptor-controlled stimulation of phosphatidylinositol turnover. *Biochem. J.* **148**, 479–485.

Jones, L. M. and Michell, R. H. (1978) Enhanced phosphatidylinositol breakdown as a calcium independent response of rat parotid fragments to substance P. *Biochem. Soc. Trans.* **6**, 1035–1037.

Kirk, C. J., Verrinder, T. R. and Hems, D. A. (1978) The influence of extracellular calcium concentration on the vasopressin-stimulated incorporation of inorganic phosphate into phosphatidylinositol in hepatocyte suspensions. *Biochem. Soc. Trans.* **6**, 1031–1033.

Kiselev, G., Minenko, A., Moritz, V. and Oehme, P. (1981) Polyphosphoinositide metabolism of spontaneously hypertensive rats. *Biochem. Pharm.* **30**, 833–837.

Kishimoto, A., Takai, Y., Mori, T., Kikkawa, U. and Nishizuka, Y. (1980) Activation of calcium and phospholipid-dependent protein kinase by diacylglycerol, its possible relation to phosphatidylinositol turnover. *J. biol. Chem.* **255**, 2237–2276.

Kotoh, N., Wrenn, R. W., Wise, B. C., Shoji, M. and Kuo, J. F. (1981) Substrate proteins for calmodulin-sensitive and phospholipid-sensitive Ca^{2+}-dependent protein kinases in heart, and inhibition of their phosphorylation by palmitoylcarnitine. *Proc. Natl. Acad. Sci. USA* **78**, 4813–4817.

Kuo, J. F., Andersson, R. G. G., Wise, B. C., Mackerlova, L., Salomonsson, I., Brackett, N. L., Katoh, N., Shoji, M. and Wrenn, R. W. (1980) Calcium-dependent protein kinase: Widespread occurrence in various tissues and phyla of the animal kingdom and comparison of effects of phospholipids, calmodulin, and trifluoperazine. *Proc. Natl. Acad. Sci. USA* **77**, 7039–7043.

Lewis, R, A. and Austen, K. F. (1981) Mediation of local homeostasis and inflammation by leukotrienes and other mast cell-dependent compounds. *Nature* **293**, 103–108.

Margolis, R. U., Press, R., Altszuler, N. and Stewart, M. A. (1981) Inositol production by the brain in normal and alloxan-diabetic dogs. *Brain Res.* **28**, 535–539.

Michell, R. H. (1975) Inositol phospholipids and cell surface receptor function. *Biochim. Biophys. Acta* **415**, 81–147.

Michell, R. H. (1979) Inositol phospholipids in membrane function. *Trends Biochem. Sci.* **4**, 128–131.

Michell, R. H. and Kirk, C. J. (1981) Why is phosphatidylinositol degraded in response to stimulation of certain receptors? *Trends Pharmacol. Sci.* **2**, 86–89.

Michell, R. H., Jones, L. M. and Jafferji, S. S. (1977) A possible role for phosphatidylinositol breakdown in muscarinic cholinergic stimulus-response coupling. *Biochem. Soc. Trans.* **5**, 77–81.

Michell, R. H., Kirk, C. J., Jones, L. M., Downes, C. P. and Creba, J. A. (1981) The stimulation of inositol lipid metabolism that accompanies calcium mobilization in stimulated cells: defined characteristics and unanswered questions. *Phil Trans. R. Soc. Lond. B.* **296**, 123–137.

Moncada, S. and Vane, J. R. (1979) The role of prostacyclin in vascular tissue. *Fed. Proc.* **38**, 66–71.

Ohsaka, S. and Deguchi, T. (1981) Stimulation by phosphatidic acid of calcium influx and cyclic GMP synthesis in neuroblastoma cells. *J. Biol.Chem.* **256**, 10945–10948.

Panagia, V., Michiel, D. F., Dhalla, K. S., Nijjar, M. S. and Dhalla, N. S. (1981) Role of phosphatidylinositol in basal adenylate cyclase activity of rat heart sarcolemma. *Biochim. Biophys. Acta* **676**, 395–400.

Pike, M. C., Kredich, N. M. and Snyderman, R. (1979) Phospholipid methylation in macrophages is inhibited by chemotactic factors. *Proc. Natl. Acad. Sci. USA* **76**, 2922–2926.

Prasad, C. and Edwards, R. M. (1981) Stimulation of rat pituitary phospholipid methyltransferase by vasopressin but not oxytocin. *Bioch. Biophys. Res Commun.* **103**, 559–564.

Prescott, S. M. and Majerus, S. P. W. (1981) The fatty acid composition of phosphatidylinositol from thrombin-stimulated human platelets. *J. Biol. Chem.* **256**, 579–582.

Putney, J. W. (1981) Recent hypotheses regarding the phosphatidylinositol effect. *Life Sci.* **29**, 1183–1194.

Putney, J. W., Weiss, S. J. Van De Walle, C. M. and Haddas, R. A. (1980) Is phosphatidic acid a calcium ionophore under neurohormonal control? *Nature* **284**, 345–347.

Putney, J. W., Poggioli, J. and Weiss, S. J. (1981) Receptor regulation of calcium release and calcium permeability in parotid gland cells. *Phil Trans. R. Soc. Lond. B.* **296**, 37–45.

Randon, J., Lecompte, T., Chignard, M., Siess, W., Marlas, G., Dray, F. and Vargaftig, B. B. (1981) Dissociation of platelet activation from transmethylation of their membrane phospholipids. *Nature* **293**, 660–662.

Ristow, H.-J., Hoffmann, R., Pachowsky, H. and Frank W. (1975) Influence of potassium and calcium ions on the phosphatidylinositol and phosphatidylcholine metabolism in rat fibroblasts after growth stimulation by calf serum. *Eur. J. Biochem.* **56**, 413–420.

Ristow, H.-J., Messmer, T. O., Walter, S. and Paul, D. (1980) Stimulation of DNA synthesis and *myo*-inositol incorporation in mammalian cells. *J. Cell Physiol.* **103**, 263–269.

Rittenhouse-Simmons, S. (1979) Production of diglyceride from phosphatidylinositol in activated human platelets. *J. Clin. Invest.* **63**, 580–587.

Rittenhouse-Simmons, S. (1980) Indomethacin-induced accumulation of diglyceride in activated human platelets. *J. Biol Chem.* **255**, 2259–2262.

Roelofsen, B. (1981) The (non)specificity in the lipid-requirement of calcium- and (sodium plus potassium)-transporting adenosine triphosphatase. *Life Sci.* **29**, 2235–2247.

Rounds, P. S., Jepson, A. B., McAllister, D. J. and Howland, J. L. (1980) Stimulated turnover of phosphatidylinositol and phosphatidate in normal and Duchenne-dystrophic human skin fibroblasts. *Bioch. Biophys. Res. Commun.* **97**, 1384–1390.

Salmon, D. M. and Honeyman, T. W. (1980) Proposed mechanism of cholinergic action in smooth muscle. *Nature* **284**, 344–347.

Sawyer, S. T. and Cohen, S. (1981) Enhancement of calcium uptake and phosphatidylinositol turnover by epidermal growth factor in A-431 cells. *Biochemistry* **20**, 6280–6286.

Shearer, W. T. and Richards, J. E. (1981) Rapid turnover of arachidonyl-phosphatidylinositol in L cells stimulated by antibody. *Biochem. Biophys. Res. Commun.* **101**, 800–806.

Sherman, W. R., Leavitt, A. L., Honchar, M. P., Hallcher, L. M. and Phillips, B. E. (1981) Evidence that lithium alters phosphoinositide metabolism.

Chronic administration elevates primarily D-*myo*-inositol-1-phosphate in cerebral cortex of the rat. *J. Neurochem.* **36**, 1947–1951.

Shier, W. T. (1980) Serum stimulation of phospholipase A$_2$ and prostaglandin release in 3T3 cells is associated with platelet-derived growth-promoting activity. *Proc. Natl. Acad. Sci. USA* **77**, 137–141.

Smith, T. L. and Hauser, G. (1981) Effects of changes in calcium concentration on basal and stimulated ^{32}P incorporation into phospholipids in rat pineal cells. *J. Neurochem.* **37**, 427–435.

Subramanian, N., Whitmore, W. L., Seidler, F. J. and Slotkin, T. A. (1981) Ontogeny of histaminergic neurotransmission in the rat brain: concomitant development of neuronal histamine, H-1 receptors, and H-1 receptor-mediated stimulation of phospholipid turnover. *J. Neurochem.* **36**, 1137–1141.

Takai, Y., Kishimoto, A., Kikkawa, J., Mori, T. and Nishizuka, Y. (1979) Unsaturated diacylglycerol as a possible messenger for the activation of calcium-activated, phospholipid-dependent protein kinase system. *Biochem, Biophys. Res. Commun.* **91**, 1218–1224.

Takai, Y., Kishimoto, A. and Nishizuka, Y. (1981) Calcium and phospholipid turnover as transmembrane signalling for protein phosphorylation, in *Calcium and Cell Function*, Vol. 2, (ed. W. J. Cheung) Academic Press, London.

Vance, D. E. and de Kruijff, B. (1980) The possible functional significance of phosphatidylethanolamine methylation. *Nature* **288**, 277–278.

Vanhoutte, P. M., Verbeuren, T. J. and Webb, R. C. (1981) Local modulation of adrenergic neuroeffector interaction in the blood vessel wall. *Physiol. Rev.* **61**, 151–247.

Walenga, R. W., Opas, E. E. and Feinstein, M. B. (1981) Differential effects of calmodulin antagonists on phospholipase A$_2$ and C in thrombin-stimulated platelets. *J. Biol. Chem.* **256**, 12523–12528.

Warefield, A. S. and Segal, S. (1978) *Myo*-inositol and phosphatidylinositol metabolism in synaptosomes from galactose-fed rats. *Proc. Natl. Acad. Sci. USA* **75**, 4568–4572.

Westfall, T. C. (1977) Local regulation of adrenergic neurotransmission. *Physiol. Rev.* **57**, 659–728.

Whittle, B. J. R., Kauffman, G. L. and Moncada, S. (1981) Vasoconstriction with thromboxane A$_2$ induces ulceration of the gastric mucosa. *Nature* **292**, 472–474.

Wolfe, L. S. (1982) Eicosanoids: prostaglandins, thromboxanes, leukotrienes, and other derivatives of carbon-20 unsaturated fatty acids. *J. Neurochem.* **38**, 1–14.

Wrenn, R. W., Katoh, N., Wise, B. C. and Kuo, J. F. (1980) Stimulation by phosphatidylserine and calmodulin of calcium-dependent phosphorylation of endogenous proteins from cerebral cortex. *J. Biol. Chem.* **255**, 12042–12046.

12

The complex structure and regulation of adenylate cyclase

Martin Rodbell

The National Institutes of Health, Bethesda, Maryland 20205, U.S.A.

Abbreviations used

- ACTH — corticotropin
- C — catalytic unit (adenylate cyclase)
- Gpp(NH)p — guanylylimidodiphosphate
- M — metal ion component
- N — guanine nucleotide regulatory component
- R — recognition component (receptor)

INTRODUCTION

For over twenty years the adenylate cyclase system has been investigated as a model for hormone action. Situated in the plasma membrane, the enzyme is regulated by a large number of hormones, neurotransmitters, and such 'local' hormones as prostaglandins and purinergic compounds. Initially considered to be a two component system consisting of the catalytic unit (C) and the recognition components (R) for the various hormones and hormone-like substances, a large body of evidence has accumulated that these systems are multi-subunit systems composed minimally of R,C, and nucleotide regulatory components (N) that are responsible for the regulation of adenylate cyclase activity by guanine nucleotides (Rodbell, 1980, Ross and Gilman, 1980). The enzyme is also regulated in opposing manner by sets of R and N units that either stimulate or inhibit the enzyme. A functional stimulatory receptor unit has been designed as RN_s whereas the opposing inhibitory receptor unit has been designated as the RN_i unit or complex (Rodbell, 1980). Systems displaying dual regulation contain, by definition, both types of receptor complexes. In addition, different R units are often linked structurally to both N_s and N_i in the same cell membrane.

To accommodate such regulatory and structural complexity, the hypothesis has been put forth that RN_s and RN_i complexes are present in membranes in the form of oligomeric structures in which mixtures of R,N_s, and N_i units are aggregated (Rodbell, 1980). It was further postulated that the concerted actions of hormones and GTP on these oligomeric structures disaggregated the oligomers to 'monomeric' structures that were capable of forming complexes with the catalytic unit.

Here I will expand and modify somewhat this theory by including recent findings on the structures of the components and on the role of ions in the regulation of adenylate cyclase activity.

1. THE ROLE OF CATIONS

Early studies of adenylate cyclase systems had shown that Mg^{2+} ions stimulate activity to levels approaching that observed with hormones; hormones reduced the concentrations of Mg^{2+} ions required to achieve maximal activation (Birnbaumer et al., 1969). These findings suggested that part of the transduction process involved in hormonal activation consisted of sites for Mg^{2+} ions. Although this idea was disputed, it is now clear that regulatory sites for Mg^{2+} ions exist in a variety of cyclase systems (Londos and Rodbell, 1975; Londos and Preston, 1977b). Mn^{2+} ions have generally higher affinity than Mg^{2+} ions for these sites. The idea that the divalent cation sites are linked functionally to R and N units stems in part from findings that the actions of hormones and guanine nucleotides require the divalent cations for expression. An interesting aspect of glucagon action on the hepatic adenylate cyclase system is that it promotes inhibition of the enzyme by adenosine acting at the so-called 'P-site' which is thought to be proximal to the catalytic component; promotion is mediated by Mn^{2+} or Mg^{2+} ions with affinities resembling those involved in the activation of adenylate cyclase by glucagon and GTP (Londos and Preston, 1977a). Based on findings that Mn^{2+} and Mg^{2+} ions affect the binding of agonists to R units, it has also been suggested that the divalent cation sites are linked to the R units (Stadel et al., 1982). The effects are generally an enhanced affinity of R for agonists, which is opposite to the well-known effects of GTP on agonist binding to R (Hanski et al., 1981). Whether this effect of divalent cations is due to a direct action on R or to the removal of bound guanine nucleotides from N (which would result in enhanced affinity of R for agonists) is unclear. Nonetheless, these findings reinforce the idea that divalent cations play an important role in the transduction process.

The component(s) through which Mg^{2+} and Mn^{2+} act to stimulate adenylate cyclase remains unknown. However, the recent findings (Hanski et al., 1981; Sternweis et al., 1981) that the N_s component purified from rabbit liver and turkey erythrocytes is a heterodimer raises the intriguing possibility that one of the components is the site of divalent cation binding. N_s contains a subunit ranging in size from 45K to 55K daltons; these are related structurally and functionally, and appear to be the major components through which hormones (through R), guanine nucleotides, cholera toxin, and fluoride ion exert their stimulatory effects (Ross and Gilman, 1980). The other subunit is a 35K dalton component which appears to be equally essential as a transduction element but which is not ADP-ribosylated by the action of cholera toxin. As illustrated schematically in Fig. 1, the 35K dalton component may contain both the sites for divalent cations as well as the P-site. Forskolin, a steroid-like compound that stimulates the activity of all animal cell cyclase systems, also acts on a component distinct from the 45K or 55K unit through which GTP acts (Seamon and Daly, 1981). This is based on the finding that forskolin stimulates cyclase activity in membranes from a variant of S49 murine lymphoma cells that lack a functional N_s unit. Similarly, it has been reported that adenosine analogues that act through the P-site also inhibit the activity of cyclase in this variant (Florio and Ross, 1981). However, in the light of recent information (Ferguson et al., 1982) that this variant contains the 35K dalton unit, the question remains open whether the P-site, forskolin site, and sites for Mg^{2+} and Mn^{2+} ions are on the 35K unit or possibly on another component more closely associated with the catalytic unit (indicated in Fig. 1 by the dash-lined region on C).

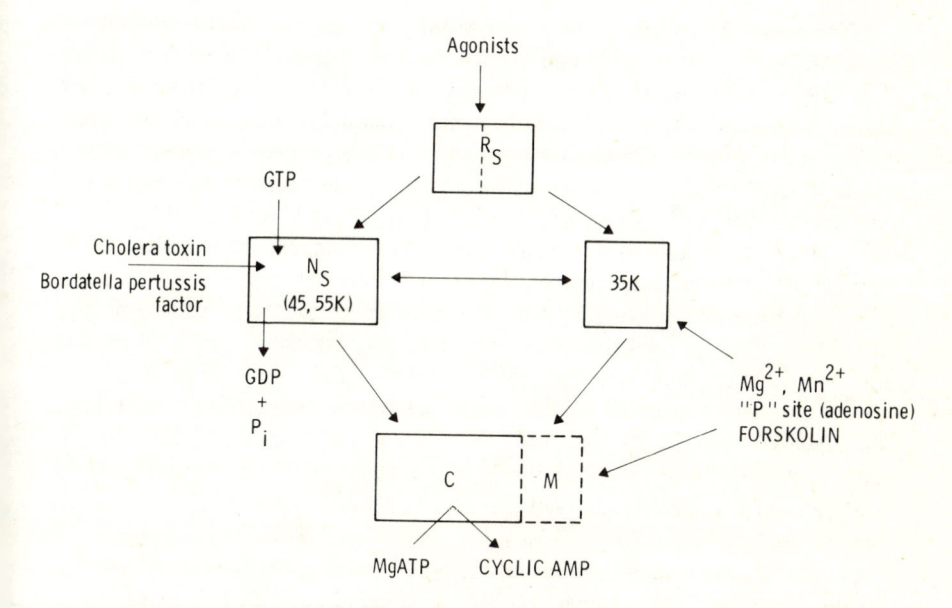

Fig. 1 – Schematic representation of the known components of systems that are stimulated by hormones and GTP. The enclosed areas are those for the hormone recognition proteins (R); the GTP-binding component (N_s) that hydrolyses GTP to GDP and reacts with cholera toxin and a factor from Bordatella pertussis (Katada et al., 1982); the 35K dalton component co-purifies with N_s and may be the site for divalent cation regulation, for adenosine inhibition (P-site) and for the stimulatory action of forskolin. It is also possible that the latter sites are situated on a component (M) more proximal to the catalytic unit (C). For more details, see text.

In addition to regulatory sites for Mg^{2+} and Mn^{2+} ions, it would appear that Ca^{2+} ions play a stimulatory role in certain cyclase systems. A role for Ca^{2+} ions in the stimulatory actions of ACTH was first suggested from studies on the rat adipocyte system (Birnbaumer and Rodbell, 1969). ACTH action was abolished in the presence of EGTA, a chelator that binds Mg^{2+} ions much less effectively than Ca^{2+} ions. Action by ACTH was restored by the addition of as little as 10^{-8} M free calcium ions. By contrast, the stimulatory actions of glucagon, epinephrine and several other stimulatory hormones were unaffected by the chelator or the addition of Ca^{2+} ions. Similar effects of the chelator and Ca^{2+} ions have been reported with cyclase systems stimulated by oxytocin and vasopressin (Bar and Hechter, 1969; Campbell et al., 1972). In view of the selective effects of Ca^{2+} ion regulation, it seems unlikely that the R unit is the site of calcium ion regulation. ACTH, in addition to its involvement with Ca^{2+} ions, promotes activation by Mg^{2+} ions. Therefore, it is possible that certain hormones promote activation through sites or components for GTP, Mg^{2+} or Mn^{2+} ions, and for Ca^{2+} ions. Perhaps two or more metal ion components (M) operate in tandem.

The idea that distinct metal ion or M units are associated with stimulation of adenylate cyclase is supported by the findings that brain adenylate cyclase is activated by calmodulin, a calcium binding protein that regulates a number of enzymes (for review, see Bradham and Cheung, 1980). Brain adenylate cyclase exists in both calmodulin-dependent and independent forms; the former accounts for 20% to 50% of the total adenylate cyclase activity found in detergent extracts of brain (Westcott et al., 1979). In such extracts, addition of calmodulin and calcium ions activates the enzyme independently of the effects of guanine nucleotides exerted through the N_s unit (Heideman et al., 1982). It has not been established that hormones or neurotransmitters regulate adenylate cyclase through the calmodulin-stimulated process. By the same token, it is not evident that calmodulin is involved in the stimulatory actions of ACTH, vasopressin, and oxytocin. Clearly, these are problems worthy of resolution. Nonetheless, there is sufficient evidence to warrant the suggestion that M components exist independently of the N_s components and that divalent cations regulate adenylate cyclase through at least two distinct types of M units, one for Mg^{2+} or Mn^{2+}, the other for Ca^{2+} ions.

A role for monovalent cations in the regulation of adenylate cyclase activity was first suggested from findings that Na^+ ion affected both the binding of opiate agonists to receptors and the ability of opiates to inhibit cyclase activity in neuroblastoma cells (reviewed in Jakobs et al., 1981). GTP, through N_i, inhibits cyclase by a process that is independent of Na^+ ion but like sodium it affects the binding of agonists to opiate receptors as well as other receptors known to effect inhibition of cyclase systems (Rodbell, 1980). Arguments that N_i and sodium-regulation are independent processes stem in part from findings that GTP and Na^+ ion act synergistically but not dependently. Recent findings (Simonds et al.,

1980; Gavish *et al.*, 1982) point to different structural components for Na$^+$-regulation. RN$_i$ complexes have been extracted with detergents in forms that retain the ability of agonist binding to be sensitive to GTP but not to Na$^+$ ion, suggesting that the Na$^+$-component was released during detergent solubilisation of RN$_i$.

The available information suggests that regulation of cyclase activity by monovalent cations (Li$^+$ is also effective) is restricted to systems that are inhibitory. By analogy to the stimulatory systems, perhaps the systems containing RN$_i$ units also contain an M unit that uniquely reacts with Na$^+$ or Li$^+$ ions, as is schematically represented in Fig. 2. Mn^{2+} and Mg^{2+} ions may also react with this M component since in systems displaying inhibition by GTP, the divalent cations (Mn^{2+} > Mg^{2+}) block inhibition. However, it is also possible that the blocking effects are related to actions at the M sites through which the cations stimulate adenylate cyclase (the N$_s$ system depicted in Fig. 1).

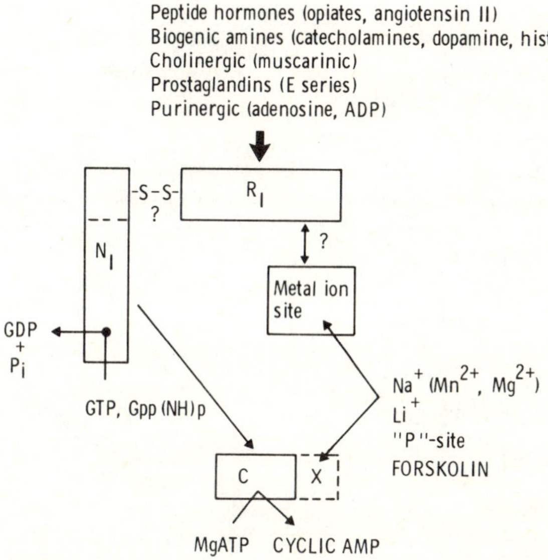

Fig. 2 — Schematic representation of components in hormone-inhibitable forms of adenylate cyclase systems. Enclosed areas are R$_I$ for hormones and neurotransmitters; a GTP-regulatory unit that inhibits the enzyme (N$_i$) and which hydrolyses GTP but not Gpp(NH)p. The putative metal ion unit (M) that reacts with the Na$^+$ or Li$^+$ ions is possibly linked to R$_I$ and/or to a component (X) associated with the catalytic unit (C). For more details, see text.

In addition to sites for monovalent cations, there are major differences in the structures of N$_i$ and N$_s$ systems. An important distinction is that the GTP-inhibitory process is abolished when cells containing both N$_s$ and N$_i$ components are treated with proteases; N$_s$ survives whereas the functions of N$_i$ are abolished

(Evain and Anderson, 1979). Another difference is that the binding of opiates and adenosine to inhibitory R units and the effects of GTP on agonist binding are preserved following detergent extraction of receptors. These findings imply that N_i remains associated with the R units during solubilization and forms stable RN_i complexes. This is not the case for the glucagon and beta adrenergic RN_s systems. Stabilisation of RN_s complexes requires binding of agonists to R prior to solubilisation with detergents (Rodbell, 1980; Stadel et al., 1982). Based on these differences, it would appear from the few examples in hand that inhibitory R units form stable, possibly covalent attachments with N_i (perhaps through disulphide bridges). The attachments may be at the outer surface of the membrane through components susceptible to trypsin degradation (hence the dashed area on N_i). This would imply that N_i units span the plasma membrane, whereas it is thought that the N_s unit is associated only with the inner aspect of

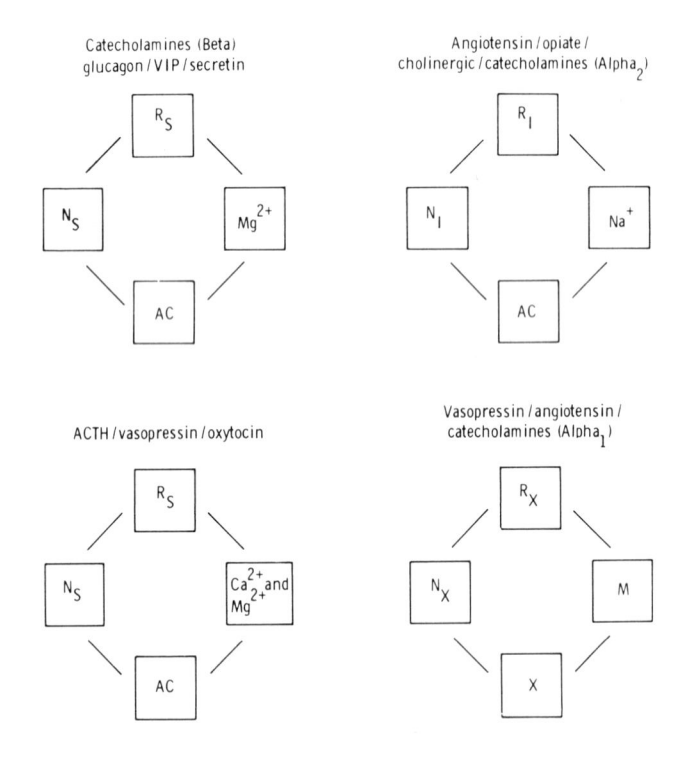

Fig. 3 — Schematic representation of different regulatory forms of adenylate cyclase systems. Each type of system contains hormone recognition units (R), a GTP-regulatory protein (N) a metal ion component for monovalent or divalent cations (M), and the catalytic unit. A hypothetical signal generating system responding to various hormones that do not regulate adenylate cyclase activity directly is also shown.

the membrane and is not an intrinsic membrane protein. Structural differences between RN_s and RN_i systems have also been indicated from differences in sizes, as revealed by target analysis (Rodbell et al., 1981). In addition, N_i is considerably more sensitive to sulfhydryl alkylating agents than is N_s (Jakobs et al., 1982).

Collectively, the information gained thus far on the role of cations argues for selective types of metal ion components (M) associated differentially with adenylate cyclase systems depending on the types of R and N units comprising the receptors. Three such 'families' of cyclase systems are depicted schematically in Fig. 3. Also included is a fourth category of signal generating systems that may contain R, N, and M units that regulate enzymes or processes other than adenylate cyclase. Hints that such systems may exist stem from findings that guanine nucleotides affect the binding of hormones to receptors thought not to be involved in cyclic AMP production (Rodbell, 1980).

2. STRUCTURE OF MULTI-RECEPTOR SYSTEMS

It is rare to find a cell which contains only one type of receptor for the regulation of adenylate cyclase. Most cells contain several types of R units associated with both RN_s and RN_i units in the same membrane, the classical example being the rat adipocyte. In this system, addition of maximally stimulating concentrations of the various stimulatory hormones does not lead to additive responses, indicating that each receptor type has access to a common C unit. The M units must vary since ACTH and vasopressin action involved Ca^{2+} ions whereas the actions of glucagon, VIP, secretin, and catecholamines do not. Moreover, the inhibitory agents (adenosine, prostaglandins, and nicotinic acid) that act through RN_i require Na^+ ion and thus a separate type of M unit. The latter agents also do not act in additive fashion. Moreover, these agents whether combined or not fail to inhibit adenylate cyclase to an extent greater than about 50% (Jakobs et al., 1981), a phenomenon seen with all systems displaying both stimulatory and inhibitory circuits. One explanation is that adenylate cyclase is a family of enzymes, some linked to inhibitory RN_i units, others not. If correct, then the various populations of the enzyme could be in discrete, non-overlapping domains that have quite different regulatory features. Data taken from studies of the actions of Gpp(NH)p on the adipocyte system (Rodbell, 1975) are consistent with discrete populations of cyclase systems that are differentially susceptible to inhibition and stimulation by guanine nucleotides.

Shown in Fig. 4 is the time course of basal activity of the adipocyte enzyme in the absence and presence of 0.1 mM Gpp(NH)p. Note that the initial basal rate is high and then declines to a much lower rate after 1 min incubation. Gpp(NH)p causes an immediate, profound (greater than 80%) inhibition of the initial basal rate. Inhibition is followed by progressive increase in rate until finally the rate is essentially equivalent to the high rate given by the basal state.

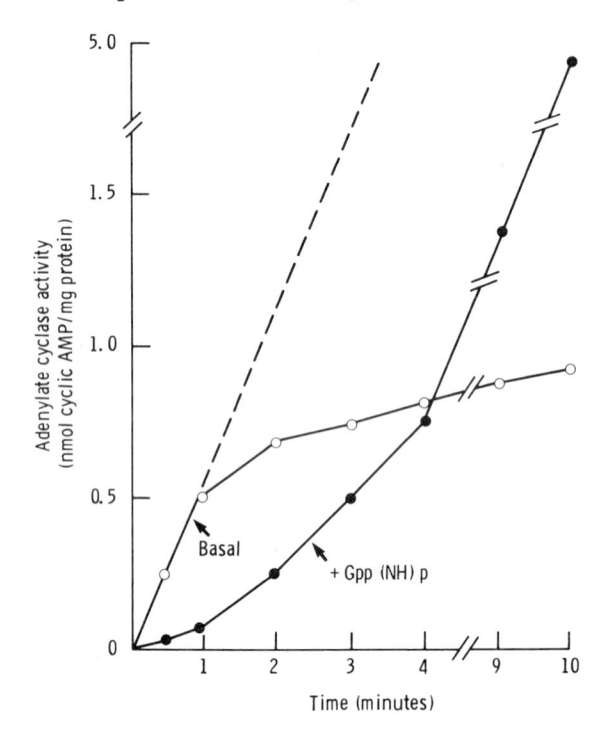

Fig. 4 — Time course of cyclic AMP production by adenylate cyclase in rat adipocyte membranes in the absence (○) and presence (●) of Gpp(NH)p. Membranes (5 μg protein) were incubated at 30 °C in 0.35 ml medium consisting of 25 mM Tris-HCl, pH 7.4/5 mM MgCl$_2$/1 mM dithiothreitol/an ATP-regenerating system/0.5 mM ATP/0.1% bovine serum albumin and, when present, 0.1 mM Gpp(NH)p. Data are from Rodbell (1975).

These findings are consistent with the existence of two non-interconvertible populations of the adipocyte enzyme, one containing an N$_i$ unit through which Gpp(NH)p inhibits, the other containing N$_s$ through which nucleotide activates. The recent report that forskolin acts in additive fashion with Gpp(NH)p on the adipocyte system (Seamon and Daly, 1981), is also consistent with two or more populations of the enzyme co-existing in the same cell membrane.

Given the large and varied number of components involved in the regulation of adenylate cyclase systems, the organisation of these components in the membrane seems a formidable problem. It has been suggested that the components are initially separate in the membrane and that hormones and guanine nucleotodes, in step-wise fashion, induce physical coupling of the units (Stadel et al., 1982). An alternative (Rodbell, 1980) is an integrated, multi-component structure in which the various effector ligands induce structural re-arrangements of the organised components. In the light of what is known of many complex regulatory systems in biology, including that of the recently purified acetylcholine receptor (Wastek and Yamamura, 1981), an organised structure would seem the most efficacious for prompt and highly coordinate regulation. In any event, without prior knowledge of structure, it is not possible to differentiate clearly between uncoupled and precoupled forms of the enzyme, as some have argued.

When applied judiciously (Schlegel et al., 1979), analysis of functional size by high energy irradiation (target analysis) is the only procedure currently available for evaluating some aspects of membrane-bound structures. This procedure has been applied to the ground-state and activated states of the enzyme in rat hepatocytes and adipocytes (Rodbell et al., 1981). The ground-state or precursor sizes in both cell membranes approximated 1.2×10^6 daltons. Analysis of the size of the hormone-activated states yielded about 350,000 daltons, or about 1/4 that of the ground-state (Table 1). Assuming that both the precursor and final activated structures contain R, N_s, and C, then the precursor is an oligomer consisting of four units of the complex formed between R, N and C. Such assumptions are, of course, subject to debate. Nonetheless, at least this methodology supplies data for interpretation. Moreover, target analysis has given some insight, particularly when taken in conjunction with reported sizes of the subunits of R, N, and C, regarding the structures of the units that are required for activation of adenylate cyclase in its membrane environment.

Table 1 — Subunit and functional sizes of adenylate cyclase components in rat hepatocyte.

Component	Subunit Analysis M_r kdaltons	Reference
Catalytic (C)	50	Stengel et al. (1982)
GTP-regulatory (N)	35, 45 & 55[a]	Sternweis et al. (1981)
Glucagon-receptor (R)	58	Johnson et al. (1981)

	Functional sizes (target analysis)[b]		
Function assayed	Target size kdaltons	Possible target	Target based on subunit sizes
MnATP as substrate	~ 100	C	C_2
F^- or Gpp(NH)p-activated	~ 250	NC	$[NC]_2$
Glucagon + GTP-activated	~ 350	RNC	$[RNC]_2$
Ground-state precursor	~ 1300	$[RNC]_n$ or $[RN]_n$	$([RNC]_2)_4$ or $[RN]_{10}$

[a] 45K and 55K dalton units are structurally and functionally homologous.
[b] Data taken from Schlegel et al. (1979).

Table 1 includes recent data on the subunit sizes reported for the glucagon receptor (Johnson et al., 1981), the purified N_s unit (Sternweis et al., 1981), and the estimated size of C when liberated in soluble form from membranes by proteases (Stengel et al., 1982). Note that the target sizes of the units are approximately twice that of the subunits for R, N, and C. Again assuming a complex between these units, such findings suggest that the R unit for glucagon is linked to C via a dimer of N_s which, as discussed previously, is a heterodimer. From the standpoint of the known regulatory properties of the glucagon-sensitive system, this is an interesting structure. For example, guanine nucleotides induce different states of glucagon receptor from that of the resting state and, under the same conditions, promote activation of C in the presence of the hormone. Analysis of the actions of guanine nucleotides on hormone binding and activation of C gave rise to the conclusion that the sites of of guanine nucleotide action on these processes were different, and were designated N_1 and N_2 (Lad et al., 1977). If indeed, a dimer of N_s links R with C, a relatively straightforward explanation for N_1 and N_2 can be surmised, namely, that one unit of the dimer interacts with R whereas the other interacts with C. In effect, each unit of the assembled N_s dimer can be considered structurally and functionally as distinct N units with differing interacting forces for R and C. With the possibility of an assymetrical arrangement of the 45K and 35K subunits in the dimeric structure and of differing articulation with the dimers of R and C, the holoenzyme could take a number of configurations. These could satisfy differing states of R for agonist binding, different activity states of N for the binding and actions of GTP and divalent cations, and different states of the catalytic unit that converts MgATP to cyclic AMP. Finally, even though the various subunits comprising the holoenzyme are structurally 'leashed' through intersubunit forces, the system could display kinetics as though they functioned as separate units and then coupled, in step-wise fashion, in response to the effector ligands.

3. CONCLUSIONS

It is evident from this brief discussion of adenylate cyclase systems that regulation of cyclic AMP production by hormones, neurotransmitters, and local hormones can take different pathways, some convergent, others divergent. From the standpoint of structural features, notable achievements have been made in purifying N_s, in identifying and isolating R units, and in gaining some information on structure with the target size technique. The nature of N_i and its attachments to R need to be further clarified. There are a number of functional similarities in the behaviour of N_s and N_i to warrant the suggestion that they are likely to be structural homologues. Since so many hormones and neurotransmitters act in common through N_s or N_i units, one might also argue that the R units for each hormone are structural homologues, the variant regions of each being the sites of binding of the hormones or neurotransmitters. However, since the R units appear to interact in rather selective manner with N_s, N_i, and the putative M units for cations, one must also consider that R units diverge in structure and function not only with respect to the specific hormone binding regions, but also with respect to the regions for interaction with the N and M components. Target analysis suggests that R units function as dimers. These could consist both of homologous and heterologous subunits, each type designed for selective interactions with the other subunits that comprise what I term the functional receptor (the RN complexes). The major point to be emphasised is that receptor specificity, long held to be primarily the function of the hormone binding sites, is actually the combined result of all of the subunits comprising the receptor, including, R, N, M and whatever other subunits are eventually discovered to be part of the regulatory process. It is already evident from studies with the purified acetylcholine receptor that much greater insights are to be gained by understanding the structure of all components, not simply those responsible for binding hormones.

REFERENCES

Bar, H. P. and Hechter, O. (1969) Adenyl cyclase and hormone action III. Calcium requirement for ACTH stimulation of adenylate cyclase. *Biochem. Biophys. Res. Commun.* **35**, 681–686.

Birnbaumer, L. and Rodbell, M. (1969) Adenyl cyclase in fat cells II. Hormone receptors. *J. Biol. Chem.* **244**, 3477–3482.

Birnbaumer. L., Pohl, S. L. and Rodbell, M. (1969) Adenyl cyclase in fat cells I. Properties and the effects of adrenocorticotropin and fluoride. *J. Biol. Chem.* **244**, 3468–3476.

Bradham, L. S. and Cheung, W. Y. (1980) Calmodulin-dependent adenylate cyclase, in *Calcium and Cell Function* (Cheung, W. Y., ed.) pp. 109–126. Academic Press, New York.

Campbell, B. J., Woodward, G. and Borber, V. (1972) Calcium-mediated interactions between the antidiuretic hormone and renal plasma membranes. *J. Biol. Chem.* **247**, 6167–6173.

Evain, D. and Anderson, W. B. (1979) Inhibitory effect of guanyl nucleotides towards adenylate cyclase of chinese hamster ovary cell membranes activated in vitro by cholera toxin. *J. Biol. Chem.* **254**, 8726–8729.

Ferguson, K. M., Northrup, J. K. and Gilman, A. G. (1982) Goat antibodies to the regulatory component of adenylate cyclase. Fed. Proc. (abstr.) p. 1407.

Florio, V. A. and Ross, E. M. (1982) Direct inhibition of the catalytic protein of adenylate cyclase at the adenosine P site. *Fed. Proc. (abstr.)* p. 1408.

Gavish, M., Goodman, R. R. and Snyder, S. H. (1982) Solubilized adenosine receptors in the brain: Regulation by guanine nucleotides. *Science* **215**, 1633–1634.

Hanski, E., Sternweis, P. C., Northrup, J. K., Dromerick, A. W. and Gilman, A. G. (1981) The regulatory component of adenylate cyclase: purification and properties of the turkey erythrocyte protein. *J. Biol. Chem.* **256**, 12911–12919.

Heideman, W., Wierman, B. M. and Storm, D. R. (1982) GTP is not required for calmodulin stimulation of bovine brain adenylate cyclase. *Proc. Natl. Acad. Sci. USA* **79**, 1462–1465.

Jakobs, K. H., Aktories, K. and Schultz, G. (1981) Inhibition of adenylate cyclase by hormones and neurotransmitters. *Adv. Cyc. Nucleotide Res.* **14**, 173–187.

Jakobs, K. H., Lasch, P., Minuth, M., Aktories, K. and Schultz, G. (1982) Uncoupling of alpha-adrenoceptor-mediated inhibition of human platelet adenylate cyclase by N-ethylmeleimide. *J. Biol. Chem.* **257**, 2829–2833.

Johnson, G. L., MacAndrew, V. I. and Pilch, P. F. (1981) Identification of the glucagon receptor in rat liver membranes by photo-affinity crosslinking. *Proc. Natl. Acad. Sci. USA* **78**, 875–878.

Katada, T., Amano, T. and Ui, M. (1982) Modulation by islet-activating protein of adenylate cyclase activity in C_6 Glioma cells. *J. Biol. Chem.* **257**, 3739–3746.

Lad, P. M., Welton, A. F. and Rodbell, M. (1977) Evidence for distinct guanine nucleotide sites in the regulation of the glucagon receptor and of adenylate cyclase activity. *J. Biol. Chem.* **252**, 5942–5946.

Londos, C. and Preston, M. S. (1977a) Regulation by glucagon and divalent cations of inhibition of hepatic adenylate cyclase by adenosine. *J. Biol. Chem.* **252**, 5951–5956.

Londos, C. and Preston, M. S. (1977b) Activation of the hepatic adenylate cyclase system by divalent cations: a reassessment. *J. Biol. Chem.* **252**, 5957–5961.

Londos, C. and Rodbell, M. (1975) Multiple inhibitory and activating effects of nucleotides and magnesium on adrenal adenylate cyclase. *J. Biol. Chem.* **250**, 3459–3465.

Rodbell, M. (1975) On the mechanism of activation of fat cell adenylate cyclase by guanine nucleotides: An explanation for biphasic inhibitory and stimulatory effects of the nucleotides and the role of hormones. *J. Biol. Chem.* **250**, 5826–5834.

Rodbell, M. (1980) The role of hormone receptors and GTP-regulatory proteins in membrane transduction. *Nature* **284**, 17–22.

Rodbell, M., Lad, P. M., Nielsen, T. B., Cooper, D. M. F., Schlegel, W., Preston, M S., Londos, C. and Kempner, E. S. (1981) The structure of adenylate cyclase systems. *Adv. Cyc. Nucleotide Res.* **14**, 3–14.

Ross, E. M. and Gilman, A. G. (1980) Biochemical properties of hormone-sensitive adenylate cyclase. *Ann. Rev. Biochem.*, **49**, 533–564.

References

Schlegel, W., Kempner, E. S. and Rodbell, M. (1979) Activation of adenylate cyclase in hepatic membranes involves interactions of the catalytic unit with multimeric complexes of regulatory proteins. *J. Biol. Chem.* **254**, 5168–5176.

Seamon, K. B. and Daley, J. W. (1981) Activation of adenylate cyclase by the diterpene forskolin does not require the guanine nucleotide regulatory protein. *J. Biol. Chem.* **256**, 9799–9801.

Simonds, W. F., Koski, G., Streaty, R. A., Hjelmeland, L. M. and Klee, W. A. (1980) Solubilization of active opiate receptors. *Proc. Natl. Acad. Sci. USA* **77**, 6423–6427.

Stadel, J. M., DeLean, A. and Lefkowitz, R. J. (1982) Molecular mechanisms of coupling in hormone-receptor-adenylate cyclase systems. *Adv. Enzymol.* **53**, 1–43.

Stengel, D., Guenet, L. and Hanoune, J. (1982) Proteolytic solubilization of adenylate cyclase from membranes deficient in regulatory component; properties of the solubilized enzyme (submitted for publication).

Sternweis, P. C., Northrup, J. K., Smigel, M. D. and Gilman, A. G. (1981) The regulatory component of adenylate cyclase: purification and properties. *J. Biol. Chem.* **256**, 11517–11526.

Wastek, G. J. and Yamamura, H. I. (1981) Acetylcholine receptors, in: *Neurotransmitter Receptors*, Part 2 (Yamamura, H. I. and Enna, S. J., eds.) pp. 103–128. Chapman and Hall, London.

Westcott, K. R., LaPorte, D. C. and Storm, D. R. (1979) Resolution of adenylate cyclase sensitive and insensitive to Ca^{2+} and calcium-dependent regulatory protein (CDR) by CDR-sepharose affinity chromatography. *Proc. Natl. Acad. Sci. USA* **76**, 204–208.

13

A human mutation affecting hormone-sensitive adenylate cyclase

Henry R. Bourne

Division of Clinical Pharmacology, Departments of Medicine and Pharmacology and the Cardiovascular Research Institute, University of California, San Francisco, California 94143, U.S.A.

Abbreviations used

ADH	—	antidiuretic hormone
AHO	—	Albright's hereditary osteodystrophy
PHP	—	pseudohypoparathyroidism
PPHP	—	pseudopseudohypoparathyroid
PTH	—	parathyroid hormone
TSH	—	thyrotropin
TRH	—	thyrotropin releasing hormone

1. INTRODUCTION

Hormone-sensitive adenylate cyclase consists of at least three distinct membrane proteins. These include hormone receptors (R), the catalytic unit (C), and a coupling component that is required for functional coupling of R and C. In the present review this coupling component will be called the N protein. As described by Rodbell in Chapter 12, several laboratories are investigating the molecular interactions of N with R and C and the regulation of these interactions by guanine nucleotides.

The hormone-sensitive cyc^- mutant cell line, derived from S49 murine lymphoma cells (Bourne et al., 1975), functionally lacks the N protein (Ross and Gilman, 1980). By observing that a protein factor extracted from normal plasma membranes can complement the phenotypic defect of cyc^- membranes, Ross and Gilman (1977, 1980) discovered the N protein. Their complementation procedure formed the basis of a sensitive assay for the N protein, which allowed its biochemical characterisation and eventual purification (Northup et al., 1980). The N protein is present in all vertebrate tissues so far examined. It appears to mediate stimulation of adenylate cyclase by a wide variety of peptide hormones, biogenic amines, and prostaglandins, and also by the enterotoxin of *Vibrio cholerae*.

Because every protein is the product of a mutable gene, and because the N protein was discovered by investigating phenotypic consequences of a mutation, it was natural to ask what hereditary disorder would result from genetic deficiency of the N protein in man. From the outset it appeared unlikely that complete absence of the N protein, analogous to the mutant lesion in cyc^-, could be compatible with life. Instead, the deficiency would have to be partial, and would presumably produce resistance to the action of any one or all of the hormones known to utilise cyclic AMP (cAMP) as an intracellular second messenger. This review will summarise biochemical and clinical evidence that N deficiency causes a well-defined hereditary disorder, pseudohypoparathyroidism.

2. PSEUDOHYPOPARATHYROIDISM, TYPE I

In 1942 Albright and his colleagues reported three patients who presented the clinical picture of hypoparathyroidism but did not respond to administration of parathyroid extract. They concluded that these cases of 'pseudohypoparathyroidism' (PHP) were caused by 'failure of the organism to respond to parathyroid hormone'. During the ensuing decades more cases of PHP were described. These patients were hypocalcemic and hyperphosphatemic, and suffered from tetany, seizures, and other consequences of lowered serum calcium. Many of these patients exhibited a typical habitus, including short stature, round face, brachydactyly, and subcutaneous calcifications, which is termed Albright's Hereditary Osteodystrophy (AHO). Some PHP patients were obese, and many exhibited mild retardation. With the advent of radioimmunoassay, PHP patients were found to have elevated serum concentrations of parathyroid hormone (PTH).

In 1969 Chase, Melson, and Aurbach devised a biochemical test of Albright's hypothesis, based on the observation that PTH stimulates synthesis of cAMP by renal adenylate cyclase and causes increased urinary cAMP excretion in normal patients. In PHP patients, PTH gave little or no increase in urinary cAMP excretion, as compared with normal or hypoparathyroid subjects. This result provided an unambiguous and useful criterion for the diagnosis of PHP. The disorder is now called PHP-I because subjects have been identified in whom resistance to the phosphaturic effect is associated with normal or even elevated urinary cAMP excretion in response to the hormone (Drezner *et al.*, 1973), a syndrome termed PHP-II.

During the past decade several reports have described endocrine abnormalities in PHP-I that cannot be explained simply on the basis of isolated resistance to PTH. Many PHP-I patients are clinically hypothyroid, with elevated serum thyrotropin (TSH) (Marx *et al.*, 1971; Werder, 1979). Some show a blunted serum prolactin response to administration of thyrotropin releasing hormone (TRH) (Carlson *et al.*, 1977), and some are resistant to the urine concentrating effect of antidiuretic hormone (ADH) (Brickman and Weitzman, 1978). Instances of resistance to gonadotrophins and glucagon have also been reported (Wolfsdorf *et al.*, 1978).

Because cAMP may act as a second messenger for all these hormones, in addition to PTH, it appeared possible that these additional endocrine abnormalities might be due to deficient cAMP synthesis in several tissues. Because each of these hormones acts via a different class of receptors, it was likely that PHP-I patients have a generalised defect in a component of adenylate cyclase that is distal to the receptor but essential for cAMP synthesis.

3. N ACTIVITY IN ERYTHROCYTES OF PHP-I PATIENTS

Human erythrocytes, unlike the target organs phenotypically affected in PHP-I, are conveniently accessible for biopsy. Erythrocytes also contain measurable amounts of N protein activity (Kaslow *et al.*, 1979), although they lack catalytic adenylate cyclase. Thus it was possible to use erythrocytes of PHP-I patients in initial tests of the hypothesis that N activity is generally deficient in cells of PHP-I patients (Farfel *et al.*, 1980, 1981; Levine *et al.*, 1980). We measured N activity in erythrocyte membranes by two methods: (1) an adaptation of the *in vitro* complementation procedure of Ross and Gilman, utilising N-deficient cyc^- membranes; (2) quantification of the cholera toxin-catalysed transfer of [^{32}P] ADP-ribose from radiolabelled NAD to the 42,000-dalton peptide subunit of the erythrocyte N protein. By both methods, we found that N activity was reduced by about 50% in erythrocytes of many PHP-I patients, as compared to measurements in erythrocytes of normal subjects, hypoparathyroid patients, and a single patient with PHP-II.

The erythrocyte N defect served as a useful biochemical marker to clarify inheritance of PHP-I. Some kindreds showed a dominant pattern of inheritance of the biochemical defect, while in other kindreds the pattern strongly indicated recessive inheritance. The dominant pattern, most often described in earlier studies, would be expected if affected patients inherited one normal and one abnormal allele coding for a peptide subunit of N; such individuals would probably express approximately 50% of normal N activity in their cells. The recessive pattern of inheritance, found in three families (Farfel *et al.*, 1981, and unpublished) implies a different genetic defect, which so far remains unexplained.

Erythrocytes of some PHP-I patients exhibited quantitatively normal N activities. For convenience, patients with reduced erythrocyte N activity are termed PHP-Ia, while patients with normal erythrocyte N activity are termed PHP-Ib. The two sets of patients are genetically and biochemically distinct: no family yet studied has included phenotypically affected individuals in both groups (Farfel *et al.*, 1980, 1981, and unpublished results). A total of 67 reported PHP-I patients have been 'typed' with respect to presence or absence of the erythrocyte N defect, in two laboratories (Table 1). It is evident that most PHP-Ia patients exhibit AHO, while most PHP-Ib patients have no skeletal abnormalities.

Table 1 — Albright's Hereditary Osteodystrophy (AHO) in subtypes of PHP-I.[a]

Skeletal phenotype	Number of reported patients		
	PHP-Ia	PHP-Ib	Total
AHO+	38	8	46
AHO−	1	20	21
Total	39	28	67

[a]Table summarises studies of 46 patients in our laboratory (Farfel et al., 1980, 1981, and unpublished) and 23 patients studied by M. A. Levine, A. Spiegel, and coworkers (1980, 1981a, 1981b). Two of the PHP-Ia patients (both AHO+) were studied in both laboratories, and are counted only once in the table.

4. N ACTIVITY IN OTHER CELLS

We have measured N activity in three additional cell types of PHP-I patients: platelets (Farfel and Bourne, 1980), cultured skin fibroblasts (Bourne et al., 1981), and Epstein-Barr virus-transformed lymphoblasts (unpublished). In every case, N activity was decreased, by about 50%, in cells from patients with reduced erythrocyte N activity (PHP-Ia), while cells from PHP-Ib patients showed quantitatively normal N activity. The observations in fibroblasts and lymphoblasts indicate that the N defect persists in cells propagated for many generations outside the endocrine and chemical environment of the patient. Thus, deficient N activity is programmed in the genome of fibroblasts and lymphoblasts of PHP-Ia patients.

Our laboratory has not succeeded in demonstrating reproducible decreases in adenylate cyclase activity or cAMP accumulation in any of these cell types. This contrasts with the prediction that N-deficient cells should exhibit deficient responses to hormones and other effectors (cholera toxin, NaF, guanine nucleotides) that stimulate cAMP synthesis by N-dependent mechanisms. This failure was related to great variability (two- to ten-fold) in cAMP and adenylate cyclase responses among cell populations (e.g. fibroblasts) obtained from different normal subjects. Such variability was sufficient to preclude detection of hyporesponsiveness in cells from PHP-Ia subjects.

Because functional abnormalities of platelets, fibroblasts, and lymphocytes have not been described in PHP-I, we cannot exclude the possibility that altered hormonal responsiveness in this disorder is confined to a small number of target organs — including kidney, thyroid, and perhaps bone — even though deficiency of N activity is generalised. Alternatively, variable hormone responsiveness of the cells tested might be due to heterogeneous expression of peptide gene products other than N protein, including hormone receptors, catalytic adenylate cyclase, and cyclic nucleotide phosphodiesterases.

Adenylate cyclase was reportedly abnormal in particulate extracts of a kidney biopsy performed on one PHP-I patient (Drezner and Burch, 1978). PTH-responsive adenylate cyclase in this extract was observed to require exogenous GTP in order to synthesise cAMP at a rate comparable to those observed in kidney biopsies of three normal subjects. While this finding might have been due to differences in removal of endogenous guanine nucleotides during preparation of particulate extracts from biopsies, it could also have reflected abnormality of the N component in membranes of the PHP-I kidney. Fortunately, particulate extracts from the PHP-I kidney and the control biopsies were preserved, and the patient was available for venipuncture. Downs, Spiegel and their co-workers measured N activity in the PHP-I kidney extract and in erythrocytes of the patient: activities were reduced by about 50% in both assays, as compared to normal controls (Allen Spiegel, personal communication).

Taken together with the previously reported measurements of adenylate cyclase in kidney of this patient, the findings of Downs, Spiegel, and their colleagues are extremely important. Their results indicate that the N defect of PHP-Ia, detectable in erythrocytes, other blood cells, and fibroblasts, is expressed biochemically as a defect of hormone-responsive adenylate cyclase in a clinically affected endocrine target tissue.

5. CLINICAL ENDOCRINE STUDIES

Resistance to the metabolic effects of PTH in PHP-I has been amply documented. More recently, several laboratories have reported target organ resistance to other hormones in PHP-I. Some of these studies were prompted by the observation of deficient N activity in erythrocytes (Table 2). From these results several generalisations can be made.

(1) Each of the hormones involved (Table 2) is thought to produce at least some of its effects via cAMP.

(2) Defective responses are often clinically silent, and are detectable only with the use of sensitive tests.

(3) Functional integrity of the endocrine response in the patient may be maintained by elevated blood concentrations of the stimulating hormone.

(4) PHP-Ia patients exhibit somewhat more severe and generalised resistance to hormones than do patients with PHP-Ib.

These results, combined with the evidence that N deficiency is present in many and perhaps all cells of PHP-Ia patients, raise an interesting question:

Table 2 — Endocrine studies in PHP-I patients.[a]

Hormone	Target organ	Response	Fraction of patients exhibiting abnormal response PHP-Ia	PHP-Ib
TSH	Thyroid	Clinical hypothyroidism	9/15	6/11
		Exaggerated TSH response to TRH	13/13	0/3
Glucagon	Liver	Plasma cAMP	12/12	0/9
		Plasma glucose	0/12	0/9
Gonadotrophins	Gonad	Oligo/amenorrhea in women	9/11	0/3
		Basal/stimulated plasma gonodotrophins	0/11	0/3
ADH	Kidney	Urine concentration and plasma ADH after water deprivation	3/3	2/3
TRH	Pituitary	Plasma prolactin	3/5	2/5

[a]Data collated from Farfel *et al.* (1980, 1981) and Levine *et al.* (1980, 1981a, 1981b).

why are most of the endocrine homeostatic mechanisms in these patients functionally intact, while deranged calcium homeostasis is a prominent clinical feature of the disorder?

Table 2 indicates that the discrepancy is not absolute, providing that sensitive tests of endocrine function are employed. In addition, even defective calcium homeostasis is not clinically obvious in all PHP-Ia patients. Some of these patients are 'pseudopseudohypoparathyroid' (PPHP), a designation Albright *et al.* (1952) applied to relatives of PHP patients who exhibited AHO but were normocalcemic. Available data suggest two possible compensatory mechanisms that could mask overt expression of the N defect in cAMP-mediated endocrine responses: (1) hyporesponsiveness of adenylate cyclase can be overcome by elevated concentrations of the stimulating hormone, as appears to be the case for PTH in PPHP, and probably for TSH and ADH in many PHP-Ia patients; (2) normal function in the face of deficient N and decreased cAMP synthesis may be preserved by excess or 'spare' capacity of a tissue to respond to the cAMP generated (e.g. the normal glycemic responses observed in response to glucagon, even when plasma cAMP responses are reduced (Levine *et al.*, 1981a, 1981b; Brickman *et al.*, 1981).

Nonetheless, these compensatory mechanisms are conspicuously ineffective in preserving calcium homeostasis in most PHP-Ia patients. Why is this so? The reasons for this failure of compensation must reside in features of the intricate physiological feedback loops that control calcium concentration. This question invites speculation, but firm evidence is not available.

6. THE PHP–Ib PHENOTYPE

The biochemical basis of PHP-Ib with normal erythrocyte N activity has not been defined. These PHP-Ib patients are resistant to the metabolic effects of PTH, including stimulation of urinary cAMP excretion. Fibroblasts (Bourne et al., 1981), platelets and lymphoblasts (unpublished) from a small number of PHP-Ib patients were found to contain quantitatively normal N activities.

It is likely that PHP-Ib will turn out to be a heterogeneous syndrome. Some of these patients may bear an isolated defect affecting membrane receptors for PTH, while others may bear defects in catalytic adenylate cyclase, cyclic nucleotide phosphodiesterases, or other gene products. Indeed, some of these patients may have a defective N protein whose activity appears normal in the relatively crude assays we have used.

ACKNOWLEDGEMENT

This work was supported in part by grants from the National Institutes of Health and The March of Dimes.

REFERENCES

Albright, R., Burnett, C. H., Smith, P. H. and Parson, W. (1942) Pseudohypoparathyroidism – an example of Seabright Bantam syndrome. *Endocrinology* **30**, 922–932.

Albright, F., Forbes, A. P. and Henneman, P. H. (1952) Pseudo-pseudohypoparathyroidism. *Trans. Assoc. Am. Phys.* **65**, 337–350.

Bourne, H. R., Coffino, P. and Tomkins, G. M. (1975) Selection of a variant lymphoma cell deficient in adenylate cyclase. *Science* **187**, 750–752.

Bourne, H. R., Kaslow, H. R., Brickman, A. S. and Farfel, Z. (1981) Fibroblast defect in psuedohypoparathyroidism, Type I: Reduced activity of receptor-cyclase coupling protein. *J. Clin. Encrinol. Metab.* **53**, 636–640.

Brickman, A. S. and Weitzman, R. E. (1978) Renal resistance to arginine vasopressin in pseudohypoparathyroidism. *Clin. Res.* **26**, 164A, abstract.

Brickman, A. S., Carlson, H., Ragavan, S., Williams, A., Katz, M. and Levine, S. (1981) Resistance to glucagon in pseudohypoparathyroidism. *Clin. Res.* **29**, 93A, abstract.

References

Carlson, H. E., Brickman, A. S. and Bottazzo, G. F. (1977) Prolactin deficiency in pseudohypoparathyroidism. *N. Engl. J. Med.* **296**, 140–144.

Chase, L. R., Melson, G. L. and Aurbach, G. D. (1969) Pseudohypoparathyroidism: Defective excretion of 3′,5′-AMP in response to parathyroid hormone. *J. Clin. Invest.* **48**, 1832–1844.

Drezner, M. K., Neelon, F. A. and Lebovitz, H. E. (1973) Pseudohypoparathyroidism type II: A possible defect in the reception of the cyclic AMP signal. *N. Engl. J. Med.* **289**, 1056–1060.

Drezner, M. K. and Burch, W. M., Jr. (1978) Altered activity of the nucleotide regulatory site in the parathyroid hormone-sensitive cyclase from the renal cortex of a patient with pseudohypoparathyroidism. *J. Clin. Invest.* **62**, 1222–1227.

Farfel, Z. and Bourne, H. R. (1980) Deficient activity of receptor-cyclase coupling protein in platelets of patients with pseudohypoparathyroidism. *J. Clin. Endocrinol. Metab.* **51**, 1202–1204.

Farfel, Z., Brickman, A. S., Kaslow, H. R., Brothers, V. M. and Bourne, H. R. (1980) Defect of receptor-cyclase coupling protein in pseudohypoparathyroidism. *N. Engl. J. Med.* **303**, 237–242.

Farfel, Z., Brothers, V. M., Brickman, A. S., Conte, F., Neer, R. and Bourne, H. R. (1981) Pseudohypoparathyroidism: Inheritance of deficient receptor-cyclase coupling activity. *Proc. Natl. Acad. Sci. USA* **78**, 3098–3102.

Kaslow, H. R., Farfel, Z., Johnson, G. L. and Bourne, H. R. (1979) Adenylate cyclase assembled *in vitro*: Cholera toxin substrates determine different patterns of regulation by isoproterenol and guanosine 5′-triphosphate. *Mol. Pharmacol.* **15**, 472–483.

Levine, M. A., Downs, R. W., Singer, M., Marx, S. J. and Aurbach, G. D. (1980) Deficient activity of guanine nucleotide regulatory protein in erythrocytes from patients with pseudohypoparathyroidism. *Biochem. Biophys. Res. Commun.* **94**, 1319–1325.

Levine, M. A., Downs, R. W., Lasker, R. D., Marx, S. J., Moses, A. M., Aurbach, G. D. and Spiegel, A. M. (1981a) Resistance to multiple hormones in pseudohypoparathyroidism and deficient guanine nucleotide regulatory protein. *Clin. Res.* **29**, 412A, abstract.

Levine, M. A., Downs, R. W., Jr., Marx, S. J., Lasker, R. D., Aurbach, G. D. and Spiegel, A. M. (1981b) Clinical and biochemical features of pseudohypoparathyroidism, in *Hormonal Control of Calcium Metabolism* (Cohn, D. V., Talmage, R. V. and Matthews, J. L., eds.) Excerpta Medica, Amsterdam, pp. 95–101

Marx, S. J., Hershman, J. M. and Aurbach, G. D. (1971) Thyroid dysfunction in pseudohypoparathyroidism. *J. Clin. Endocrinol. Metab.* **33**, 822–828.

Northup, J. K., Sternweis, P. C., Smigel, M. D., Schleifer, L. S., Ross, E. M. and Gilman, A. G. (1980) Purification of the regulatory component of adenylate cyclase. *Proc. Natl. Acad. Sci. USA* **77**, 6516–6520.

Ross, E. M. and Gilman, A. G. (1977) Reconstruction of catecholamine-sensitive adenylate cyclase activity: Interaction of solubilized components with receptor-replete membranes. *Proc. Natl. Acad. Sci. USA* **74**, 3715–3719.

Ross, E. M. and Gilman, A. G. (1980) Biochemical properties of hormone-sensitive adenylate cyclase. *Annu. Rev. Biochem.* **49**, 533–564.

Werder, E. A. (1979) Pseudohypoparathyroidism. *Ergeb. Inn. Med. Kinderheilkd.* **42**, 191–221.

Wolfsdorf, J. I., Rosenfield, R. L., Fang, V. S., Kobayashi, R., Razdan, A. K. and Kim, M. H. (1978) Partial gonadotrophin-resistance in pseudohypoparathyroidism. *Acta Endocrinol.* **88**, 321–328.

14

A study of the mRNA and genes coding for the nicotinic acetylcholine receptor

K. Sumikawa,[*] **R. Miledi,**[†] **M. Houghton**[**] **and E. A. Barnard**[‡]

Molecular Genetics Department, Searle Research & Development, P.O. Box 53, Lane End Road, High Wycombe, Buckinghamshire HP12 4HL, U.K.

[*]present address: Mitsubishi-Kasei Institute of Life Sciences, Minamiooya, Machidashi, Tokyo, Japan.

[†]Department of Biophysics, University College London, Gower Street, London WC1E 6BT, U.K.

[**]present address: Chiron Corporation, 4560 Horton Street, Emeryville, California, 94608, USA.

[‡]Department of Biochemistry, Imperial College of Science and Technology, London SW7 2AZ, U.K.

Abbreviations used
- AChR — acetylcholine receptor
- ACh — acetylcholine
- SDS — sodium dodecyl sulphate
- DTT — dithiothreitol
- α-BuTX — α-bungarotoxin
- MBTA — 4-(N-maleimido)-benzyl-trimethylammonium iodide
- AChE — acetylcholinesterase

1. INTRODUCTION

In the last decade, considerable advances have been made towards an understanding of the structure and function of a neurotransmitter receptor, namely the nicotinic acetylcholine receptor (AChR). The fact that this is the best-characterised receptor at the molecular level owes much to the use of the electric organ of the electric ray *Torpedo*, an exceptionally rich source of AChR, and α-toxins from snake venom, which bind to the AChR with very high affinity and specificity. Hence, a number of laboratories have purified and characterised the receptor from *Torpedo* electric organ, and shown it to be an oligomeric glycoprotein with 4 types of subunit (α, β, γ and δ). This oligomer contains the binding sites for the neurotransmitter, acetylcholine (ACh), and also the ion channel (selective for Na^+ and K^+) which in the membrane is opened when ACh is bound (Heidmann and Changeux, 1978; Karlin, 1980). The very powerful methods of modern molecular genetics can be applied to facilitate the analysis of structural requirements in this assembly. Thus, if mRNAs encoding the different subunits can be obtained, and if the corresponding genes can be cloned, then the requirement of each subunit in the assembly can in principle be probed, and selective changes therein can be tested, both for the formation of the ligand-binding site and for the formation and control of the receptor channel. We describe here initial studies in this approach, which requires firstly a consideration of the essential elements of the receptor, then the identification of mRNAs coding for the subunits, and then their translation to form an oligomer which binds the specific ligands and forms the receptor channel structure.

2. STRUCTURE OF TORPEDO NICOTINIC ACETYLCHOLINE RECEPTOR

2.1 Oligomeric composition

AChR can be isolated in detergent solution as a pure glycoprotein, with an isoelectric point near pH 5. In detergent solution the AChR from *Torpedo* species shows two major forms with sedimentation coefficients of about 9 S (corresponding to a molecular weight of about 260,000) and 13 S (Fig. 1A) (Karlin, 1980, Raftery, *et al.*, 1980; Heidmann and Changeux, 1978). The following evidence strongly suggests that the heavy (13 S) form is a dimer of the light (9 S) form. (a) Treatment with a reducing agent converts almost all of the heavy form to the light form (Chang and Bock, 1977). (b) The peptide maps of tryptic digests of separated heavy and light forms are identical (Fig. 1B). (c) The SDS gel patterns of the heavy form and the light form were found to be identical when these forms were reduced before electrophoresis (Suarez-Isla and Hucho, 1977; Chang and Bock, 1977; McNamee *et al.*, 1975).

The question arises, therefore, whether the existence of the receptor in the heavy or the light form is physiologically important. The inhibition of the permeability response by reduction of *Torpedo* membranes with 5 mM DTT is completely reversed by 5 μM dithiobischoline, which does not generate the dimer (Hamilton *et al.*, 1979). The implication is that the disulphide cross-link between the light oligomers plays no role in the control of the cation channels of the receptor. In addition alkylation of electric organ membranes from *Torpedo* with iodoacetamide results in isolation of only the heavy oligomer following solubilisation, but this has no discernible effect on ion efflux mediated by carbamylcholine (Miller *et al.*, 1978). Moreover, when each form is artificially reconstituted in a lipid membrane, the ACh-controlled channel is formed in both cases (Anholt, *et al.*, 1980; Wu and Raftery, 1981). The functional significance of the heavy oligomer in native membranes is not yet clear.

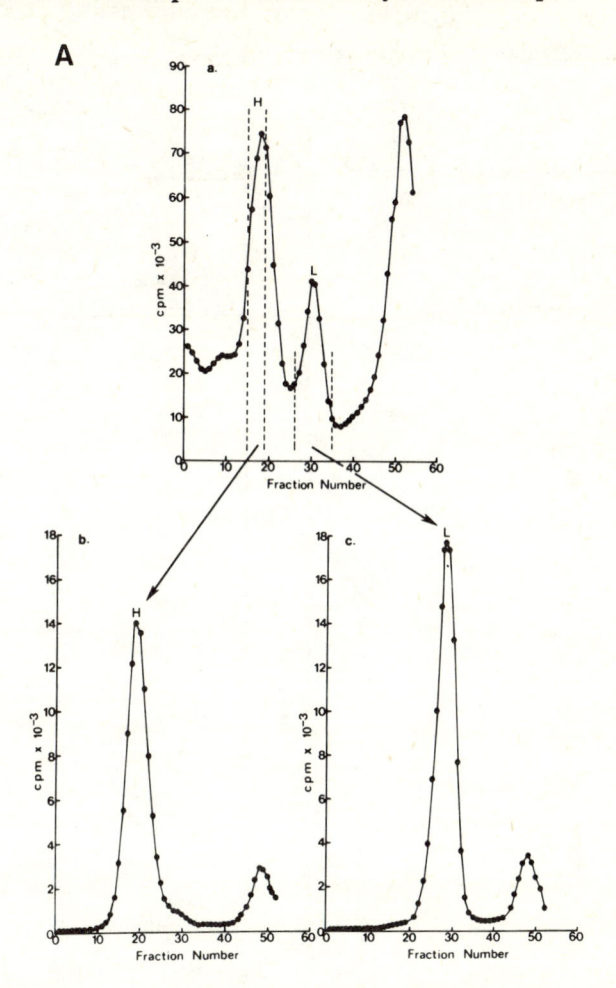

Fig. 1 – A. Isolation of two forms of AChR. (a) Purified AChR labelled with [^{3}H] α-BuTX was sedimented through a sucrose gradient to determine which fractions contain the light form and the heavy form. (b), (c) Re-centrifugation of separately harvested heavy and light forms from parallel gradients on which purified unlabelled AChR was layered. (Dotted lines in (a) show the fractions pooled from parallel gradients for re-centrifugation. The two forms were labelled with [^{3}H] α-BuTX, before run.) B. Peptide maps of tryptic digests (carried out using high-pressure liquid chromatography combined with a fluorescent assay with o-phthalaldehyde). (a) heavy form (b) light form. Eight stepwise buffers 0.2 M in sodium, were used to elute the peptides. Their pH values were: 3.25 (citrate), 4.15 (citrate) 4.60 (citrate), 5.00 (citrate), 5.45 (citrate), 6.25 (citrate) 7.20 (phosphate), 9.50 (borate). The schedule of buffers, expressed in minutes, was: 60, 60; 180 120, 120; 60; 60; 60. The peaks eluted within the first 60 min are probably amino acids. From Sumikawa and O'Brien (unpublished).

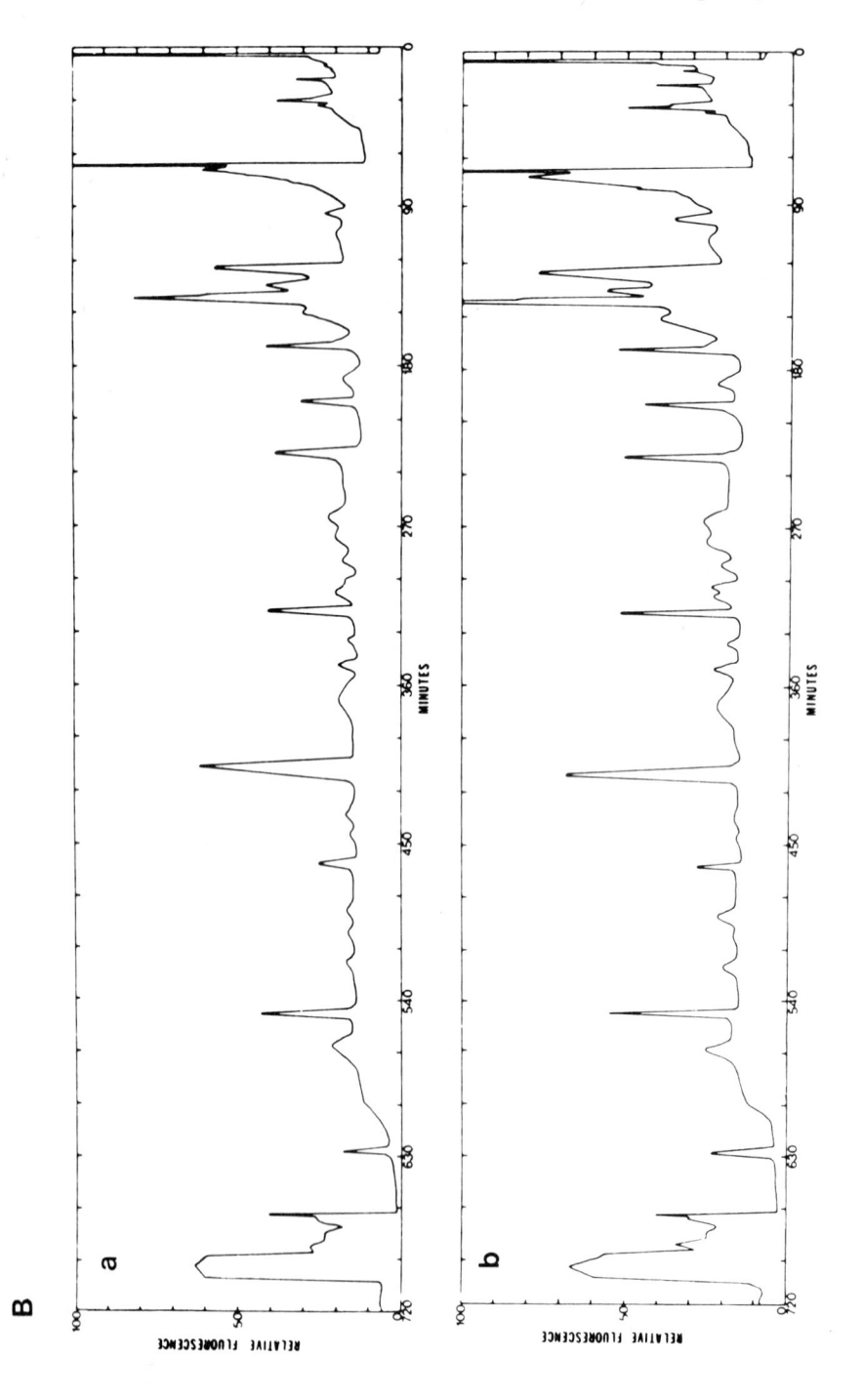

252 mRNA/Genes Coding for the Nicotinic Acetylcholine Receptor [Ch. 14

2.2 Subunit structure

There is now general agreement that *Torpedo* AChR consists of four different polypeptides (α, β, γ and δ) of apparent molecular weights of 40,000, 49,000, 57,000 and 65,000 (Fig. 2A) (the values observed depending on the species of *Torpedo* and the experimental conditions) in the molar ratio 2:1:1:1 (Karlin, 1980; Raftery et al., 1980). The individual subunits have quite different peptide maps (Fig. 2B). However, Raftery et al. (1980) have reported the first partial amino acid sequence data in these polypeptides, for up to 54 residues at the N-terminus, and these show that the 4 subunits have distinct but homologous structures. At 11 of these positions, all 4 subunits have the same amino acid. The sequence homologies in this region range, for the 4 subunits, between 35% and 50%. The peptide analysis failed to detect this subunit homology. It was, however, reflected in antibody cross-reactivities of the 4 subunits (Mehraban et al., 1982).

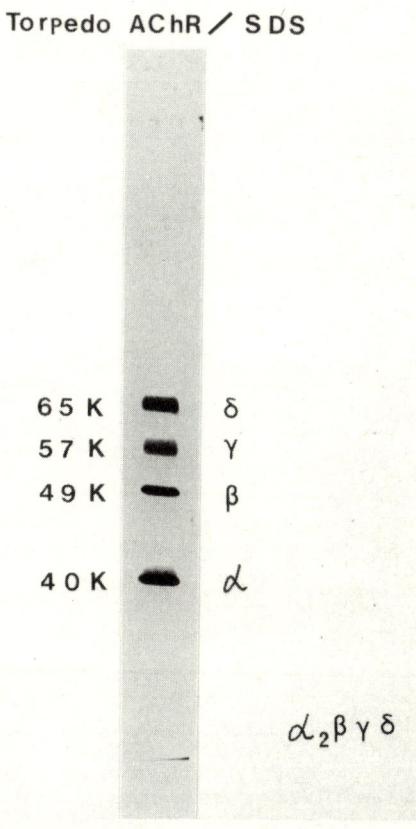

Fig. 2A — SDS gel electrophoresis of purified *Torpedo* AChR. The gel was stained with silver after electrophoresis.

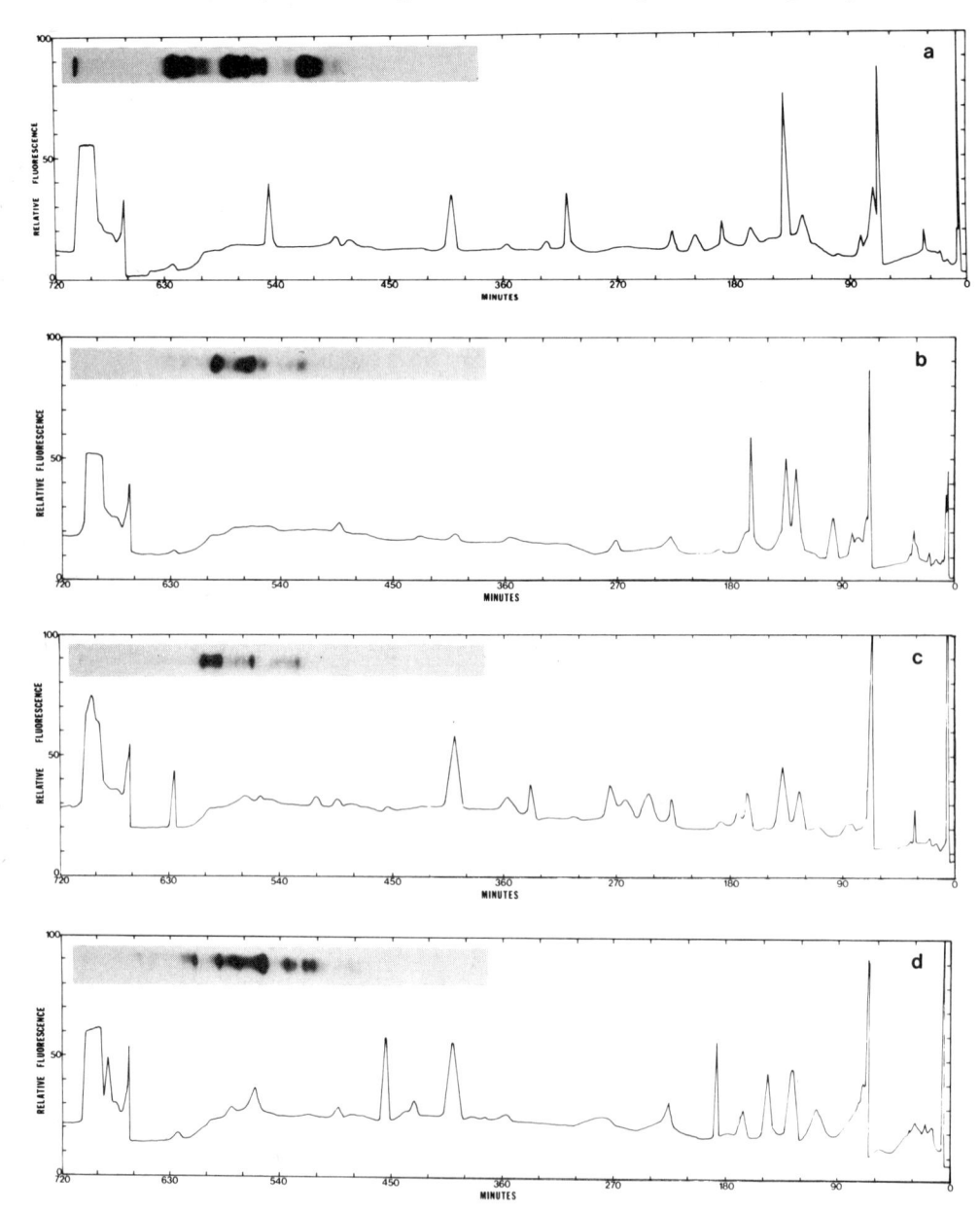

Fig. 2B — Peptide maps of tryptic digests of the separated subunits by high-pressure liquid chromatography and peptide maps of the separated subunits by limited proteolysis in SDS gels using *S. aureus* protease V8 (inserts). (a) α subunit (b) β subunit (c) γ subunit (d) δ subunit From Sumikawa and O'Brien (unpublished).

The smallest subunit (α) has been shown by affinity labelling to carry the ACh binding site (Karlin, 1980; Lyddiatt et al., 1979). However, the functional contribution of the other subunits remains to be determined. A most important question outstanding is the identity of the essential molecular elements of the ion channel.

3. STUDIES ON THE BIOSYNTHESIS OF THE ACETYLCHOLINE RECEPTOR

3.1 Biosynthesis of the acetylcholine receptor in a cell-free system

This is necessarily the first stage in the analysis, i.e. production and recognition of receptor polypeptides. Translation of the appropriate mRNA in a cell-free system followed by immunoprecipation with anti-AChR antibody constitutes a simple assay for the identification of AChR mRNA. We have isolated mRNA from the electric organ of *Torpedo marmorata* and translated it in an mRNA-dependent rabbit reticulocyte lysate in the presence of [^{35}S] methionine. Following immunoprecipitation with a rabbit antiserum raised against the native AChR and gel electrophoresis in the presence of SDS, three of the specific AChR polypeptides could be clearly identified (Sumikawa et al., 1981), with apparent molecular weights of 40,000, 51,000, and 59,000 (Fig. 3). Although these sizes differ from those of the subunits of native AChR, this result demonstrates that the antigenic sites in native receptor already exist in the separate non-processed chains. The presence of a 49,000 polypeptide found by others (Anderson and Blobel, 1981; Mendez et al., 1980) is not clearly identified (Fig. 3), since the non-AChR polypeptide with a molecular weight of 49,000 was seen in control samples. Using AChR-subunit specific antisera, Anderson and Blobel (1981) identified in mRNA translation in a wheat germ system four polypeptides with apparent molecular weights of 38,000, 49,000, 50,000 and 60,000 as the α, β, γ and δ non-processed chains, respectively. Treatment with any one of the subunit-specific antisera did not bring down all four polypeptides, suggesting that AChR polypeptides newly synthesised in a cell-free system are not assembled. In addition, the lysate translation products lack the ability to bind α-bungarotoxin (α-BuTX) (Sumikawa et al., 1981) (Fig. 3), a specific property of the native AChR, in agreement with the findings of others (Anderson and Blobel, 1981; Mendez et al., 1980) in cell-free systems. These results suggest that the α subunit requires interactions with other subunits to form the toxin-binding site. This is consistent, also, with the observation that in cultured myotubes (Anderson and Blobel, 1981) or the clonal mouse cell line BC3H-1 (Merlie and Sebbane, 1981) the formation of the toxin-binding site occurs post-synthetically over a period of about 15 min. However, interestingly, the α subunit isolated by SDS gel electrophoresis followed by renaturation binds to α-BuTX (Haggerty and Froehner, 1981), although the binding affinity is about 10^4-fold lower than that observed for native AChR. This shows that some

256   mRNA/Genes Coding for the Nicotinic Acetylcholine Receptor [Ch. 14

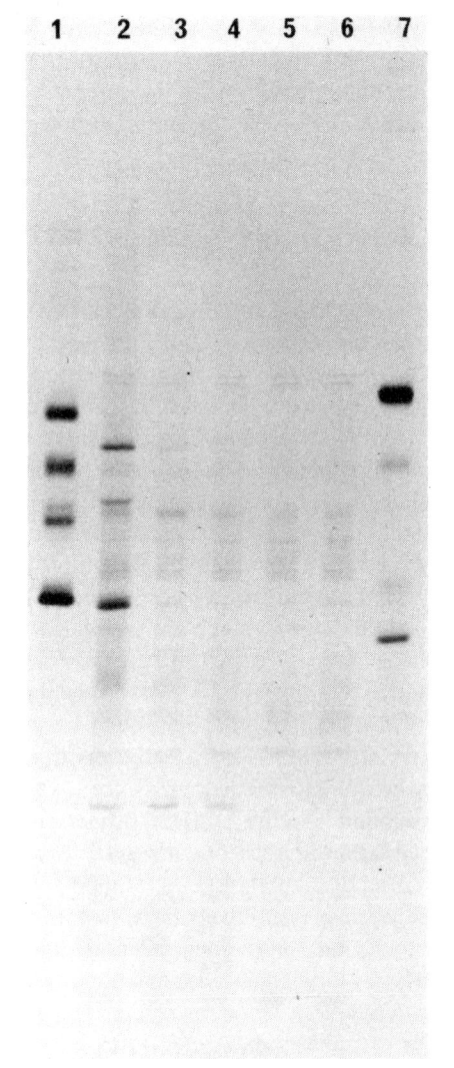

Fig. 3   Electrophoretic analysis of AChR polypeptides synthesised in a cell-free system. *Torpedo* mRNA was translated in a nuclease-treated rabbit reticulocyte lysate in the presence of [^{35}S] methionine and then the translation products were immunoprecipitated. (1) Native *Torpedo* AChR purified and radioactively-labelled with succinimidyl [2,3-^{3}H] propionate. (2) Immunoprecipitate obtained with antibody against native *Torpedo* AChR. (3) Immunoprecipitate obtained with antibody against native *Torpedo* AChR, but in the presence of a large excess of native *Torpedo* AChR. (4) Immunoprecipitate obtained with normal rabbit serum. (5) Immunoprecipitate produced with antibody against α-BuTX (control for (6)). (6) Immunoprecipitate obtained by pre-incubation with α-BuTX, before addition of antibody against α-BuTX. (7) [^{125}I] labelled standard proteins, bovine serum albumin (68K), catalase (58K), ovalbumin (43K) and lactate dehydrogenase (35K). Reproduced from Sumikawa *et al.* (1981) with permission from *Nature*.

part of the α-BuTX binding site is on the α subunit, and can be compared with the much lower but specific binding to AChR of a fragment of α-toxin comprising about half of its structure (Juillerat et al., 1982) and the fact that the binding site for ACh, as monitored by affinity labelling with MBTA (Karlin, 1980) and bromoacetylcholine, (Lyddiatt et al., 1979) is on the α subunit alone.

It seems likely therefore, that the cell-free system is incapable of faithfully synthesising the intact receptor molecule. The properties found do not go beyond those of the isolated subunits, and assembly to the native structure does not occur.

3.2 Biosynthesis of the AChR in *Xenopus* oocytes

The next stage in the analysis must involve translation to yield products which assemble, and this was achieved in an *in vivo* system, the *Xenopus* oocyte. The *Xenopus* oocyte has been shown by Gurdon and his colleagues (Gurdon, 1974; Gurdon et al., 1971. Lane et al., 1979; Lane, 1981) to constitute a highly efficient translation system for microinjected exogenous mRNA, extracted from a large variety of heterologous cells, and also to be capable of faithfully executing post-translational processing such as glycosylation, sequestration, cleavage of the pre-peptide and passage into or through all membranes of appropriate products. Since the AChR is thought to be glycosylated and sequestered in the plasma membrane, this system appears to be ideal for studying the biosynthesis of this molecule. When *Torpedo* mRNA was microinjected into oocytes, α-BuTX binding activity could be detected in the translation products. In the range of 0–15 ng mRNA injected per oocyte, the toxin binding activity increased approximately linearly with the amount of heterologous mRNA injected. Above this level, apparent saturation of the oocytes translational system was evident (Sumikawa et al., 1981). As in the case of the native *Torpedo* receptor (Vandlen et al., 1979; Lindstrom et al., 1979), the AChR synthesised in microinjected oocytes is also glycosylated, as it can all be bound to concanavalin A immobilised on Sepharose gels (Sumikawa et al., 1981). Further, tunicamycin (which inhibits protein N-glycosylation (Tkacz and Lampen, 1975)) blocked active receptor production (K. Sumikawa, unpublished results).

When [^{35}S] methionine-labelled translation products were produced in the injected oocytes and purified by affinity chromatography on α-BuTX-Sepharose, analysis by SDS/polyacrylamide gel electrophoresis disclosed the production of four toxin-specific polypeptides (Sumikawa et al., 1981) of apparent molecular weights of 40,000, 49,000, 58,000 and 66,000 (Fig. 4A). These sizes are very similar to those of the native *Torpedo* AChR. A comparison of the peptide maps for the 40,000 polypeptide synthesised in oocytes and the α subunit of *Torpedo* AChR shows that the two are very similar proteins, if not identical (Fig. 4B). These results indicate that the oocyte is, indeed, able to execute faithfully post-translational processes in the case of the mRNAs coding for this integral membrane protein.

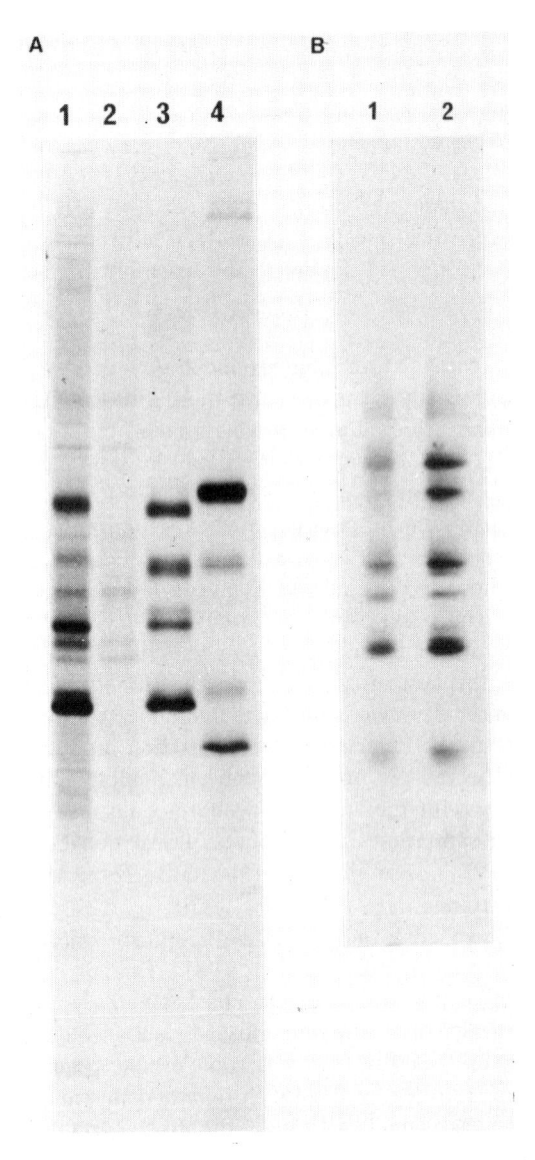

Fig. 4 – A. Electrophoretic analysis of the AChR purified from oocytes microinjected with *Torpedo* mRNA by affinity chromatography on α-BuTX-Sepharose. (1) AChR eluted from the α-BuTX-Sepharose. (2) A control for (1). (3) [^{3}H] propionylated native *Torpedo* AChR. (4) [^{125}I] labelled standard proteins as in Fig. 3. Reproduced from Sumikawa *et al.* (1981) with permission from *Nature*. B. Comparison of peptide maps in SDS gel by limited proteolysis using *S. aureus* V8 protease. From Sumikawa (unpublished). (1) From [^{125}I] α-subunit of *Torpedo* AChR. (2) From [^{35}S] 40K polypeptide of the AChR synthesised in oocytes. The polypeptides were purified by SDS gel electrophoresis prior to peptide mapping as described by Sumikawa *et al.* (1982a).

Sec. 3]   **Studies in the Biosynthesis of the Acetylcholine Receptor**   259

When the size of the AChR molecule synthesised in the oocyte was analysed by sucrose gradient centrifugation, the translation products produced a peak of α-BuTX binding activity at the same position (9 S) (Sumikawa et al., 1981) as the native monomer (Fig. 5). This result indicates the ability of the oocyte to assemble the newly synthesised subunits into intact receptor molecules.

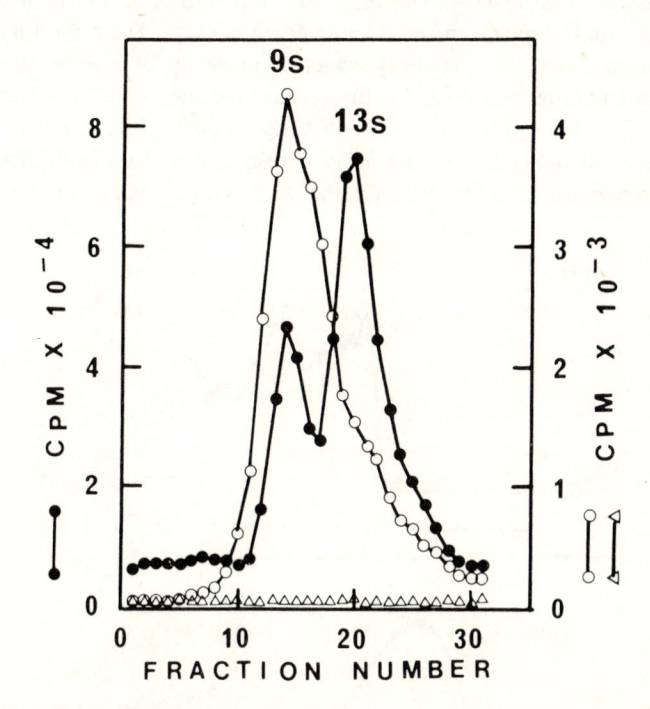

Fig. 5 – Sucrose gradient centrifugation of the AChR newly synthesised in oocytes. After centrifugation, fractions were collected and analysed for α-BuTX binding activity using [^{125}I] α-BuTX. ●, Purified AChR from *Torpedo*, △, extract from uninjected oocytes; ○, extract from oocytes injected with *Torpedo* mRNA. Reproduced from Sumikawa *et al.* (1981) with permission from *Nature*.

In vivo, newly synthesised AChR are transported to the plasma membrane and inserted into it to form the functional AChR. *Xenopus* oocytes are able to faithfully assemble this multi-subunit AChR, which binds α-BuTX. Further, after microinjection with *Torpedo* mRNA and incubation of oocytes with [^{125}I] α-BuTX, we found that α-BuTX binds to these intact oocytes (Barnard *et al.*, 1982), i.e. to their external surface. This suggests that some of the AChR molecules produced are inserted into the oocytes' surface membranes. However, we cannot state that this assembled structure is a functional AChR molecule without providing evidence that it possesses the ACh-controlled ion channel.

To address this question, the microinjected oocytes were studied electrophysiologically (Barnard et al., 1982). When ACh was applied to the surface membranes of oocytes previously injected with *Torpedo* mRNA, a nicotinic type of response to ACh (absent in the uninjected oocyte) was now present, as shown in Fig. 6. Miledi and co-workers have earlier shown that the membrane of some *Xenopus* oocytes contains a native muscarinic type of AChR (Kusano et al., 1977; Kusano et al., 1982). The response of this to ACh differs in several respects from that observed in the microinjected oocytes. The response in normal oocytes occurs after a long delay which is made up of a series of potential (or current) fluctuations (Fig. 6). In contrast the membrane potential change in microinjected oocytes consists of an early smooth phase followed by a delayed phase. Atropine blocks the delayed response to ACh in normal oocytes (and microinjected oocytes) but not the early smooth phase in microinjected

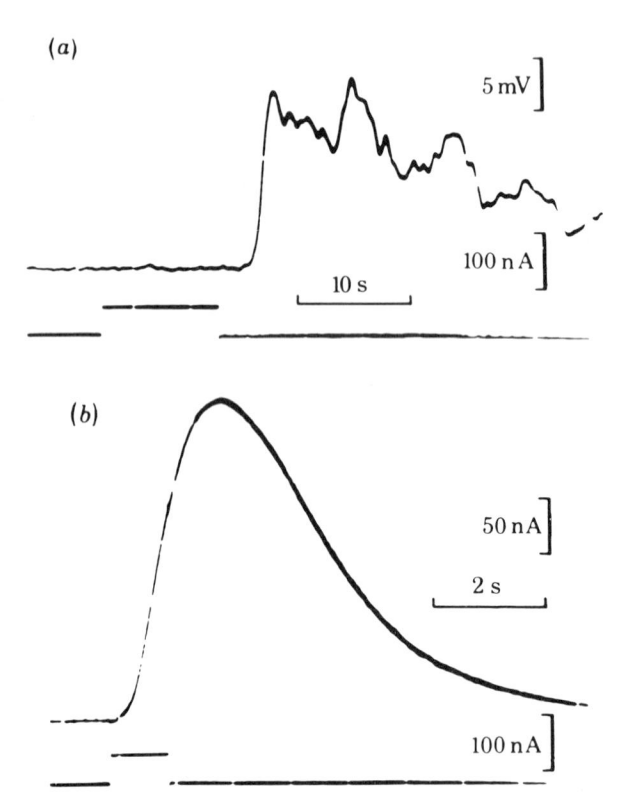

Fig. 6 — Responses to ACh applied ionophoretically to *Xenopus* oocytes. (a) Muscarinic response to ACh applied to a control oocyte. (b) Nicotinic response to ACh applied to an oocyte microinjected with *Torpedo* mRNA; atropine (2.5 × 10^{-7} M) was used to block the muscarinic response. Reproduced from Barnard, Miledi and Sumikawa (1982) with permission from *Proc. R. Soc. (Lond.)*.

Sec. 3] Studies in the Biosynthesis of the Acetylcholine Receptor 261

oocytes. The native, delayed response is unaffected by d-tubocurarine or α-BuTX, but the early response of microinjected oocytes is eventually blocked by d-tubocurarine or α-BuTX (Fig. 7). Thus, the AChR induced only in microinjected oocytes shows typical nicotinic characteristics. Intracellular (instead of the usual external) application of ACh, did not evoke a response, suggesting that many *Torpedo* AChR molecules have been inserted in the cell membrane in the correct orientation (Barnard *et al.*, 1982). We conclude, therefore, that microinjection of mRNA extracted from the *Torpedo* electric organ into oocytes leads to the synthesis and incorporation of functional *Torpedo* AChR into the oocyte cell membranes.

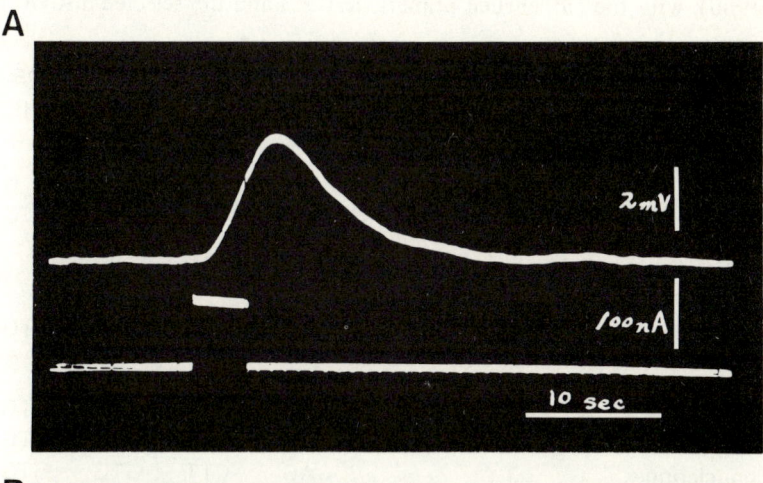

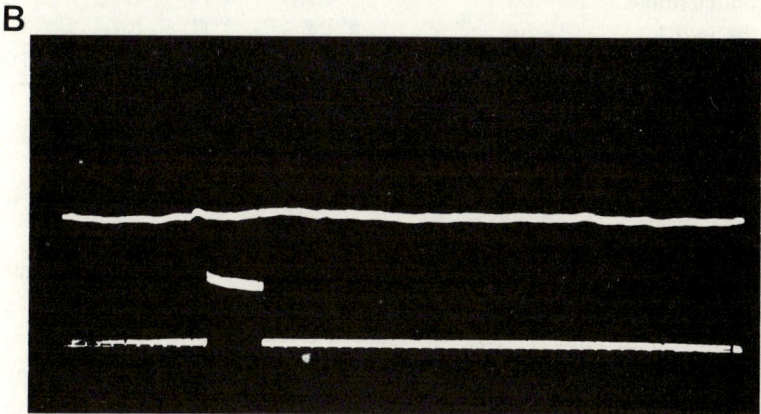

Fig. 7 – Blockade, by α-BuTX, of the response to ACh applied ionophoretically to an oocyte microinjected with *Torpedo* mRNA. A. Before α-BuTX. B. The response obtained 4 minutes after addition of α-BuTX to the bathing fluid (final α-BuTX concentration 1 μg/ml). Resting potential −45 mV. From Miledi and Sumikawa (unpublished).

4. CLONING OF AN AChR cDNA

If we could isolate the mRNA species coding for each subunit, the oocyte system would provide a new approach to studying the functional contribution of each subunit. We have initiated work aimed at cloning the AChR genes which can be used for the purification of a complementary mRNA.

The amino-terminal sequences of the subunits comprising the *Torpedo* AChR are known (Raftery *et al.*, 1980) and we have used this knowledge to synthesise oligodeoxyribonucleotides by the method of Ito *et al.* (1982) which could be used to prime specific reverse transcription of α-subunit mRNA (Fig. 8). We carried out priming experiments, following the method of Houghton *et al.* (1980), with the ^{32}P labelled primers (α1–α6) and the selected mRNA from

		22	23	24	25
Amino acid sequence		VAL	GLU	HIS	HIS
Possible codons		5' GUU (A,C,G)	GA(A,G)	CA(U,C)	CA(U,C) 3'
Possible primers		3' CA(T,G,A,C)	CT(T,C)	GT(A,G)	GTG 5'
Oligonucleotides Synthesised	α1	3' CAT	CTT	GTG	GTG 5'
	α2	3' CAT	CTC	GTG	GTG 5'
	α3	3' CAG	CTT	GTG	GTG 5'
	α4	3' CAG	CTC	GTG	GTG 5'
	α5	3' CAA	CTT	GTG	GTG 5'
	α6	3' CAA	CTC	GTG	GTG 5'

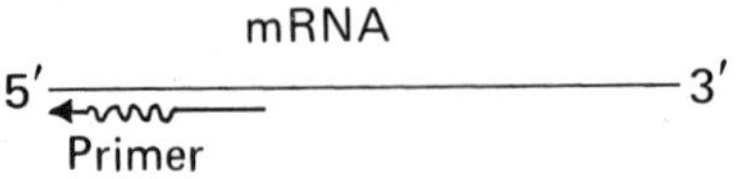

```
              mRNA
      5'_____3'
         ←∽∽
         Primer
```

Fig. 8 — Deduction of potential oligonucleotide primers for the AChR α subunit mRNA.

sucrose gradient fractions. We found that the cDNA products using α2 primer gave a band at ~275 b.p. (Fig. 9) which is in the range of the predicted size of the primed cDNA complementary to the α subunit mRNA and which appears to correlate with the content of AChR mRNA in the gradient fractions (Fig. 9). The predicted amino acid sequence (Sumikawa, *et. al.*, 1982b) from the nucleotide sequence of the primed cDNA coincides with the published N-terminal

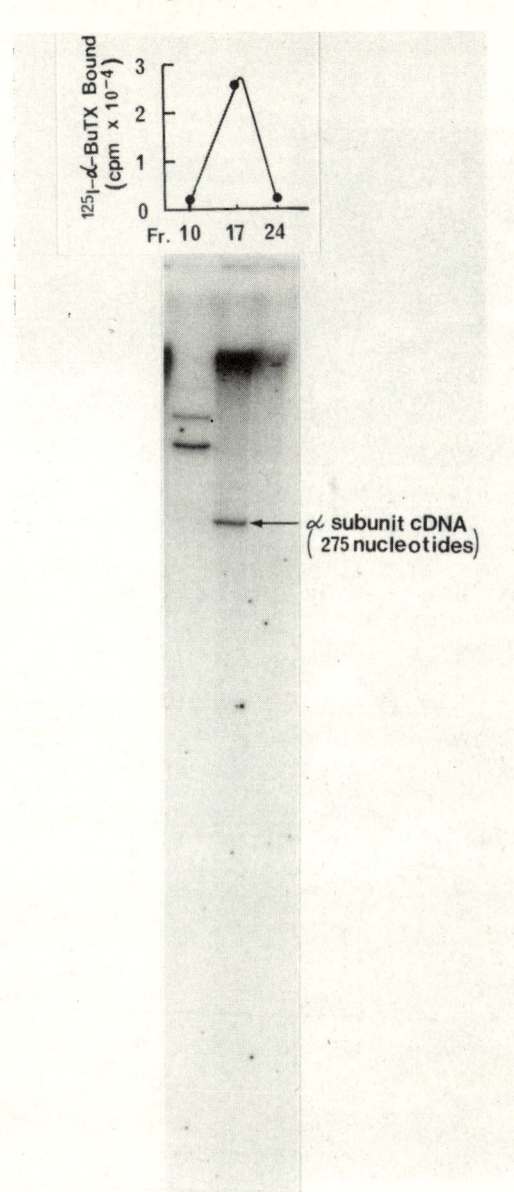

Fig. 9 — Electrophoretic analysis of the cDNA products obtained using the α2 primer, gradient-purified AChR mRNA and reverse transcriptase. AChR mRNA was purified by sucrose gradient centrifugation. Fractions 10, 17 and 24 were reverse transcribed with α2 primer and the transcripts electrophoresed through a 12.5% acrylamide gel containing 7 M urea. The α-BuTX binding activity found in the translation products of oocytes microinjected with each of these gradient fractions is shown. From Sumikawa and Houghton (Unpublished).

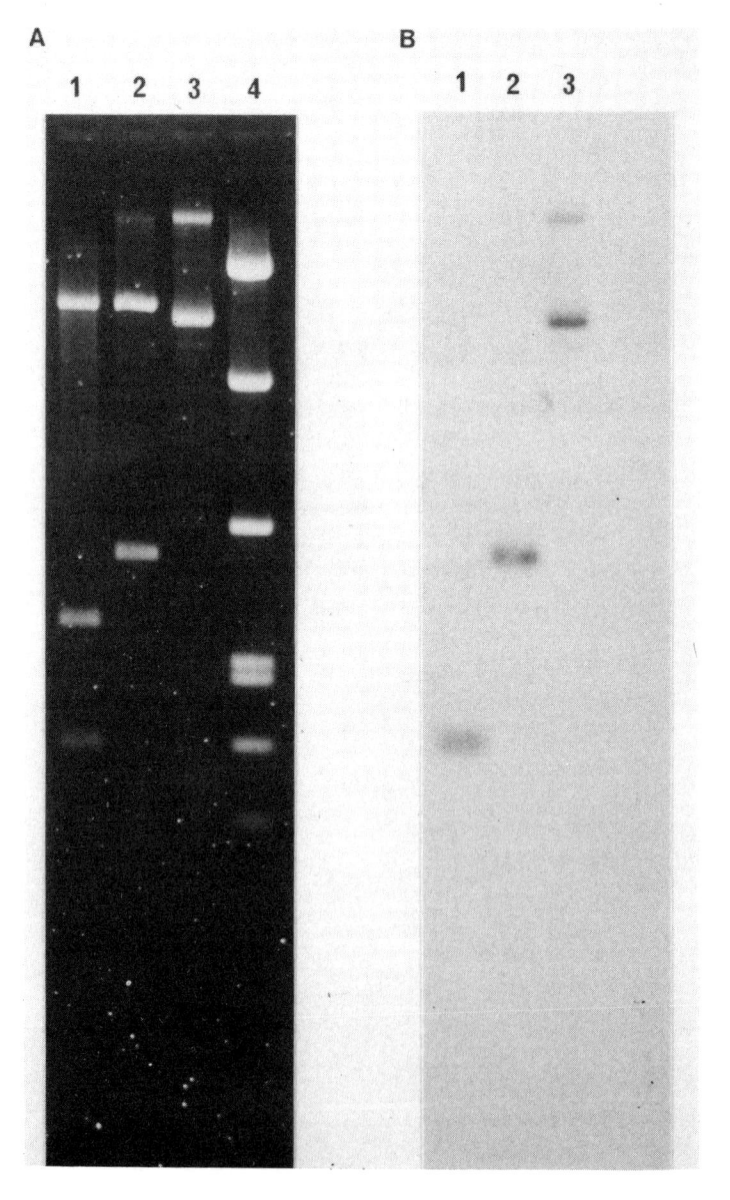

264 mRNA/Genes Coding for the Nicotinic Acetylcholine Receptor [Ch. 14

Fig. 10 — A. Analysis of the cloned cDNA by electrophoresis on 1.4% agarose. (1) Plasmid DNA cut with PvuII/Hind III. (2) Plasmid DNA cut with Hind III. (3) Plasmid DNA. (4) Hind III restriction fragments of phage PM2 as standards. B. Southern blot analysis of the cloned cDNA. The DNA fragments separated on a 1.4% agarose gel (A) were transferred to nitrocellulose paper and the blot was hybridised with [^{32}P] labelled 19-mer (i.e. α-subunit N-terminal probe) by the method of Southern (1975). (1) PvuII/Hind III fragments. (2) Hind III fragments. (3) Plasmid DNA (non-cut). From Sumikawa and Houghton (unpublished).

amino acid sequence of the α-subunit (Raftery et al., 1980). From this result we were able to synthesise an oligonucleotide (3'CTTTGTCAAACCAACGAT 5') to act as a specific hybridisation probe for recombinant bacterial clones containing the α subunit cDNA. Six positive clones were obtained (Sumikawa, et al., 1982b) from a cDNA library prepared from *Torpedo* mRNA by the method of Emtage et al. (1980) and plasmid DNA was prepared from them as described by Ish-Horowicz and Burke (1981). The largest cDNA insert (900 b.p.) (Fig. 10) was selected and restriction sites were mapped. We found that the smaller fragment obtained by restriction with Pvu II endonuclease hybridised with [^{32}P] labelled probe (Fig. 10), therefore, this fragment should contain the nucleotide sequence coding for the N-terminal amino acid sequence of the α subunit. In fact the amino acid sequence predicted from the nucleotide sequence of this fragment (Fig. 11) did coincide with the published N-terminal amino acid sequence of the mature α subunit. Beginning with the translation initiation codon ATG, the first 24 amino acids consist of many hydrophobic amino acids, particularly Leu (Fig. 11), characteristic of a signal peptide found in secretory proteins (Habener et al., 1978) and viral membrane glycoproteins (Irving et al., 1979). Since the route of intracellular transport of AChR has several properties in common with the secretion of AChE (Rotundo and Fambrough, 1980), it is likely that the association of the signal peptide with the endoplasmic reticulum would be the first step in their insertion into the plasma membrane.

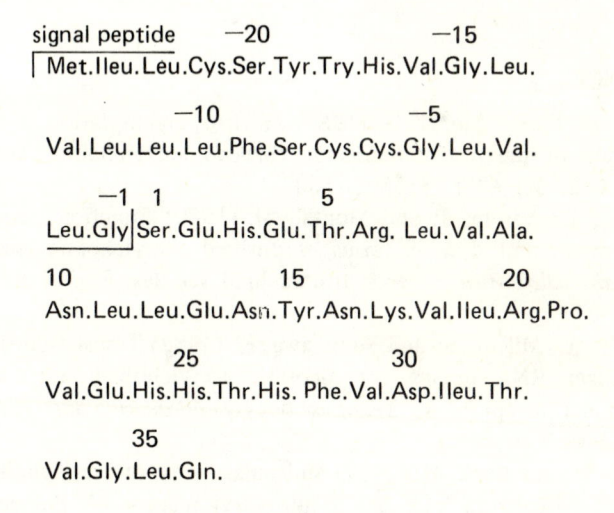

signal peptide —20   —15
| Met.Ileu.Leu.Cys.Ser.Tyr.Try.His.Val.Gly.Leu.

           —10   —5
Val.Leu.Leu.Leu.Phe.Ser.Cys.Cys.Gly.Leu.Val.

      —1 1   5
Leu.Gly|Ser.Glu.His.Glu.Thr.Arg. Leu.Val.Ala.

 10   15   20
Asn.Leu.Leu.Glu.Asn.Tyr.Asn.Lys.Val.Ileu.Arg.Pro.

            25   30
Val.Glu.His.His.Thr.His. Phe.Val.Asp.Ileu.Thr.

       35
Val.Gly.Leu.Gln.

Fig. 11 – The amino acid sequence predicted from the nucleotide sequence of the smaller fragment produced by Hind III/Pvu II cleavage of the cloned α subunit cDNA. The encoded amino acid sequence is numbered from the amino-terminus of the apparent mature α subunit polypeptide. From Sumikawa et al., 1982b.

5. CONCLUSIONS

Xenopus oocytes microinjected with mRNA extracted from *Torpedo* electric organ acquire functional *Torpedo* AChR in their surface membranes. Clearly, microinjection of mRNA coding for any particular neurotransmitter receptor into oocytes would provide a new approach for identification and study of the associated ion channel. Cloned cDNA copies of mRNA coding for the α subunit of AChR could not only provide the primary structure but also be used for the purification of the α subunit mRNA. Injection into oocytes of the purified mRNA encoding the subunit could determine whether the mature α subunit alone, in the total absence of any denaturant, is capable of binding appreciably to α-BuTX. As indicated in the introduction, experiments using the purified mRNA species coding for each subunit are likely to be an important means of elucidating the functional contribution of each subunit.

ACKNOWLEDGEMENTS

We would like to thank Dr J. S. Emtage, Dr. B. M. Richards and D. Beeson for help and advice. We also thank Dr L. Bell and co-workers for synthesising oligonucleotides. K. Sumikawa is a Research Fellow of G. D. Searle Company and thanks Dr R. D. O'Brien for advice and help, in whose laboratory peptide mapping analyses were conducted. Part of this work was supported by the M.R.C. and the Royal Society.

REFERENCES

Anderson, D. J. and Blobel, G. (1981) *In vitro*, glycosylation, and membrane insertion of the four subunits of *Torpedo* acetylcholine receptor. *Proc. Natl. Acad. Sci. USA* **78**, 5598–5602.

Anholt, R., Lindstrom, J. and Montal, M. (1980) Functional equivalence of monomeric and dimeric forms of purified acetylcholine receptors from *Torpedo californica* in reconstituted lipid vesicles. *Eur. J. Biochem.* **109**, 481–487.

Barnard, E. A., Miledi, R. and Sumikawa, K. (1982) Translation of exogenous messenger RNA coding for nicotinic acetylcholine receptors produces functional receptors in *Xenopus* oocytes. *Proc. R. Soc. (Lond.)* **B215**, 241–246.

Chang, H. W. and Bock, E. (1977) Molecular forms of acetylcholine receptor. Effects of calcium ions and a sulfhydryl reagent on the occurrence of oligomers. *Biochemistry* **16**, 4513–4520.

Emtage, J. S., Tacon, W. C. A., Catlin, G. H., Jenkins, B., Porter, A. G. and Carey, N. H. (1980) Influenza antigenic determinants are expressed from haemagglutinin genes cloned in *Escherichia coli*. *Nature* **283**, 171–174.

Gurdon, J. B. (1974) *The Control of Gene Expression in Animal Development.* Clarendon, Oxford.

Gurdon, J. B., Lane, C. D., Woodland, H. R. and Marbaix, G. (1971) Use of frog eggs and oocytes for the study of messenger RNA and its translation in living cells. *Nature (Lond.)* 233, 177–182

Habener, J. F., Rosenblatt, M., Kemper, B., Kronenberg, H. M., Rich, A. and Potts, J. T. (1978) Pre-proparathyroid hormone: Amino acid sequence, chemical synthesis, and some biological studies of the precursor region. *Proc. Natl. Acad. Sci. USA* 75, 2616–2620.

Haggerty, J. G. and Froehner, S. C. (1981) Restoration of ^{125}I-α-bungarotoxin binding activity to the α subunit of *Torpedo* acetylcholine receptor isolated by gel electrophoresis in sodium dodecyl sulfate. *J. Biol Chem.* 256, 8294–8297.

Hamilton, S., McLaughlin, M. and Karlin, A. (1979) Formation of disulfide-linked oligomers of acetylcholine receptor in membrane from *Torpedo* electric tissue. *Biochemistry* 18, 155–163.

Heidmann, T. and Changeux, J.-P. (1978) Structural and functional properties of the acetylcholine receptor protein in its purified and membrane-bound states. *Ann. Rev. Biochem.* 47, 317–357.

Houghton, M., Stewart, A. G., Doel, S. M., Emtage, J. S., Eaton, M. A. W., Smith, J. C., Patel, T. P., Lewis, H. M., Porter, A. G., Birch, J. R., Cartwright, T. and Carey, N. H. (1980) The amino-terminal sequence of human fibroblast interferon as deduced from reverse transcripts obtained using synthetic oligonucleotide primers. *Nucleic Acids Res.* 8, 1913–1931.

Irving, R. A., Toneguzzo, F., Rhee, S. H., Hofmann, T., and Ghosh, H. P. (1979) Synthesis and assembly of membrane glycoproteins: Presence of leader peptide in nonglycosylated precursor of membrane glycoprotein of vesicular stomatitis virus. *Proc. Natl. Acad. Sci. USA* 76, 570–574.

Ish-Horowicz, D. and Burke, J. F. (1981) Rapid and efficient cosmid cloning. *Nucleic Acids Res.* 9, 2989–2998.

Ito, H., Ike, Y., Ikuta, S. and Itakura, K. (1982) Solid phase synthesis of polynucleotides. VI. Further studies on polystyrene copolymers for the solid support. *Nucleic Acids Res.* 10, 1755–1769.

Juillerat, M. A., Schwendimann, B., Hauert, J., Fulpius, B. W. and Bargetzi, J. P. (1982) Specific binding to isolated acetylcholine receptor of a synthetic peptide duplicating the sequence of the presumed active center of a lethal toxin from snake venom. *J. Biol. Chem.* 257, 2901–2907.

Karlin, A. (1980) Molecular properties of nicotinic acetylcholine receptors, in *The Cell Surface and Neuronal Function* (Cotman, C. W., Poste, G. and Nicolson, G. L., eds.) pp. 191–260. Elsevier–North Holland Biomedical Press, New York.

Kusano, K., Miledi, R. and Stinnarkre, J. (1977) Acetylcholine receptors in the oocyte membrane. *Nature* 270, 739–741.

Kusano, K., Miledi, R. and Stinnakre, J. (1982) Cholinergic and catecholaminergic receptors in the *Xenopus* oocyte membrane. *J. Physiol. (Lond.)* **328**, 143–170.

Lane, C. D. (1981) The fate of foreign proteins introduced into *Xenopus* oocytes. *Cell* **24**, 281–282.

Lane, C. D., Shannon, S. and Craig, R. (1979) Sequestration and turnover of guinea-pig milk proteins and chicken ovalbumin in *Xenopus* oocytes. *Eur. J. Biochem.* **101**, 485–495.

Lindstrom, J., Merlie, J. and Yogeeswaran, G. (1979) Biochemical properties of acetylcholine receptor subunits from *Torpedo californica*. *Biochemistry* **18**, 4465–4469.

Lyddiatt, A. Sumikawa, K., Wolosin, J. M., Dolly, J. O. and Barnard, E. A. (1979) Affinity labelling by bromoacetylcholine of a characteristic subunit in the acetylcholine receptor from muscle and *Torpedo* electric organ. *FEBS Lett.* **108**, 20–24.

McNamee, M. G., Weill, C. L., and Karlin, A. (1975) Purification of acetylcholine receptor from *Torpedo californica* and its incorporation into phospholipid vesicles. *Ann. N.Y. Acad. Sci.* **264**, 175–182.

Mehraban, F., Dolly, J. O. and Barnard, E. A. (1982) Antigenic similarities between the subunits of acetylcholine receptor from *Torpedo marmorata*. *FEBS Lett.* **141**, 1–5.

Mendez, B., Valenzuela, P., Martial, J. A. and Baxter, J. D. (1980) Cell-free synthesis of acetylcholine receptor polypeptides. *Science* **209**, 695–697.

Merlie, J. P. and Sebbane, R. (1981) Acetylcholine receptor subunits transit a precursor pool before acquiring α-bungarotoxin binding activity. *J. Biol. Chem.* **256**, 3605–3608.

Miller, D. L., Moore, H.-P. H., Hartig, P. R. and Raftery, M. A. (1978) Fast cation flux from *Torpedo californica* membrane preparation: Implications for a functional role for acetylcholine receptor dimers. *Biochem. Biophys. Res. Commun.* **85**, 632–640.

Raftery, M. A., Hunkapiller, M. W., Strader, C. D. and Hood, L. E. (1980) Acetylcholine receptor: Complex of homologous subunits. *Science* **208**, 1454–1456.

Rptundo, R. L. and Fambrough, D. M. (1980) Secretion of acetylcholinesterase: Relation to acetylcholine receptor metabolism. *Cell* **22**, 595–602.

Southern, E. M. (1975) Detection of specific sequences among DNA fragments separated by agarose gel electrophoresis. *J. Mol. Biol.* **98**, 503–515.

Suarez-Isla, B. A. and Huch, F. (1977) Acetylcholine receptor: SH group reactivity as indicator of conformational changes and functional states. *FEBS Lett.* **75**, 65–69.

Sumikawa, K., Houghton, M., Emtage, J. S., Richards, B. M. and Barnard, E. A. (1981) Active multi-subunit ACh receptor assembled by translation of heterologous mRNA in *Xenopus* oocytes. *Nature* **292**, 862–864.

References

Sumikawa, K., Barnard, E. A. and Dolly, J. O. (1982a) Similarity of acetylcholine receptors of denervated, innervated and embryonic chicken muscles. Subunit compositions. *Eur. J. Biochem.*. **126**, 473–479.

Sumikawa, K., Houghton, M., Smith, J. C., Bell, L., Richards, B. M. and Barnard, E. A. (1982b) The molecular cloning and characterisation of the cDNA coding for the α subunit of the acetylcholine receptor. *Nucleic Acids Research*, **10**, 5809–5822.

Tracz, J. S. and Lampen, J. O. (1975) Tunicamycin inhibition of polyisoprenyl N-acetylglucosaminyl pyrophosphate formation in calf-liver microsomes. *Biochem. Biophys. Res. Commun.* **65**, 248–257.

Vandlen, R. L., Wu, W. C.-S., Eisenach, J. C. and Raftery, M. A. (1979) Studies of the composition of purified *Torpedo californica* acetylcholine receptor and its subunits. *Biochemistry* **18**, 1845–1854.

Wu, W. C.-S. and Raftery, M. A. (1981) Functional properties of acetylcholine receptor monomeric and dimeric forms in reconstituted membranes. *Biochem. Biophys. Res. Commun.* **99**, 436–444.

15

Quantitative drug assays using radioreceptor techniques

S. R. Nahorski and **D. B. Barnett**

Department of Pharmacology and Therapeutics, Medical Sciences Building, University of Leicester, University Road, Leicester LE1 7RH, U.K.

Abbreviations used

ACTH	—	adrenocorticotropic hormone (corticotropin)
CNS	—	central nervous system
CSF	—	cerebrospinal fluid
DHA	—	dihydroalprenolol
DP	—	diazepam
GLC	—	gas-liquid chromatography
HPLC	—	high-performance liquid chromatography
5-HT	—	5-hydroxytryptamine
K_i	—	inhibition constant
K_D	—	dissociation constant
QNB	—	quinuclidinyl benzilate
RIA	—	radioimmunoassay
RRA	—	radioreceptor assay
TAD	—	tricyclic antidepressant

1. INTRODUCTION

The last decade has seen dramatic developments in our understanding of hormone and neurotransmitter receptors. Much of this has come from technical advances made at the biochemical level, not least of which has been the development of specific radioligand receptor binding assays. The availability of a large number of neurotransmitters and receptor-active drugs labelled to very high specific activity coupled with rapid filtration or centrifugation techniques now allows many receptors to be examined directly using membrane or, in some cases, whole cell preparations. These approaches have undoubtedly revealed much new information on the pharmacological and biochemical characterisation of these recognition sites. Moreover, since they provide a quantitative estimate of receptor density, they have provided new information on the regulation of receptors in tissues in many different situations. Adequate proof of these developments is highlighted in several chapters of this monograph.

A further application of these techniques was the development of quantitative assays for hormones, neurotransmitters and drugs in body fluids and tissues. The first example of a specific radioreceptor assay (RRA) was described by Lefkowitz et al. (1970) who demonstrated that ACTH can be assayed by competition with [^{125}I] ACTH for specific sites in homogenates of adrenal cortical tissue. There followed a number of other similar assays for hormones and neurotransmitters (Enna, 1978), though in the case of drug assays these techniques have yet to be fully exploited. These RRAs are, of course, directly analogous to radioimmunoassays (RIA) in that the presence of a drug in a specimen can be quantitated by competition between that drug and a labelled ligand for specific binding sites. In contrast to the RIA, however, the binding site is very often a specific biological receptor rather than an immunoglobulin, and in many cases is the site at which the drug binds with a unique affinity to produce its particular action. Thus, it can be readily recognised that a major advantage of a RRA in drug estimations is that it is essentially a 'bioassay' sensitive to all active drug species, metabolites as well as the parent compound. Further, in the case of drugs that are administered as a racemic mixture of active and inactive isomers, by definition only the active isomer will be measured. However, there are some examples of RRAs in which a drug is assayed by virtue of possessing affinity for receptors which may not be related to its therapeutic effects (e.g. tricyclic antidepressants; see later). Therefore, although we wish to emphasise here those assays in which relevant sites of action 'receptors' are used, it should be emphasised that a 'bioassay' of other properties of a drug and its metabolites could be very useful, for example, in the assessment of expected side-effects.

In the present chapter we will discuss primarily the application of RRA to the measurement of drugs in tissues and biological fluids. In particular, we will emphasise the basic principles, methodology and possible biological implication of drug RRAs. Special reference will then be placed on those assays for beta-adrenoceptor antagonists and neuroleptics since it is these systems that have, to date, been relatively extensively studied with reference to their clinical effectiveness.

2. RADIORECEPTOR ASSAYS: PRINCIPLES AND METHODOLOGY

2.1 Principles

By definition all competitive radioreceptor assays relate to the ability of an unlabelled ligand to compete with a labelled ligand for a specific binding site. Thus, at a single radioligand concentration, the amount bound will directly relate to the quantity and affinity of the labelled and unlabelled ligands present. It follows that under optimal conditions, half maximal inhibition of radioligand binding will occur at a concentration of unlabelled competing ligand that approximates the latter's affinity constant for the receptor. Hence, by performing a receptor binding assay in the presence of an unknown quantity of competing drug, the amount of that unlabelled material can be calculated by determining the percentage inhibition of radioligand binding and comparing this to the inhibition produced by known quantities of compound in question. It naturally follows that the higher the affinity of the competing drug for the receptor, the smaller the quantity that can be accurately assayed.

2.2 Methods
It is, of course, essential when setting up an RRA that optimal conditions for binding are established. These relate not only to incubation buffer, temperature and reaction time, but also to receptor preparation, methods used to separate bound from free radioligand and assessment of specific and non-specific binding components. It is not appropriate here to provide precise details for each RRA, but it is sufficient to emphasise that unless optimal conditions are first established, then the inherent difficulties that are present when assaying drugs in biological fluids (e.g. non-specific interference) will often be amplified.

In order to cover practical details we describe here assays we have ourselves developed for the assay of beta-adrenoreceptor antagonists, using both membrane or soluble receptor preparations (Nahorski *et al.*, 1978; Barnett *et al.*, 1980; Barnett *et al.*, 1981). These assays depend upon the ability of beta-adrenreceptor antagonists to compete with a radiolabelled ligand: $(-)[^{3}H]$ dihydroalprenolol ($[^{3}H]$DHA) for beta-adrenoreceptor binding sites in particulate or soluble preparations of mammalian lung.

2.3 Membrane assay
Rat or bovine lung is freed of major bronchi, minced with scissors and homogenised in 50 vols. of 50 mM Tris-HCl buffer (pH 7.8) at room temperature. The homogenate is then passed through a single layer of cheesecloth to remove connective tissue which hinders the rapid filtration in the binding assay. The supernatant is recentrifuged at 30,000 g for 10 min and the final pellet washed three times before resuspension is assay buffer (Tris-HCl 50 mM, pH 7.8). Membranes prepared in this way can be stored for at least two months at $-40\,°C$ without significant loss of binding activity.

All constituents in the binding assay are diluted in 50 mM Tris-HCl (pH 7.8) and each tube contains 100–300 μg of membrane protein, 1–2 nM $[^{3}H]$ DHA (102 Ci/mmole, Radiochemical Centre, Amersham) and plasma sample (2–10 μl) or standard in a final volume of 250 μl. Incubations are carried out at room temperature for 30 min and are terminated by the addition of 1 ml ice-cold assay buffer followed by rapid vacuum filtration over Whatman GF/B glass-fibre filters. The filters are then rapidly washed with 3 × 5 ml buffer and radioactivity bound to the membranes trapped by the filters measured by counting in a liquid scintillation counter. Separate incubations are carried out in each assay using 200 μM (−)-isoprenaline to assess non-specific binding. Specific binding, defined as total radioactivity bound minus non-specific binding, is generally 80–90% of the total binding at the equilibrium dissociation constant (K_D) (0.5 nM) of these preparations.

2.4 Soluble receptor assay

Solubilisation of lung tissue is performed by digitonin extraction of particulate preparations for 30 min at 4 °C in a medium containing 0.5% digitonin, 50 mM Tris-HCl (pH 7.8), 100 mM NaCl using a digitonin:protein ratio of 3:1. Following centrifugation at 50,000 g for 1 h, the supernatant is used as the solubilised receptor preparation. Solubilised beta-adrenoreceptors are assayed in an incubation volume of 500 μl containing [^{3}H] DHA, 50 mM Tris-HCl (pH 7.8) and competing drugs. Bound and free [^{3}H] DHA are separated by addition of 0.25 ml 1.8% Norit GSX charcoal, 0.4% BSA in 50 mM Tris-HCl (pH 7.8) followed by centrifugation at 2,000 g for 10 min. Aliquots of supernatant containing bound ligand are counted by liquid scintillation spectrometry. Specific binding is defined as the binding displaced by 200 μM (−)-isoprenaline and represents >95% in these solubilised preparations at the K_D of [^{3}H] DHA.

We have used receptor preparations from rat or bovine lung since this tissue possesses a high density of beta-adrenoceptors (Nahorski *et al.*, 1978; Barnett *et al.*, 1978; Rugg *et al.*, 1978) and under conditions described above it displays an excellent specific to non-specific binding ratio. It should, however, be emphasised here that quite different conditions of receptor preparation, incubation and separation of free and bound ligand may be more appropriate for the assay of other receptor active drugs.

2.5 Sample preparation

One inherent advantage of RRAs lies with their simplicity and in many cases assays can be performed on unextracted plasma or other biological fluids. However, many ligands bind avidly to plasma proteins and thus plasma can produce significant interference when relatively large volumes are assayed without extraction. This appears to be a variable phenomena with respect to the receptor preparations used and requires careful preliminary evaluations for each assay. In many cases supplementing the standard curve with appropriate volumes of drug-free plasma is all that is required to account for this interference, though in some cases it may be necessary to extract or dialyse plasma samples (Innis *et al.*, 1978; Bilezikian *et al.*, 1979). The latter procedure will, of course, provide an estimate of the free fraction of active drug and this may be useful in the evaluation of drug kinetics. Differences in plasma interference between different assay systems are discussed in later sections. In the case of tissue levels of drugs, extraction may be necessary though again a reliable quantitative assay may be achieved simply by homogenisation of tissue in aqueous media and assaying appropriately diluted homogenates (Cohen et al., 1980).

2.6 Specificity and sensitivity

The specificity of RRAs obviously relates to the specificity of the receptor used. By its very nature, these assays will detect any material that will interact with the receptor. Thus, in the case of the beta-adrenoceptor, the specificity is such that only beta-adrenoceptor antagonists (and any active metabolites) possess sufficient receptor affinity to be assayed though, in theory, catecholamines, if they were present at very high levels (e.g. in patients with phaeochromocytomas), could interfere. However, with some other RRA systems (e.g. neuroleptic assays) drugs other than those directly under assay could interfere with direct estimations. However, as emphasised earlier, RRAs do not provide a quantitative assay for a specific chemical entity but provide an important assay of the total biological activity present in an extract for a particular receptor system. Thus, as discussed later, antidepressants can be assayed by virtue of their relatively high affinity for muscarinic or 5-HT receptors.

Similarly, the sensitivity of a RRA depends upon the relative receptor affinity of the drug being assayed. Thus, for example, with a very potent beta-blocker such as propranolol that exhibits an inhibition constant (K_i) for beta-receptors of about 1×10^{-9} M, under optimal conditions using a solubilised receptor assay, 1–2 pmoles propranolol/ml of plasma can be reliably detected. Alternatively, a weak beta-blocker such as sotalol ($K_i = 1 \times 10^{-7}$ M) would have a limit of sensitivity of about 100 pmoles/ml plasma. In this example, the absolute sensitivity for a specific drug is not particularly relevant since the index of the assay relates to the ability of the drug to interact with the beta-adrenoceptor, which directly correlates with biological activity. However, in other RRAs (e.g. for tricyclic antidepressants) this relationship does not hold (see Section 4.3).

Optimisation of the assay *per se* is of course very important and although it is not appropriate to discuss this in detail here, factors such as assay volume, specific activity of the labelled ligand and the relative non-specific interference of assay samples are clearly critical. Theoretical aspects for the optimisation of competitive binding assays are discussed in detail by Brown *et al.* (1972).

2.7 Calculation of results

The amount of unlabelled receptor-active compound in a sample is simply calculated by measuring the inhibition of specifically labelled radioligand binding observed in the presence of the sample and comparing this value to a standard curve of known amounts of unlabelled drug. There are a number of ways to transform the typical dose–response curve generated by such a standard curve into a linear form suitable for computation. The data may be simply transformed by dividing the specific radioactivity bound in the absence of competing cold drug by the radioactivity bound in the presence of a known quantity of standard and plotting this ratio against concentrations of standard (see Nahorski et al., 1978). Alternatively, a logit-log plot (Rodbard and Lewald, 1970) or computer-assisted iterative curve fitting of the raw data (e.g. Munson and Rodbard, 1980) may be used.

As discussed previously, since an RRA measures biological activity at the appropriate receptor rather than quantitation of the chemical drug, it is not really appropriate to use the same drug in the standard curve, but rather to use a single standard for a particular receptor and to express levels in samples as units of drug equivalents with respect to this standard.

3. *IN VIVO* AND *EX VIVO* RRAs

RRAs for drug levels in plasma have been highlighted here since they have gained immediate application in man under various clinical situations. However, it is also clearly desirable to have information concerning the *in vivo* distribution of drugs since it is likely that different inter- and perhaps intra-tissue distribution of a compound relate to its physico-chemical properties and could influence its potency and selectivity independently of receptor interaction. With these aims in mind, a number of groups have examined the *in vivo* tissue binding of drugs in animals, assessed by the ability of the drugs administered to the intact animal to compete with the specific *in vivo* binding of a radioligand (Bylund et al., 1977; Laduron et al., 1978). Theoretically, such approaches could provide useful information on the relative distribution of drugs and the kinetics of *in vivo* receptor occupation. However, a major limitation of this approach could lie with unequal distribution of the radioactive drug between tissues, thus precluding a complete examination of the disposition of competing cold drugs.

These problems can be largely overcome in *ex vivo* experiments. Here, animals are first treated with different doses of non-radioactive drug and the amount of drug present in a particular tissue is assessed by performing binding assays with specific radioligand under *in vitro* conditions (Laduron et al., 1978; Sriwatanakul and Nahorski, 1980). Comparison of the inhibition of specific radioligand binding produced by the cold drug present in crude homogenates of the tissue in question with that produced by appropriate standards, provides a quantitative estimate of the concentration of 'bioactive' drug within that tissue. This approach possesses several advantages over the direct *in vivo* binding assay. Firstly, since the labelling of the receptor is performed *in vitro*, the conditions of assay can be carefully controlled to allow a precise estimation of the competing drug in the tissue. Secondly, unlike the *in vivo* assay, the kinetics of cold drug distribution are not complicated by the kinetics of the administered radiolabelled ligand.

The *ex vivo* RRA may also provide certain advantages over a simple tissue

RRA in which the drug is extracted from a tissue and then assayed with a receptor preparation from another source. If it is assumed that there is little redistribution of the drug during homogenisation, then the *ex vivo* assay may provide an estimate of the degree of receptor occupation by the administered cold drug. Evidence for this has been provided for the beta-adrenoceptor antagonist propranolol in rat lung and heart (Sriwatanakul and Nahorski, 1980).

In conclusion, therefore, the use of *ex vivo* receptor binding assays provides a novel approach to the analysis of the tissue disposition and 'bioactivity' of receptor directed drugs. The use of this approach in combination with conventional plasma RRAs should provide important new information concerning pharmacodynamic–pharmacokinetic relationships in different tissues and plasma.

4. EXAMPLES OF RADIORECEPTOR ASSAYS

In the following sections we will concentrate on RRAs that have been developed to measure specific drugs in plasma and examine their possible role in relationship to human therapeutics.

4.1 Beta-adrenoceptor antagonists

Over the last ten years there has been increasing usage of beta-adrenoceptor antagonist drugs (beta-blockers) in the treatment of a wide variety of cardiovascular disease states, including angina and hypertension. Measurement of tissue levels in animals and plasma levels in humans of these drugs has proved useful in studying dose-response relationships *in vivo* as well as evaluating the complex pharmacokinetics of these compounds. The general clinical applicability of plasma level measurements of beta-blockers is debatable as direct correlation between clinical effect and plasma level is not universal for all actions of these drugs (see later). However, the widespread use of such measurements as might be indicated, has been, until recently, hampered by the technical complexity of some of the assay systems available. Thus, most previously described assays universally required prior extraction of plasma followed by gas liquid (Walle, 1974) or high-pressure liquid chromatographic techniques (Nation *et al.*, 1978). Assays which ostensibly did not require prior extraction of plasma such as the fluorimetric assay described by Shand *et al.* (1970), were relatively insensitive and prone to considerable interference from plasma constituents with the production of a high and variable assay blank.

Recently, we and others (Nahorski *et al.*, 1978; Innis *et al.*, 1978; Bilezikian *et al.*, 1979) have developed an RRA for beta-blocking drugs based on the principles described above. The RRAs described have used different tissue membrane preparations including mammalian lung (Nahorski *et al.*, 1978), rat cerebral cortex (Innis *et al.*, 1978) and turkey erythrocytes (Bilezikian *et al.*, 1979). Membranes are prepared by conventional methods and binding assays performed as previously described using rapid vacuum filtration for separation of bound from free radioactive ligand. Assays use high affinity ligands, [^{3}H] DHA with lung and brain membranes or [^{125}I] hydroxybenzylpindolol for the turkey erythrocytes. All RRAs for beta-blockers are highly sensitive, detecting final concentrations in plasma of propranolol as low as 0.5–2 ng/ml. By definition these assay systems will measure total beta-blocking activity of all compounds in the plasma in relationship to their affinity at the beta-adrenoceptor. Thus, for propranolol this will include not only the parent drug but contributions from the presence of active metabolites, such as 4-OH-propranolol, in proportion to their beta-blocking potency. In theory the presence of variable levels of plasma catecholamines might also be expected to contribute to overall beta-adrenoceptor activity in the RRA and constitute a source of interference for direct measurement of levels of beta-blocking drugs. In practice, however, this effect is of little importance as the circulating catecholamines have much lower affinity for the beta-adrenoceptor (approximately 1,000–10,000 nM) than the radioactive ligands used (0.2–2 nM). Thus, even the highest circulating concentrations of catecholamines encountered, for example, in phaeochromocytoma (50–100 nM) would produce less than 10% inhibition of radioligand binding.

The affinity that the beta-blocking drugs possess for the beta-adrenoceptor also varies and this is most relevant in relationship to the selectivity of the drugs for the beta$_1$ adrenoceptor subtype, the so called 'cardioselective' beta-blockers. The ability of the RRA to measure beta$_1$ selective compounds will depend upon the presence of beta$_1$ adrenoceptors in the membrane system used. If a tissue with a high proportion of the beta$_2$ subtype, e.g. rat lung (80% beta$_2$), is used, then the cardioselective drugs with low affinity for this subtype will not be measurable at usual clinical plasma concentrations and, more importantly, levels estimated will not relate directly to the drugs activity at its major site of action, namely the beta$_1$ adrenoceptor. In theory, therefore, use of a tissue with a high proportion of beta$_1$ adrenoceptors, e.g. rat cerebral cortex (70% beta$_1$), would allow the measurement by RRA of beta$_1$ selective drugs. However, because of the heterogeneity of the receptor subtypes within this tissue, biphasic displacement of non-selective labelled ligands by the selective compounds occurs in competition experiments. This would possibly lead to inaccuracies in measurement of absolute biologically active drug concentrations of the selective drugs. Certain manipulations of this heterogeneous population of receptors are possible to reduce the interference from the beta$_2$ adrenoceptor. Thus, inclusion within the assay of a carefully calculated concentration of a cold highly selective beta$_2$ agent (e.g. ICI 118.551) will exclude binding of the radioligand to the beta$_2$ adrenoreceptors and leave effectively a pure homogeneous population of beta$_1$ adrenoceptors for use in the RRA. Despite these theoretical considerations, to date an RRA for beta$_1$ selective compounds has not been reported.

The major clinical use of the RRA for beta-blockers is in the measurement of plasma drug levels. The inherent simplicity of the assay system is attractive but interference with radioligand binding to particulate preparations by added unextracted human plasma has been a problem and has led some workers to suggest various treatments of the plasma prior to assay (Innis et al., 1978; Bilezikian et al, 1979). Ideally direct plasma measurements without prior extraction would be desirable to simplify the application of the assay to clinical situations. This plasma interference relates, in part, to binding of the radioactive ligand to plasma proteins, effectively excluding a proportion of the ligand from interaction with the receptor in the assay. This effect is seen at quite small concentrations of plasma, may vary between different donors, and may also, in part, be a function of the particulate receptor preparation used. Since plasma interference is less for any particular concentration in the RRA using lung membranes than in other assay systems, and is further reduced by using a solubilised preparation of the lung beta-adrenoceptors as described above (Table 1), it is probable that the presence of detergent in the soluble preparation reduces the binding of the labelled ligand to plasma proteins. In view of this we now routinely use the solubilised receptor for assay of beta-blocking drugs (Barnett et al., 1981). To allow for minimal remaining plasma interference we still supplement the standard curve with drug-free plasma in appropriate quantities but as interference in this assay system is constant between individuals, stock samples of pooled plasma can be used. Dilution of the plasma samples in the assay causes complete dissociation of drug bound to plasma and consequently the assay measures total drug levels.

Table 1 — Comparison of inhibition of specific ligand binding[a] by increasing volumes of drug-free plasma in different radioreceptor assay systems

	Volume of plasma added			
	10 μl	20 μl	30 μl	50 μl
Solubilised receptor (Barnett et al., 1981)	5%	5%	—	20%
Rat lung membranes (Barnett et al., 1980)	5–10%	25%	45%	—
Rat cerebral membranes (Innis et al., 1978)	25–35%	65%	—	—
Turkey erythrocyte (Bilezikian et al., 1979)	10%	35%	55%	75%

[a]Results expressed as percentage inhibition of specific binding. Values are means of a number of experiments performed by authors quoted in the table and reproduced from other published work for comparison.

Sec. 4] Examples of Radioreceptor Assays 279

The dextro (+) rotatory isomers of beta-antagonists generally have about 100-fold lower affinity at the beta-adrenoceptor than the laevo (−) isomer and consequently the RRA preferentially detects the (−) isomer in a racemic mixture. This can be shown quite easily by administering either racemic (±) or (+)-propranolol to volunteers and measuring their plasma levels of beta-blocking activity (Fig. 1). Other assay systems, including GLC, HPLC and most RIAs, do not display stereoselectivity and equally detect both isomers of beta-blocker.

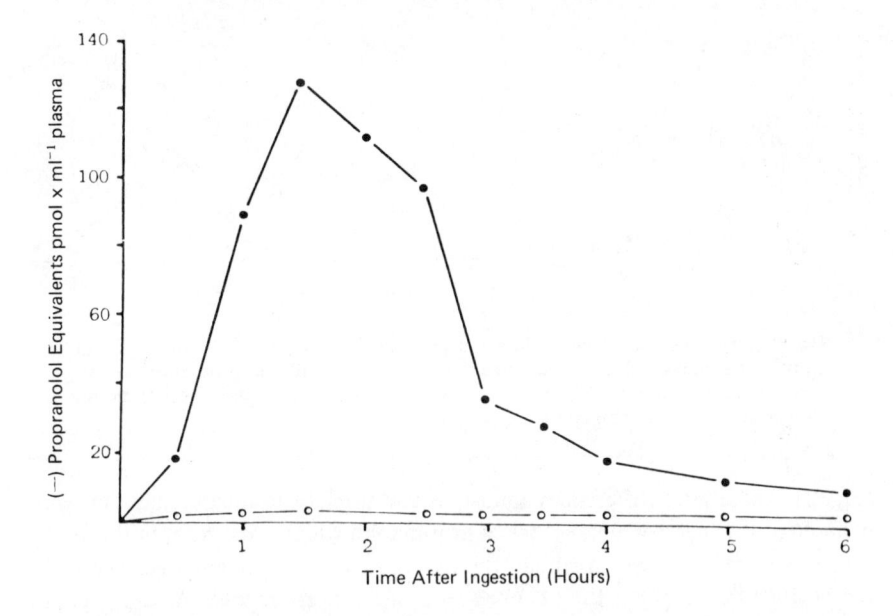

Fig. 1 — Time course of plasma levels measured in ( )-propranolol equivalents by RRA in volunteers administered orally 40 mg (±)-propranolol (●) or 40 mg (+)-propranolol (o). Each point is the mean of at least three duplicates performed on different dilutions of plasma. SEM for each point <5%. (Reproduced with permission from Nahorski *et al.* (1978).

To illustrate this difference we have made a direct comparison between RRA and RIA for propranolol. Examples of this comparison in three different volunteers are shown in Fig. 2. Peak plasma levels were observed with both assays 1.5–2.0 hours after ingestion of 40 mg (±)-propranolol, though at this time in all cases the RIA gave concentrations approximately twice those obtained by RRA. This difference undoubtedly relates to the inability of the RIA to distinguish between the stereoisomers of propranolol. In these same experiments the difference between the two assays at later time points (>3 hours) was even greater, such that at 6–7 hours after ingestion no biologically active drug could be detected by RRA, although levels estimated by RIA were still relatively high

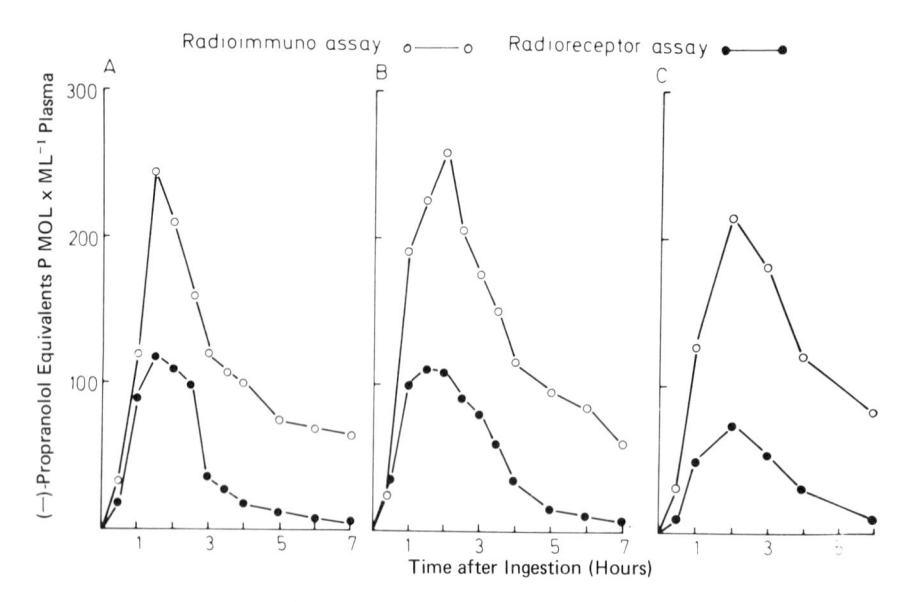

Fig. 2 — Time course of plasma levels of propranolol expressed as (−)-propranolol equivalents measured by radioimmunoassay (o) or radioreceptor assay (•) in three separate volunteers following oral administration of 40 mg (±)-propranolol. (Reproduced with permission from Barnett et al. (1980).

(Fig. 2). These later differences almost certainly relate to measurement by RIA of biologically inactive metabolites of propranolol which cross react in the assay. This was examined experimentally by assaying plasma from volunteers who had received 40 mg of either the (+) or (−) isomer of propranolol. As expected, in the volunteer who had taken the (+) isomer, plasma levels were easily measured by RIA but rarely detectable by RRA (Fig. 3A). On the other hand, in the volunteer who received the (−) isomer, as would be expected, plasma levels were similar at early times after drug administration, but a discrepancy between the methods became increasingly obvious at later times (Fig. 3B), suggesting the production during metabolism of biologically inactive, though immunoassayable, material.

Thus, the RRA so far described is applicable to the measurement of all non-selective beta-blocking drugs. The advantages of the RRA, as previously mentioned, include the ability to measure total beta-blocking activity of parent compound and metabolite. In view of this it is inappropriate to measure drug concentrations in absolute quantities but rather by comparison to a standard curve constructed concurrently in the assay to a reference compound. Thus, by using (−)-propranolol as the standard it is possible to directly compare all non-selective beta-blockers and express the level of beta-blocking activity measured in (−)-propranolol equivalents (Fig. 4).

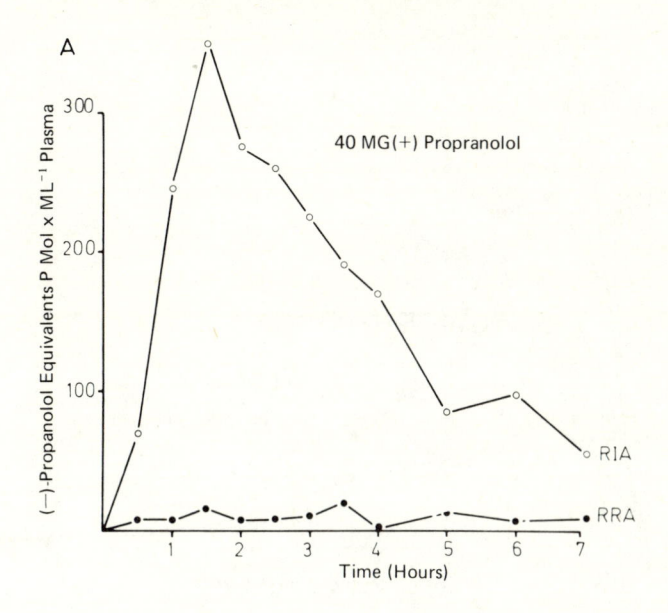

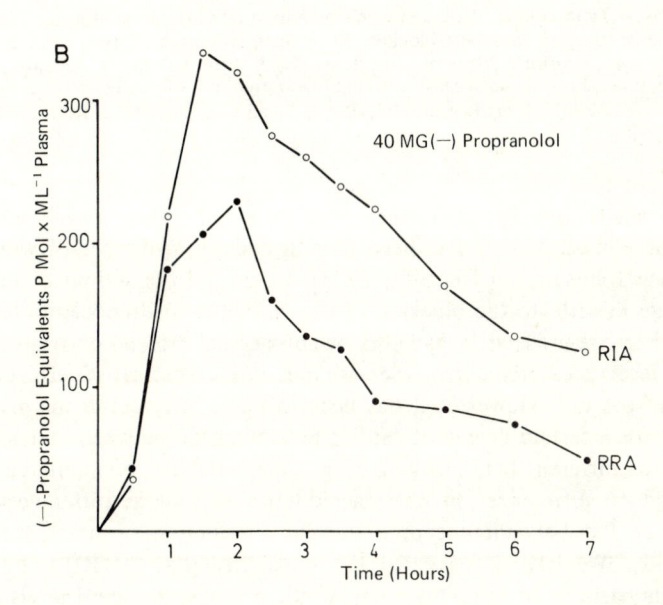

Fig. 3 — Time course of plasma levels of propranolol expressed as (−)-propranolol equivalents measured by both radioimmunoassay (○) and radioreceptor assay (●) in two volunteers, one of whom (A) had received orally 40 mg (+)-propranolol and the other (B) 40 mg (−)-propranolol. (Reproduced with permission from Barnett et al. (1980).

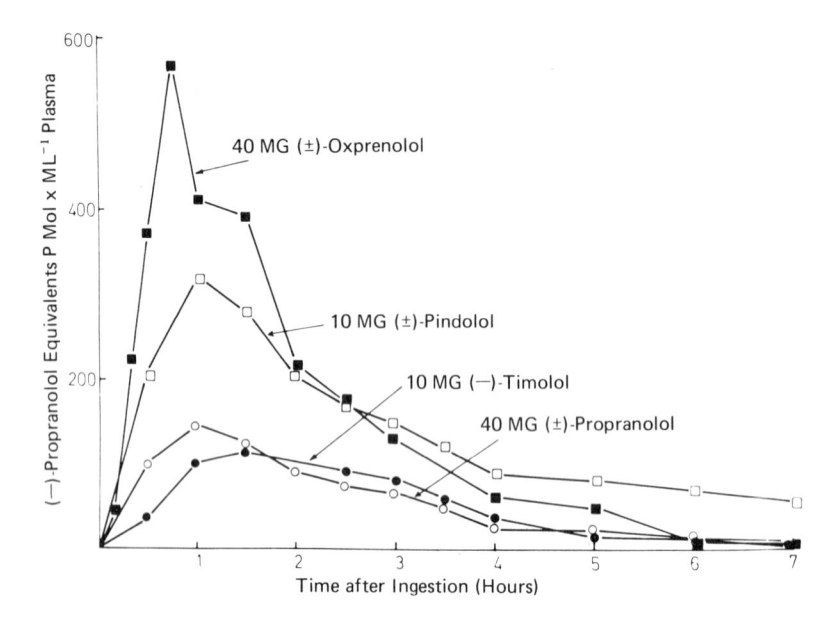

Fig. 4 — Time course of plasma levels measured by radioreceptor assay following oral ingestion of four beta-blockers by a single volunteer. Levels are expressed as (−)-propranolol equivalents (see text). Each drug was taken separately with at least weekly intervals. Drugs: 40 mg (±)-oxprenolol, 40 mg (±)-propranolol, 10 mg (±)-pindolol, 10 mg (−)-timolol. (Reproduced with permission from Barnett et al. (1980).

The relationship between plasma level measurements of beta-blocking drugs and their clinical effects has been investigated by a number of workers (for review see Johnson and Regardh, 1976). In general the actions of these drugs that relate directly to the blockade of the cardiac beta-adrenoceptor (e.g. reduction of exercise-induced tachycardia or anti-anginal effects) correlate well with plasma level measurements, whereas the antihypersensitive effect of beta-blockers does not. However, it has been difficult to establish the plasma level required for a certain degree of cardiac beta-blockade between individuals and/or between different beta-blocking drugs. The variability encountered has been attributed to differences in pharmacokinetics as well as differences in drug activity at the beta-adrenoceptor and the production of unmeasured active metabolites. We have investigated the possibility that the RRA, by allowing direct comparisons of different drugs, would establish standard levels of plasma beta-blocking activity (in (−)-propranolol equivalents) correlating with physiological cardiac beta-blockade assessed by reduction in exercise-induced tachycardia (Barnett et al., 1981). Plasma levels were measured by RRA in healthy male volunteers after oral ingestion of different beta-blocking drugs (timolol 10 mg, propranolol 40 mg, oxprenolol 40 mg) administered randomly on sepa-

Sec. 4] Examples of Radioreceptor Assays 283

rate occasions. Heart rate was measured immediately after a standardised exercise test on a bicycle ergometer prior to dosing and at intervals up to 24 hours afterwards. Blood samples for RRA were taken at each exercise time point and levels for each drug expressed as (−)-propranolol equivalents. Direct comparison was made of log total plasma beta-blocking activity and percentage reduction in exercise heart rate by graphical means (Fig. 5) using the composite data from all volunteers. This revealed significantly different but parallel dose–response relationships for each drug, indicating a distinct hierarchy of beta-blocking potency with timolol > propranolol > oxprenolol. These results indicate that even when plasma levels are normalised using a single direct assay of beta-blocking activity, significant difference in dose–response relationships remain

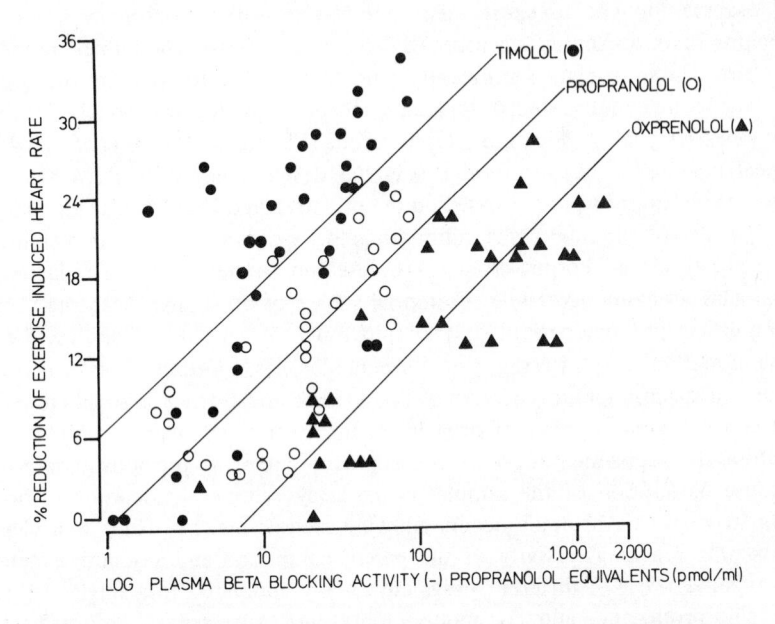

Fig. 5 – Direct comparison of total log plasma beta-blocking activity measured by RRA ((−)-propranolol equivalents) and percentage reduction in exercise-induced heart rate (%↓EHR) for three different beta-blocking drugs. The three drugs (10 mg timolol, 40 mg propranolol, 40 mg oxprenolol) were administered in a single blind cross-over study with 5 normal volunteers. Drugs were given at least one week apart at various times up to 24 hours after dosing. Blood samples for RRA were taken immediately prior to each exercise test. Composite data for all experimental points is represented in the figure. Covariance analysis within and between drug groups was used to generate the regression lines of best fit for each assay. These are significantly different ($P = <0.001$) but parallel, conforming to a standard regression equation: $y = K + 11.02x$ (where $y = \%\downarrow$ EHR, $x = $ log plasma beta-blocking activity). Interindividual variation in subject dose–response relationships is indicated by the scatter around these lines and was similar for all three drugs. Correlation coefficients (r) = timolol 0.71, propranolol 0.74, oxprenolol 0.86.

between beta-blocking drugs, most probably related to differences in tissue distribution, protein binding and other pharmacological properties. However, using the RRA it is possible to suggest average plasma levels required for a given degree of beta-blockade for individual drugs.

4.2 Neuroleptic drugs

Several groups of workers have reported increased numbers of dopamine receptors, identified by radioligand binding, in post-mortem brains of schizophrenic patients (Chapter 9, Owen *et al.*, 1978; Lee and Seeman, 1980). From these and other observations it has been postulated that brain dopamine systems are hyperactive in schizophrenics and may relate to disease activity. Despite reservations about the general significance of this hypothesis, neuroleptic drugs in clinical use are thought to exert their therapeutic anti-schizophrenic actions by blocking brain dopamine receptors (Snyder *et al.*, 1974). The clinical potencies of these drugs correlate significantly with their ability to compete for specific binding to mammalian brain dopamine receptors by the radiolabelled butyrophenones [^{3}H] haloperidol and [^{3}H] spiperone (Chapter 4, Creese *et al.*, 1976). A logical conclusion of this finding was the development of a RRA assay for neuroleptic drugs in plasma based on the principle that these drugs will compete for the specific binding of the radioligands to dopamine receptors in membranes prepared from calf corpus striatum (Creese and Snyder, 1977). The latter was chosen as an easily accessible substantial source of dopamine receptors. Assays using either [^{3}H] haloperidol or [^{3}H] spiperone are carried out as described in general methods. As with the RRA for beta-blockers, addition of plasma reduces total and specific binding. However, this plasma interference is small enough to allow direct measurement of drug levels in unextracted plasma samples, with appropriate supplementation of the standard curve, as previously described. Because of dilution of the samples in the assay volume, dissociation of bound drug from plasma proteins occurs allowing measurement of total 'neuroleptic' (dopamine blocking) activity of the parent compound and any active metabolite. Plasma levels of different drugs are usually compared to a single standard, e.g. chlorpromazine, and commonly, therefore, expressed as 'chlorpromazine equivalents'. It should be noted that [^{3}H] spiperone, though not [^{3}H] haloperidol, can label 5-HT as well as dopamine receptors even within the striatum (Chapter 4, Howlett and Nahorski, 1980; Withy *et al.*, 1981). This could complicate the interpretation of neuroleptic RRAs and the use of selective agents to suppress binding to the 5-HT sites (Withy *et al.*, 1981) is clearly necessary when [^{3}H]-spiperone is used as a radioligand.

Creese and Snyder (1977) have compared the RRA measurements of serum haloperidol levels with an RIA for this drug and, as expected, close correlation was found between the two methods since haloperidol is minimally metabolised and then only to inactive compounds (see similar comparisons for beta-blocking drugs above). However, other neuroleptics such as chlorpromazine or thioridazine may exert their therapeutic actions in part through active metabolites (7-hydroxy-chlorpromazine and mesoridazine) not measured by most chemical assays, but their contribution to total plasma 'neuroleptic' activity is readily assessed by the RRA.

Investigation of the relationship between clinical response and plasma neuroleptic drug levels (using standard chemical assays) has been undertaken by several groups, some reporting good correlations, whereas others have been less successful (Phillipson *et al.*, 1977; Wiles *et al.*, 1976). Most studies have attempted to correlate response with either levels of parent drug or its metabolites separately. However, the inherent advantage of the RRA, allowing synchronous estimation of the total contribution of all active compounds, has been exploited in more recent work. Despite the intrinsic problem in assessing clinical status and response to therapy in schizophrenic patients (as opposed to assessment of beta-blockade discussed previously) good general correlation has been shown between RRA measured plasma 'neuroleptic' activity and disease activity. Thus, Tune *et al.* (1980) studying serum neuroleptic levels by RRA in 30 schizophrenic patients receiving a variety of neuroleptic drugs, showed that poor therapeutic response was associated with levels under 50 ng/ml (chlorpromazine equivalents) but all patients with levels of 100–200 ng/ml (chlorpromazine equivalents) had exhibited good clinical improvement. Furthermore, there was no correlation between neuroleptic dosage and response or between dose and serum neuroleptic levels. These observations were extended by the same workers (Tune *et al.*, 1981) who examined serial neuroleptic levels and clinical response in 10 newly treated schizophrenic patients. Results were similar to the earlier study with the exception of thioridazine which produced much higher serum levels than the other neuroleptics, confirming previous observations that thioridazine crosses the blood brain barrier less readily and the ratio of serum to CSF levels for this drug are higher than for other neuroleptics (Rosenblatt *et al.*, 1981). Excluding thioridazine, neuroleptic levels measured by RRA were very similar for the other drugs and correlated across the board with clinical response.

In summary, therefore, the RRA for neuroleptic drugs affords a simple and sensitive technique for measuring blood levels of these compounds and their active metabolites which is potentially useful in monitoring or predicting clinical response in schizophrenic patients.

4.3 Tricyclic antidepressants
Blood levels of tricyclic antidepressants (TAD) measured by chemical techniques have been employed extensively in attempts to correlate levels with clinical status and improve therapeutic management of depressed patients. Response to TAD is variable and interindividual variation in plasma steady state levels of these drugs as a consequence of differences in metabolism has been claimed to be the reason for lack of effects in some patients (Glossmann *et al.*, 1979). A

simple sensitive, easily reproducible assay for TAD is, therefore, desirable to allow routine clinical monitoring of plasma levels. The major problem in producing an RRA for TADs has been that in contrast to beta-blocking drugs (Section 4.1), neuroleptics (Section 4.2) and benzodiazepines (Section 4.4), there is no known receptor interaction which is unequivocally responsible for the mood elevating properties of the TADs. Three RRAs for these compounds have, however, been described.

Most TADs have anticholinergic (muscarinic) effects probably unrelated to their mode of action but relevant to the production of variable degrees of unwanted side-effects. Innis et al. (1979) have described an RRA which exploits this side-effect in order to quantitate TAD in plasma. The ability of unknown quantities of TAD in plasma to compete for the binding of [^{3}H] quinuclidinyl benzilate ([^{3}H]QNB) to muscarinic cholinergic receptors in mammalian brain membranes is compared to similar displacement by known quantities of drug in a concurrently constructed standard curve. Considerable plasma interference is reported such that small quantities of plasma containing no TAD added directly to the RRA cause variable (10-60%) reduction of specific binding. Therefore, the authors suggest that in routine assays the blood samples must be subjected to an extraction procedure. Despite this drawback, good correlation of plasma levels of nortriptyline in patients receiving this drug are reported between the RRA and GLC methods (Innis et al., 1979). However, two major drawbacks of this RRA are in the estimation of total blood activity of TAD including active metabolites, and comparisons in one assay system of different drugs within the TAD group. Both of these aspects of measurement, as previously mentioned, are considered important advantages of other RRAs. Thus, following administration of the tertiary amines, amitriptyline or imipramine, blood levels of their active metabolites, nortriptyline and desimipramine are comparable to the parent drug and contribute significantly to the total antidepressant activity of the drugs. However, parent drug and metabolite differ in their ability to inhibit [^{3}H]QNB binding (Snyder and Yamamura, 1977) which may bear no relationship, as previously mentioned, to antidepressant action. Thus, it is impossible to determine directly total biologically active TAD levels in patients treated with these drugs using the RRA. In addition, the assay will also detect any other substance which potently inhibits [^{3}H]QNB binding. This will include any compound that may be co-administered with the TAD that also acts directly or indirectly as a muscarinic cholinergic antagonist, for example, neuroleptic drugs (e.g. thioridazine), anticholinergic drugs (e.g. atropine) and anti-Parkinsonian agents (e.g. benztropine).

Another RRA described for measurement of TAD employs the ability of these drugs to compete for specific [^{3}H] 5-hydroxytryptamine (5-HT) binding to whole platelets prepared from rat blood (De Filipe et al., 1982). In this assay, unknown levels are estimated as with other RRAs by comparison with a standard inhibition curve constructed with known amounts of the drug. The authors claim little or no interference with [^{3}H] 5-HT binding by added drug-free plasma up to 0.2 ml, thus obviating the need for prior extraction procedures. As with the [^{3}H] QNB assay, levels of drug are measured indirectly dependent on their ability to bind to the 5-HT recognition sites in platelets, which may not be linked with antidepressant activity. Assay sensitivity varies between drugs in this group from a low limit of 0.1 ng/ml for clomipramine to 20 ng/ml for nortriptyline, making it impossible to directly compare plasma activity between drugs in the same assay. Finally, other compounds (e.g. chlorpromazine) cross-react with TAD in the assay by amounts related to their affinity for the 5-HT receptor.

In contrast to the first two RRAs described, the approach adopted by Paul *et al.* (1980a) seems at first sight more rational and pharmacologically relevant. These workers have used the presence of high affinity binding sites for [^{3}H]-imipramine on human platelet membranes (Paul *et al.*, 1980b) to develop a RRA. Platelet membranes obtained from normal healthy donors are used in a competition assay using [^{3}H] imipramine along similar lines to those described for beta-blocking and neuroleptic drugs. Prior extraction of samples is not required as dilution of plasma 1 in 4 produces negligible effects on binding in volumes up to 25 µl added directly to the assay tubes (final volume 250 µl). Inhibition of [^{3}H] imipramine binding was observed at nanomolar concentrations for most TADs. However, tertiary amine antidepressants such as imipramine and amitriptyline, were all potent inhibitors of binding, whereas their corresponding secondary amines (desipramine and nortriptyline) were much less potent. The only non-TAD psychotropic drug which might cross-react in the assay was chlorpromazine, but even this drug was approximately 50-fold less potent than the tertiary amines at displacing specific [^{3}H] imipramine binding.

Application of this RRA to clinical samples from patients maintained on steady state doses of imipramine was shown to correlate well with levels measured by combined gas chromatography/mass spectrometry. In addition, measurement of total TAD plasma 'biological' activity by RRA correlated very closely with combined levels of imipramine and its active metabolite 2-OH-imipramine, estimated separately by HPLC.

The major drawback of this RRA as presently described, is that it is only capable of quantitating the parent tertiary amine TAD. The relatively low affinity of the secondary amine derivatives for [^{3}H] imipramine binding precludes routine measurement in clinical samples, despite their important contribution to overall antidepressant activity.

Because of the various problems with the RRAs for TAD outlined, routine use of these methods seems less promising. To date no clinical studies with any of the assays described, have been reported.

4.4 Benzodiazepines

The benzodiazepine tranquillisers comprise the most widely prescribed class of compounds in current therapeutic use (see Chapter 3). Evidence exists that there is a relationship between therapeutic and/or side-effects of these drugs and blood levels, suggesting that the clinical use of the benzodiazepines might be improved by blood level monitoring (Greenblatt and Shader, 1974). Methods available for drug assay have not achieved routine clinical use because of overall complexity and the need for sample extraction (Zingales, 1975; Linnoila and Dorrity, 1977). The identification of high affinity, stereospecific binding sites for [^{3}H] diazepam ([^{3}H] DP) in mammalian CNS (Braestrup et al., 1977; Mohler and Okada, 1977) has provided the basis for the development of an RRA for benzodiazepines. The assay has been described by two groups of workers and essentially depends on the competition for [^{3}H] DP binding in rat or calf cerebral cortex membranes (Skolnick et al., 1979; Owen et al., 1979). In these methods the plasma samples are firstly deproteinised with perchloric acid and neutralised prior to assay. The supernatant after centrifugation is added directly to the assay and little or no interference with [^{3}H] DP binding of drug-free samples is reported. Standard curves are constructed using known quantities of cold diazepam added to drug-free plasma in varying concentrations and then treated as for the unknown samples. Final quantities of plasma activity are, therefore, expressed in 'diazepam equivalents' and include relative contributions from parent drugs and active metabolites. No significant cross-reactivity was observed in the assay with a number of other drugs tested including: neuroleptics, tricyclic antidepressants, anticholinergics, anticonvulsants and beta-adrenoceptor antagonists. Measurement of blood levels of diazepam and N-demethyl-diazepam (nordiazepam) have been compared in the same samples using the RRA and GLC techniques (Skolnick et al., 1979; Owen et al., 1979). As diazepam and nordiazepam have very similar affinities at the diazepam receptor (7–8 nM), separately estimated plasma concentrations of each compound using GLC can be combined to produce a calculated composite level expressed as 'diazepam equivalents', which compares very closely with measurements using the RRA and similarly expressed. Other diazepam metabolites, although active, have lower affinities for the receptor (20–500 nM) and, therefore, contribute to overall plasma biological activity to a much lesser extent. Despite the relative simplicity and ease of use of this RRA, no clinical studies employing it have been reported to date. Evaluation of the pharmocokinetics and correlation of plasma levels of benzodiazepines with clinical effect using the RRA would be of great interest.

5. CONCLUSIONS

In the preceding sections we have described and emphasised the advantages of RRAs for the quantitative analysis of several drugs in plasma. The present availability of specifically labelled receptor-directed ligands coupled to well-established conditions for ligand–receptor binding assays now allows a technically simple though very sensitive assay to be quickly developed at relatively low cost. As discussed, these techniques possess considerable advantages over chemical drug analysis since they provide an assay of the 'bioactivity' of that drug and possibly its metabolites at a particular receptor. Although these techniques have yet to be fully exploited, preliminary pharmacokinetic/pharmacodynamic studies in man emphasise their considerable potential. It should also be emphasised that with the recent availability of specific ligands for alpha$_1$, alpha$_2$, histamine H$_1$ and multiple opiate receptors to name a few, we can anticipate rapid proliferation of RRAs in a variety of clinical situations.

ACKNOWLEDGEMENTS

The authors would like to thank Jenny Bell for her excellent manuscript preparation.

REFERENCES

Barnett, D. B., Rugg, E. and Nahorski, S. R. (1978) Direct evidence for two types of beta-adrenoceptor binding site in lung tissue. *Nature* **273**, 166—168.

Barnett, D. B., Batta, M., Davies, B. and Nahorski, S. R. (1980) Evaluation of a radioreceptor assay for beta-adrenoceptor antagonists. *Eur. J. Clin. Pharmacol.* **17**, 349—354.

Barnett, D. B., Cook, N., Dickinson, K. E. J. and Nahorski, S. R. (1981) Radioreceptor assay for beta-adrenoceptor antagonists using solubilised receptor protein. *Br. J. Clin. Pharmacol.* **13**, 284P.

Barnett, D. B., Heller, B. and Jagger, C. (1982) Comparison of beta-blocking activity measured by radioreceptor assay and physiological cardiac beta-blockade. *Br. J. Clin. Pharmacol.* **13**, 613P.

Bilezikian, J. P., Gammon, D. E., Lee Rochester, A. B. C. and Shand, D. G. (1979) A radioreceptor assay for propranolol. *Clin. Pharmacol & Therap.* **26**, 173—180.

Braestrup, C., Albrechsten, A. and Squires, R. F. (1977) High densities of benzodiazepine receptors in human cortical areas. *Nature* **269**, 702—704.

Brown, B. L., Ekins, R. P. and Albano, J. D. M. (1972) Saturation assay for cyclic AMP using endogenous binding protein. *Adv. Cycl. Nucl. Res.* **2**, 25—40.

Bylund, D. B., Charness, M. E. and Snyder, S. H. (1977) Beta-adrenergic receptor labelling in intact animals with [^{125}I] hydroxybenzylpindolol. *J. Pharmacol. Exp. Ther.* **201**, 644—652.

Cohen, B. M., Herschel, M., Miller, E., Mayberg, H. and Baldessarini, R. J. (1980) Radioreceptor assay of haloperidol tissue levels in the rat. *Neuropharmacology* **19**, 663—668.

Creese, I., Burt, D. R. and Snyder, S. H. (1976) Dopamine receptor binding predicts clinical and pharmacological potencies of antischizophrenic drugs. *Science* **192**, 481—483.

Creese, I. and Snyder, S. H. (1977) A simple and sensitive radioreceptor assay for antischizophrenic drugs in blood. *Nature* **270**, 180–182.

De Felipe, M. C., Fuentes, J. A. and Drummond, A. H. (1982) Specific 5-Hydroxytryptamine binding to rat platelets as a system to evaluate tricyclic antidepressants in plasma. *Biochem. Pharmacol.* **31** (8), 1661–1663.

Enna, S. J. (1978) Radioreceptor assay techniques for neurotransmitters and drugs, in *Neurotransmitter Receptor Binding* (Yamamura, H. I., Enna, S. J. and Kuhar, M. J., eds.). Raven Press, New York.

Glossmann, A. H., Perel, J. M., Shostak, J., Trautor, S. J. and Fleiss, J. L. (1979) Clinical implications of imipramine plasma levels for depressive illness. *Arch. Gen. Psychiatry* **34**, 197–204.

Greenblatt, D. J. and Sjader, R. I. (1974) *Benzodiazepines in Clinical Practice.* Raven Press, New York.

Howlett, D. R. and Nahorski, S. R. (1980) Quantitative assessment of heterogeneous [^{3}H]spiperone binding to rat neostriatum and frontal cortex. *Life Sci.* **26**, 511–517.

Innis, R. B., Bylund, S. B. and Snyder, S. H. (1978) A simple, sensitive and specific radioreceptor assay for beta-adrenergic antagonist drugs. *Life Sci.* **23**, 2031–2038.

Innis, R. B., Tune, L., Rock, R., De Paulo, R., U'Prichard, D. C. and Snyder, S. H. (1979) Tricyclic antidepressant radioreceptor assay. *Eur. J. Pharmacol.* **58**, 473–479.

Johnsson, G. and Regardh, C. G. (1976) Clinical pharmacokinetics of beta-adrenoceptor blocking drugs. *Clin. Pharmacokinetics* **1**, 233–263.

Laduron, P. M., Janssen, P. F. M. and Leysen, J. E. (1978) Characterisation of specific *in vivo* binding of neuroleptic drugs in rat brain. *Life Sci.* **23**, 581–586.

Lee, T. and Seeman, P. (1980) Elevation of brain neuroleptic/dopamine receptors in schizophrenia. *Am. J. Psychiatry* **137**, 191–197.

Lefkowitz, R., Roth, J. and Pastan, I. (1970) Radioreceptor assay of adrenocorticotrophic hormone: new approach to assay of polypeptide hormones in plasma. *Science* **170**, 633–635.

Linnoila, M. and Dorrity, F. (1977) Rapid gas chromatographic assay of serum diazepam, N-desmethyldiazepam and N-desalkylflurazepam. *Acta Pharmacol. Toxicol.* **41**, 458–464.

Mohler, H. and Okada, T. (1977) Benzodiazepine receptor: demonstration in the central nervous system. *Science* **198**, 849–851.

Munson, P. and Rodbard, D. (1980) Ligand: a versatile computerised approach for the characterisation of ligand binding systems. *Anal. Biochem.* **107**, 220–239.

Nahorski, S. R., Batta, M. I. and Barnett, D. B. (1978) Measurement of beta-adrenoceptor antagonists in biological fluids using a radioreceptor assay. *Eur. J. Pharmacol.* **52**, 393–396.

Nation, R. L., Pens, G. W. and Chion, W. L. (1978) High pressure liquid chromatographic method for the simultaneous quantitative analysis of propranolol and 4-hydroxypropranolol in plasma. *J. Chromatogr.* **145**, 429–436.

Owen, F., Crow, T. J. and Poulter, M. (1978) Increased dopamine receptor sensitivity in schizophrenia. *Lancet* **i**, 223–225.

Owen, F., Lofthouse, R. and Bourne, R. C. (1979) A radioreceptor assay for diazepam and its metabolites in serum. *Clinica Chimica Acta* **93**, 305–310.

Paul, S. M., Rehavi, M., Hulihan, B., Skolnick, P. and Goodwin, F. K. (1980a) A rapid sensitive radioreceptor assay for tertiary amine tricyclic antidepressants. *Commun. in Psychopharmacol.* **4**, 487–494.

Paul, S. M., Rehavi, M., Skolnick, P. and Goodwin, F. K. (1980b) Demonstration of specific high affinity binding sites for [^{3}H] imipramine on human platelets. *Life Sci.* **26**, 953–959.

Phillipson, O. T., McKeown, J. M., Baker, J. and Healey, A. F. (1977) Correlation between plasma chlorpromazine and its metabolites and clinical ratings in patients with actual relapse of schizophrenic and paranoid psychoses. *Br. J. Psychiatry* **131**, 172–185.

Rodbard, D. and Lewald, J. E. (1970) Computer analysis of radioligand assay and radioimmunoassay data. *Acta Endocrinol. (Suppl.)* **147**, 79–103.

Rosenblatt, J. E., Pary, R. J. and Bigenew, L. B. (1981) Measurement of serum neuroleptic concentrations by radioreceptor assay: concurrent assessment of clinical response and toxicity, in *Proceedings of the International Phenothiazine Congress* (Usdin, E. ed.). Raven Press, New York.

Rugg, E. L., Barnett, D. B. and Narhorski, S. R. (1978) Co-existence of beta$_1$ and beta$_2$ adrenoceptors in mammalian lung: evidence from binding studies. *Mol. Pharmacol.* **14**, 996–1005.

Shand, D. G., Nuckolls, E. M. and Oates, J. A. (1970) Plasma propranolol levels in adults with observations in four children. *Clin Pharmacol. & Therap.* **11**, 112–120.

Skolnick, P., Goodwin, F. K. and Paul, S. M. (1979) A rapid and sensitive radioreceptor assay for benzodiazepines in plasma. *Arch. Gen. Psychiatry* **36**, 78–80.

Snyder, S. H. and Yamamura, H. I. (1977) Antidepressants and the muscarinic acetylcholine receptor. *Arch. Gen. Psychiatry* **34**, 236–239.

Snyder, S. H., Banerjee, S. P., Yamamura, H. I. and Greenberg, D. (1974) Drugs, neurotransmitters and schizophrenia. *Science* **184**, 1243–1253.

Sriwatanakul, K. and Nahorski, S. R. (1980) Disposition and activity of beta-adrenoceptor antagonists in the rat using an *ex vivo* receptor binding assay. *Eur. J. Pharmacol.* **66**, 169–178.

Tune. L. E., Creese, I., De Paulo, J. R., Slavney, P. R., Coyle, J. T. and Snyder, S. H. (1980) Clinical state and serum neuroleptic levels measured by radioreceptor assay in schizophrenia. *Am. J. Psychiatry* **137**, 187–190.

Tune, L. E., Creese, I., De Paulo, J. R., Slavney, P. R. and Snyder, S. H. (1981) Neuroleptic serum levels measured by radioreceptor assay and clinical response in schizophrenic patients. *J. Nervous & Mental Dis.* **196**, 60–63.

Walle, T. (1974) GLC determination of propranolol, other beta-blocking drugs and metabolites in biological fuids and tissues, *J. Pharmaceut. Sci.* **63**, 1885–1891.

Wiles, D. H., Kolakowska, T. and McNally, A. S. (1976) Clinical significance of plasma chlorpromazine levels. I. Plasma levels of the drugs, some of its metabolites and prolactin during acute treatment. *Psychol. Med.* **6**, 407–415.

Withy, R. M., Mayer, R. J. and Strange, P. G. (1981) Use of [^{3}H] spiperone for labelling dopaminergic and serotonergic receptors in bovine caudate nucleus. *J. Neurochem.* **37**, 1144–1154.

Zingales, J. A. (1975) Diazepam metabolism during chronic medication, unbound fraction in plasma, erythrocytes and urine. *J. Chromatogr.* **75**, 55–78.

Editor's Conclusions – Future strategies

From the foregoing chapters it is clear that research into cell surface receptors is very active yielding results important for basic science and which lead to clinical applications. In considering prospects for the future I intend only to consider a few key areas where I feel progress may be made.

Undoubtedly progress in understanding cell surface receptor structure will be rapid as more receptors are isolated and characterised and modern immunological and molecular genetic techniques are used. It will be of great interest to see whether there are similarities between different receptors – possibly groups of receptors will be identified with similar structural and functional determinants, the groups of receptors possibly having similar mechanisms of action. This structural knowledge should clarify the current confusion in some areas over multiple sub-classes of receptors/binding sites and should allow the general relationships between binding sites identified by *in vitro* assays and true receptors to be determined. Thus it should be possible to use the structural information to design more selective drugs and it cannot be long before information about receptor binding site structure is available, in some cases enabling rational design of specific receptor-directed ligands to be achieved.

Structural work on receptors will be greatly aided by the monoclonal antibody technique. Using this technique it should be possible to obtain monoclonal antibodies against specific receptors for use in receptor purification. The monoclonal technique combined with a specific receptor assay should circumvent a major problem in receptor purification, namely the low number of receptors and the problems this could raise in conventional purification schemes. Monoclonal antibodies against receptors may also enable immunoassays for receptors to be set up which might eventually supercede ligand-binding assays. It is likely that monoclonal antibodies directed against determinants other than the receptor sites may recognise putative common structures in several receptors as suggested above. Immunological approaches are likely to be used widely and the use of anti-idiotypic antibodies (antibodies against anti-antagonist or anti-agonist antibodies) may offer an alternative approach to receptor purification.

The anti-idiotypic antibodies may in favourable circumstances recognise the specific receptor thus offering a new affinity reagent for receptor purification (see for example Schreiber *et al.* (1980)).

Immunological approaches are likely to be important clinically in that our knowledge of autoimmune diseases, especially those where receptors are involved, should progress and perhaps further receptor-linked autoimmune diseases will be discovered (see Chapters 4 and 7). Knowledge about receptors may aid in diagnosis in these diseases and anti-receptor antibodies or the related anti-idiotypes may be of real practical use in the treatment of these disorders.

Immunological approaches will be important in the application of the modern techniques of molecular genetics to receptor studies. As has been described in Chapter 14 for the nicotinic acetylcholine receptor these techniques enable information on receptor polypeptide structure and function to be obtained without the extensive use of conventional sequencing methods on purified proteins.

Another area of current and future interest concerns the mechanism of action of receptors. Although a great deal of information is available on this, the precise mechanism of action is ill-defined in many cases and it is not clear whether receptors can really be grouped according to their mechanisms of action or whether each receptor must be considered separately in its own environment. Points of great interest that are likely to be clarified in the near future are the role of phospholipid alterations in receptor mechanisms (Chapter 11) and the possible multiplicity of N-proteins (Chapter 12) and their relation to adenylate cyclase and other non-cyclase effector mechanisms.

A further area of interest concerns the organisation and interdependence of receptors where coexistence of neurotransmitters has been demonstrated. It will be of great interest to understand in detail the cases where receptor interaction has been described (Chapter 6) and such studies may reveal other similar interactions. Such receptor interaction is of interest clinically because it may offer alternative sites for drug intervention, for example in schizophrenia.

Knowledge about the true clinical function of receptors, particularly in the brain, has been difficult to obtain. Positron emitting ligands with high receptor affinity should allow positron emission tomography to be applied to human clinical problems. This may enable a non-invasive study of receptor function to be made *in vivo* in disorders such as schizophrenia and depression.

Related to the question of the mechanism of action of receptors is the question of receptor turnover via internalisation. Changes in receptor number may be important in some diseases, e.g. schizophrenia (Chapter 9) and in drug treatments although the effects of drugs on receptors often seem to be slow indirect effects. It is far from clear at present whether the receptosome mechanism may be applied to all receptors or whether the receptor-turnover mechanism is related to the particular receptor itself. Also the importance of internalisation of the ligand for its action is unknown. Knowledge in this area is likely to progress rapidly and may lead to the development of drugs that can interfere directly with receptor turnover rather than indirectly, thus provoking rapid changes in receptor numbers and offering more rapid treatments for certain diseases.

It seems likely therefore that progress in the field of cell surface receptors will be exciting leading to rapid advancement of existing knowledge and offering new clinical insights.

REFERENCE

Schreiber, A. B., Couraud, P. O., Andre, C., Vray, B. and Strosberg, A. D. (1980) Anti-alprenolol anti-idiotypic antibodies bind to β-adrenergic receptors and modulate catecholamine sensitive adenylate cyclase. *Proc. Natl. Acad. Sci. U.S.A.* **77**, 7385–7389.

Index

A
acetylcholine, 118, 146, 210, 249
acetylcholine receptor, muscarinic, 105, 119, 166, 210, 231, 260, 286
acetylcholine receptor, nicotinic, 14, 16, 126, 210, 249
adenylate cyclase, 14, 15, 82, 118, 144, 163, 209, 210, 227, 241, 294
α-adrenergic receptor, 21, 105, 166, 210, 217, 231, 289
β-adrenergic receptor, 35, 105, 138, 166, 208, 210, 232, 273
$α_2$-adrenoreceptor agonists, therapeutic potential, 24
$α_2$-adrenoreceptor antagonists, therapeutic potential, 28
Albright's Hereditary Osteodystrophy, 241
allergic reaction, 216
allergic rhinitis, 138, 216
Alzheimer's disease, 164, 178
γ-amino butyric acid receptor, 53, 105, 122, 166
anaphylactic shock, 216
angiotensin II, 208, 231
antibodies, 16, 64, 96, 112, 127, 197, 255, 271, 293
antidepressants, 33
anti-idiotypic antibodies, 127, 293
anxiety, 51
arachidonic acid, 214
asthma, 16, 138, 216
atherosclerosis, 217
autoimmune diseases, 16, 127, 138, 294
autoradiography, 46, 59

B
barbiturates, 65
benzodiazepine receptor, 50, 54, 105, 122, 166, 287
benzodiazepines, 50, 287
beta blocking drugs, 274, 276
Bordatella pertussis factor, 229
α-bungarotoxin, 126, 249

C
calcium ions, 15, 101, 209, 210, 230, 241
cellular proliferation, 219
cholecystokinin, 116
cholera toxin, 229, 241
clonidine, antihypertensive effect, 24
coexistence of neurotransmitters, 110, 294
concanavalin A, 96, 208
corticotropin (ACTH), 41, 232, 270
cyclic AMP, 14, 15, 121, 143, 210, 217, 227, 241
cyclic GMP, 54, 215

D
depression, 33, 285, 294
desensitisation, 16, 17, 61, 199
diabetes, 16, 138
diazepam, 55, 288
dihydropyridine receptor, 101
diptheria toxin, 215
cDNA, 41, 262
dopamine, 116, 142, 231
dopamine receptor, 82, 105, 118, 142, 163, 231, 284
down-regulation, 18, 34, 199
dynorphin, 42

E
endogenous ligand, 17, 41, 62, 122
β-endorphin, 42, 111
enkephalin, 41, 111, 171
estrous cycle, 187
ethyl-β-carboline carboxylate, 50

Index

F
fluid mosaic model, 14, 16
flupenthixol, 85, 146, 165
forskolin, 229, 231

G
gene cloning, 250
glucagon, 208, 228, 242
glucagon receptor, 232, 235
gonadotrophin releasing hormone, 184
gonadotrophin releasing hormone receptor, 184
Graves's disease, 16, 138
guanine nucleotide regulatory protein, 15, 93, 207, 227, 240, 294
guanosine triphosphate, 15, 17, 46, 86, 93, 207, 227, 240
guanylate cyclase, 215

H
hay fever, 138, 216
histamine receptor, 105, 210, 217, 231, 289
Huntington's chorea, 56, 164, 178
5-hydroxytryptamine, 112
hypertension, 24, 218, 276

I
immunoglobulin E, 208
immunohistochemistry, 112
inflammation, 217
insulin receptor, 138

L
leukotrienes, 214, 217
Librium, 50
ligand-binding assay, 16, 30, 43, 55, 83, 101, 118, 142, 163, 184, 270, 293
lithium ions, 211, 219
lymphoma cell (S49), 229, 240
lysophosphatidylcholine, 89

M
manic-depressive illness, 220
membrane, 15, 207, 259, 294
messenger RNA, 41, 250
metastasis, 217
mianserin, 85
microinjection, 259
molecular genetics, 41, 249, 295
monoclonal antibodies, 96, 127, 293
myasthenia gravis, 16, 88, 127

N
α-neo-endorphin, 42
neuroleptic drugs, 83, 142, 164, 284
neurotransmitter coexistence, 110, 294
nimodipine, 102

O
opiate receptor, 17, 41, 105, 166, 231, 289
opioid peptides, 41

P
parathyroid hormone, 241
Parkinson's disease, 88
phosphatidylcholine, 16, 208
phosphatidylethanolamine, 16, 208
phosphatidylinositol, 15, 207, 294
phosphoinositides, 207, 210
phospholipid methylation, 16, 208, 294
plasma levels of drugs, 271
platelet aggregation, 217
positron emission tomography, 59, 180, 294
postjunctional, 21, 111
prejunctional receptors, 21, 111
pre-proenkephalin, 41
pre-proopiocortin, 41
prostaglandins, 214, 217, 227
protein kinase, 215
pseudohypoparathyroidism, 241
purinergic receptor, 231

R
radioimmunoassay, 270, 293
radioreceptor assay, 270
receptor
  α-adrenergic, 21, 105, 166, 210, 217, 231, 289
  β-adrenergic, 35, 105, 138, 166, 208, 210, 273
  γ-amino butyric acid, 53, 105, 122, 166
  benzodiazepine, 50, 54, 105, 122, 166, 287
  dihydropyridine, 101
  dopamine, 82, 105, 118, 142, 163, 231, 284
  glucagon, 232, 235
  gonadotrophin releasing hormone, 184
  histamine, 105, 210, 217, 231, 289
  insulin, 138
  muscarinic acetylcholine, 105, 119, 166, 210, 231, 260, 286
  nicotinic acetylcholine, 14, 16, 126, 210, 249
  opiate, 17, 41, 105, 166, 231, 289
  postjunctional, 21, 111
  prejunctional, 21, 111
  purinergic, 231
  serotonin, 86, 105, 114, 166, 287
  thyrotropin, 138
  V_1-vasopressin, 210
  V_2-vasopressin, 210
receptor concept, 13, 16, 293
receptor purification, 95
receptor reconstitution, 250

receptor regulation, 17, 33, 60, 142, 164, 184, 294
receptor solubilisation, 66, 88, 127, 250, 273
receptosome, 18, 294
restriction enzymes, 265
reticulocyte lysate, 255

S

schizophrenia, 87, 88, 142, 163, 178, 284, 294
second messenger, 14, 209
secretin, 232
serotonin, 112
serotonin receptor, 86, 105, 114, 166, 287
signal peptide, 265
solubilisation, 66, 88, 127, 250, 273
spiperone, 85, 150, 165, 284
spirodecanone site, 89
stereotype, 87, 144

steroid hormones, 17, 185
substance P, 112

T

tardive dyskinesia, 143, 158
target size analysis, 235
thromboxanes, 214, 217
thyrotropin receptor, 138
thyrotropin releasing hormone, 112, 241
Torpedo fish, 127, 249
transmembrane signalling, 13, 18, 208
tricyclic antidepressant drugs, 34, 285

V

vasoactive intestinal polypeptide, 118, 232
vasopressin, 208
V_1-vasopressin receptor, 210
V_2-vasopressin receptor, 210

X

Xenopus oocyte, 257